This, the first full-length biography of Edward Emerson Barnard, tells the remarkable tale of endurance and achievement of one of the leading astonomers of the late nineteenth and early twentieth century. As a 'man who was never known to sleep', Barnard scoured the heavens endlessly, leaving an astonishing legacy of observations – of planets, satellites, comets, double stars, bright and dark nebulae, and globular clusters – that make him one of the greatest observers of all time.

This book traces Barnard's life from poverty to international recognition. We are told how he grew up fatherless and in hardship during the American Civil War; that he later acquired a small telescope and discovered so many comets that, despite his lack of formal education, he won a position at the Lick Observatory, California. His success as a professional astronomer then unfolds, and we are told, in particular, how he discovered the fifth satellite of Jupiter and pioneered wide-angle photography of comets and the Milky Way.

Beautifully illustrated throughout, this book includes many of Barnard's famous wide-field photographs of comets and the Milky Way. It provides a complete history of Barnard's fascinating life and work, based largely on archival material hitherto unpublished. It also offers unusual insight into the astronomers he knew and the observatories with which he was associated and will be of interest to astronomers and historians of science.

The Immortal Fire Within

The Life and Work of Edward Emerson Barnard

The Immortal Fire Within

The Life and Work of Edward Emerson Barnard

WILLIAM SHEEHAN

CAMBRIDGE
UNIVERSITY PRESS

CAMBRIDGE UNIVERSITY PRESS
Cambridge, New York, Melbourne, Madrid, Cape Town, Singapore, São Paulo

Cambridge University Press
The Edinburgh Building, Cambridge CB2 8RU, UK

Published in the United States of America by Cambridge University Press, New York

www.cambridge.org
Information on this title: www.cambridge.org/9780521444897

First published 1995
This digitally printed version (with corrections) 2007

A catalogue record for this publication is available from the British Library

Library of Congress Cataloguing in Publication data

Sheehan, William, 1954–
The immortal fire within: the life and work of Edward Emerson Barnard/William Sheehan.
p.   cm.
ISBN 0 521 44489 6
1. Barnard, Edward Emerson, d. 1923. 2. Barnard, Edward Emerson, d. 1923. 3. Astronomy – United States – History.
4. Astronomers – United States – Biography. I. Title.
QB36.B3S48 1995
520′.92–dc20   94-26742 CIP

ISBN 978-0-521-44489-7 hardback
ISBN 978-0-521-04601-5 paperback

To Don and Dale

Contents

Preface to the Paperback Edition

The publication of this full-length biography of Edward Emerson Barnard in paperback fittingly coincides with the 150th anniversary of his birth.

Barnard is one of the most famous astronomers of the late 19th and early 20th century. Among amateurs he is almost a patron saint. His early life in Nashville, Tennessee was one of enormous difficulties. During the troubled years before the American Civil War, he began without advantages—fatherless, impoverished, hungry, and harassed—but he had his lucky chance when he was hired as a lowly helper at the Photograph Gallery of John Van Stavoren, an old friend of his ailing mother. At the Photograph Gallery, he found mentors and learned a skill—photography—which served him well during the rest of his career.

Psychologists are fascinated with the concept of resiliency—the way that some individuals are able, despite adverse circumstances, to succeed in spite of them. Barnard remains a prime example, a stirring case of one who managed to find "sweet uses of adversity." He could not have survived without the adults who took an interest in him, who saw his promise and were willing to mentor him. However, no less critical was his discovery of his ruling passion. He came down hard with a case of what psychologist William E. Kelly calls Noctcaelador: "emotional attachment to, or adoration, of the night-sky." The condition dawned early, and it proved incurable; ironically, it cured him at once of almost everything else that had ailed him.

From acquisition (by chance) of a first book containing a few small star-charts to the cobbling together of a small spyglass allowing him the thrill of seeing the craters of the Moon, the satellites of Jupiter, and the phases of Venus, Barnard advanced rapidly in his astronomical craft. Photographs of him in his teens and early twenties—and he was surprisingly often photographed–show him to have been a scrawny and malnourished youth; but he was possessed of tremendous drive and the ability to get by on very little sleep. As soon as he saved up enough money to purchase a proper telescope, he rapidly distinguished himself as the discoverer of comets, and received a fellowship to Vanderbilt University where his exploits are still proudly recited. He continued on to Lick Observatory, at Mt. Hamilton, California, overlooking San Jose (today the center of Silicon Valley). Despite the miseries of working under a tyrannical director, Barnard began the photography of comets and the Milky Way with a modest and inexpensive lens; his discovery of the fifth satellite of Jupiter in September 1892 catapulted him to international fame. That satellite, Amalthea, has now been photographed at close range by spacecraft orbiting around Jupiter.

When Barnard left Mt. Hamilton in 1895, he did so with many regrets; but in the end, he was lured to the University of Chicago's Yerkes Observatory by its irresistibly energetic young director, George Ellery Hale. (Ironically, Hale himself abandoned Yerkes for the warmer climate of California soon after Barnard arrived; Hale's successor's name was, rather ominously, Frost!) Though Barnard plotted his eventual return to the West Coast—he dreamed of one day returning and planting an orchard and growing orange trees on a tract of land on Alum Rock Road in San Jose–he remained in the Midwest with its terrible winters to the bitter end. The only exception was eight blissful months in 1905 he spent on Mt. Wilson, where he took most of the remarkable photgraphs of the Milky Way that grace the *Atlas of Selected Regions of the Milky Way* – a legendary volume indeed, and a book of which every astronomical book collector dreams (copies of it routinely sell for tens of thousands of dollars!). Fortunately, all of the photographs have now been digitally scanned and made available online through a project organized by James Sowell at Georgia Tech. They can be accessed at:

www.library.gatech.edu/Barnard

In the more than eighty years since Barnard's death, his meticulous records continue to yield valuable insights to research astronomers. Since this book first appeared, James Bryan of McDonald Observatory has reevaluated Barnard's 1889 observations of the eclipse of Iapetus and sleuthed out fascinating new details about the structure of Saturn's rings. Gareth Williams of the Minor Planet Center at Cambridge, Massachusetts, has added to the list of Barnard's accomplishments the first known observation of a Trojan asteroid, which Barnard unwittingly chanced upon in 1904 as he searched near Saturn for its recently discovered satellite, Phoebe. And Tony Misch, Laurie Hatch, Rem Stone and I repeated his famous series of observations of Mars with the Great Lick Refractor in 2003, when the planet was closer than it had been since the Upper Paleolithic; we abundantly confirmed Barnard's mastery as an observer of the planets.

As with the heroes of a thousand faces Joseph Campbell wrote about, the places where Barnard toiled—where he slew astronomical dragons, or conquered mountains of doubt—stand as shrines sacred to the memory of all who love the stars. At Nashville, among the sweet and spicy fragrance of the magnolias, at Lick Observatory, where the straw-colored hills roll away from Mt. Hamilton and recall his view in the Great Lick Refractor of the "seas" of Mars, above all at Yerkes, where the geese still gather in the autumn in Lake Geneva before setting south for the winter as he must often have wished to do, the astronomically-minded person still finds his spirit everywhere. Even more so is this true of the skies, where he is memorialized in Barnard's Star, Barnard's Loop, and Barnard's dark nebulae. If you seek his monument, look around you.

His life continues to inspire as we carriers of his torch continue to follow the paths he trod among the stars. The Immortal Fire still burns. It is my hope that this paperback will find its way into the hands of many an amateur astronomer and inspire them to do as he did and to "follow their bliss."

William Sheehan
Willmar, Minnesota, U.S.A.
May 6, 2007

Key to abbreviations in notes

Journals

AA *Astronomy and Astro-Physics*
AJ *Astronomical Journal*
AN *Astronomische Nachrichten*
Ap J *Astrophysical Journal*
BMNAS *Biographical Memoirs of the National Academy of Science*
JAH *Journal for the History of Astronomy*
JTAS *Journal of the Tennessee Academy of Science*
MNRAS *Monthly Notices of the Royal Astronomical Society*
PA *Popular Astronomy*
PASP *Publications of the Astronomical Society of the Pacific*
PLO *Publications of the Lick Observatory*
PYO *Publications of the Yerkes Observatory*
SM *Sidereal Messenger*

Individuals

EMA Eugène M. Antoniadi
RGA Robert G. Aitken
WSA Walter S. Adams
EEB Edward Emerson Barnard
JAB John A. Brashear
SWB Sherburne Wesley Burnham
HC Henry Crew
MRC Mary R. Calvert
SCC Seth Carlo Chandler, Jr.
WWC William Wallace Campbell
GD George Davidson
EBF Edwin Brant Frost
RSF Richard S. Floyd
ESH Edward S. Holden
GEH George Ellery Hale
WRH William Rainey Harper

JEK James E. Keeler
SN Simon Newcomb
ECP Edward C. Pickering
TGP Timothy Guy Phelps
JMS John M. Schaeberle
LS Lewis Swift
HHT H. H. Turner
CAY Charles A. Young

Archival Sources

BL Bancroft Library, University of California, Berkeley
George Davidson Papers
CIF Carnegie Institution Files, Washington, DC
DCL Dartmouth College Library, Hanover, New Hampshire
Charles A. Young Papers
HCO Records of the Harvard College Observatory, Harvard University
Archives, Cambridge, Massachusetts
HHL Manuscripts Department, Henry E. Huntington Library, San Marino California
George Ellery Hale Collection
Mt. Wilson Observatory Collection
LC Manuscript Division, Library of Congress
Simon Newcomb Papers
LO Lick Observatory Plate Vault, Mt. Hamilton, California
LOA Lowell Observatory Archives, Flagstaff, Arizona
RAS Royal Astronomical Society Archives, Burlington House, London
RGO Royal Greenwich Observatory Archives, University Library, Cambridge
RL Special Collections, Regenstein Library, University of Chicago
Edward Emerson Barnard Observing Books
William Rainey Harper Papers
SLO Mary Lea Shane Archives of Lick Observatory, University Library, University of
California, Santa Cruz
USNO US Naval Observatory Archives, Washington, DC
VUA Special Collections, Vanderbilt University Library, Nashville, Tennessee
Edward Emerson Barnard Papers
VVO Van Vleck Observatory Archives, Wesleyan University, Middletown, Connecticut
Frederick Slocum Papers
YOA Yerkes Observatory Archives, Williams Bay, Wisconsin

1

Through rugged ways

1

One of the things that has always been most appealing about Edward Emerson Barnard is the fact that he was able to overcome so much childhood hardship. Growing up as 'the poorest of poor boys in a 3rd rate Southern town . . . just after the war which ruined the South,' as E. S. Holden later described him,[1] he was determined to make good, and in so doing became an outstanding example of what might be called the 'Horatio Alger syndrome,' after the nineteenth century Unitarian minister and author of the *Ragged Dick, Luck and Pluck,* and *Tattered Tom* series of books for boys – tales of poor boys who rose up from harsh circumstances entirely through exceptional industry and ability. All contained variants on the familiar pull-oneself-up-by-one's-own-bootstraps theme. In overcoming the disadvantages of his early life, Barnard became a sterling example of the characteristically American mythos of the self-made man. The mythos ties in with the American ideal that, despite all appearances to the contrary, everyone is in some sense 'born equal'; that whatever one's material circumstances, so long as one is clever enough or industrious enough, one can rise above them.

Olin Landreth, who knew Barnard at the beginning of his career at Vanderbilt, described him as 'a man who, without other aids than those which his own character, enthusiasm and energy commanded for him, rose from an humble origin, overcame circumstantial difficulties, and made the astronomical world his debtor.'[2] A later colleague, J. A. Parkhurst, saw him as 'an example of the possibilities which America offers its youth,' and added: 'Nowhere else in the world could a boy rise from such restricted and difficult conditions to such complete and abundant manhood.'[3] Cases like Barnard's prove the point – partially. Yet the question remains: How is it that some individuals, despite such hardship, manage to rise above it?

Barnard was born into the slums of Nashville, Tennessee, on December 16, 1857.[4] 'It would be impossible,' he wrote toward the end of his life, 'even if I were in Nashville, to more than roughly locate my birthplace':

> When a boy I was shown an old brick house which I was told by my mother was my birthplace. This was on what was known as 'Rolling Mill Hill,' so called, I believe, from the fact that there were rolling mills near. This was up on the high bluff close north of the old city reservoir, overlooking the Cumberland River. This is as near as I can come to the location of my birthplace. The house must have long ago perished.[5]

The second son of Reuben and Elizabeth Jane (née Haywood) Barnard, who at the beginning of the year 1857 were living in Cincinnati, Edward never saw his father. Reuben died three months before he was born, and nothing more is known about him. After Reuben's death, Elizabeth, originally a native of Kentucky,[6] moved to Nashville, possibly in the hope of finding work – she was skilled in modeling flowers in wax. At the time of Edward's birth, she was already forty-two years old. One gathers that she and her two sons were quite alone in Nashville; there seem to have been no relations near.

Edward's older brother, Charles, born in 1854, is almost as obscure a figure as Reuben, but what little is known of him makes abundantly clear that he did not share his brother's intellectual endowment, and there is even a faint suspicion that he may have been feeble-minded. In all the accounts Edward wrote of his early years, Charles garners nary a mention.

One guesses that Elizabeth must have been partial to her gifted son, and it seems likely that she implanted in him the desire – perhaps the expectation – of doing something extraordinary with himself. Certainly there is a belief among many peoples that a male child born after his father's death, a posthumous, is often remarkable, and Edward himself made a point of recording, at the top of a later manuscript sketch of his early life, the fact that he had been born with a caul, another supposed indication of future success which he can only have learned from his mother.[7]

Though Elizabeth's health began to fail when Edward was still very young – from the time he was about twelve she was a 'confirmed invalid,' and by then her mind seems to have been failing also – he fondly remembered that in better days she had been 'an educated woman and an artist.' The unusual middle name she chose for him, after the New England writer Ralph Waldo Emerson, attests to her taste for literature, and she taught him when he was very young to read from the Bible. Other books that were among his youthful favorites were a volume of the writings of the Jewish historian Josephus, which 'had a fascination equalled only by the story of Robinson Crusoe,' the *Arabian Nights*, and *Nicholas Nickleby*. And, he later recalled, 'an old volume of Scientific Discovery and Invention which I read among my first books perhaps had some influence in turning my mind to the wonders of science.'[8]

2

The Nashville into which Edward Barnard was born in the 1850s was a place of culture and cultivation. It was then the largest and most important American city south of the Ohio River with the exception of New Orleans. It had grown between the turn of the century and the eve of the Civil War from a frontier town consisting of some sixty or eighty families to a city of nearly 17 000, of whom some 4000 were black slaves. Wide turnpikes and busy railroad lines radiated from it, and the coiling Cumberland, draining Cumberland Gap on its way to the Mississippi, was crowded with river traffic carrying goods from the city's foundries, ironworks, and small manufactures to Memphis and New Orleans. With its Female Academy and Nashville University, it was already a center of education, and its free public school system was the first in the South – Barnard himself was briefly numbered among its pupils, though because of his mother's failing health he was able to attend for only two months. There were a theater and an opera, and one observer commented that there was as much fashion in Nashville as in New York.[9]

For all its dazzling surfaces, Nashville in the 1850s was a troubled city. The decade had begun with the Convention of Southern States, which convened in the Nashville Capitol in June 1850 with the express purpose of forming a southern sectional party to protest the attempt then being made in Washington by President Zachary Taylor's administration to exclude southern men with their slaves from the territories recently ceded by Mexico. Admittedly there was a far from unanimous viewpoint at the Convention, some seeing the movement as nothing less than a treasonable scheme concocted by the leaders of South Carolina and Mississippi to secede from the Union and to form their own southern confederacy. The Convention adjourned without reaching any definite conclusions, but its members planned to reconvene in November if Congress failed to respond to the demands of the southern states. The Convention never reconvened; Taylor died suddenly, and his successor, Millard Fillmore, was eager to work out with Congress compromise measures in which popular sovereignty was proposed as a way of deciding the issue of slavery in the territories, thereby averting a rupture at least for the time being.

The Compromise of 1850 unravelled with the election of Abraham Lincoln in 1860, when ten southern states withdrew from the Union. In April 1861, Fort Sumter fell. Tennessee threw in its lot with the seceded states in June, and almost at once became a theater of war. General Albert Sidney Johnston was placed in command of the Confederate defense line that ran from the mountains around Cumberland Gap to the Mississippi, and Fort Donelson was built to defend the vital supply lines where they crossed the Cumberland. After a prolonged siege, the fort surrendered to Ulysses S. Grant on February 16, 1862, with the capture of 15 000 men, thereby opening Nashville itself to the invading Union troops. Confederate soldiers cut the cables and burned the railroad bridge as they abandoned the city. Within two days, Nashville had fallen into Union hands, and it remained under Union occupation for the duration of the war. The Capitol was surrounded with temporary stockades made of cotton-bales, and an extensive system of fortifications was built up throughout the city.

3

The dreadful privations that fell upon the South during and just following the war only intensified the misery of the impoverished Barnard family. Barnard recalled many years later that his early youth was 'so sad and bitter that even now I cannot look back to it without a shudder.'[10]

During the war, the Barnards lived briefly at a place near the old toll-gate on the Lebanon Pike south of the city, and later moved to the bank of the Cumberland a mile or so above the old waterworks. During these terrible years, the family was often on the verge of starvation. Once, as Barnard recounted to his niece Mary Calvert, a steamer loaded with provisions was sunk as it was trying to get into the city. He and many others, gaunt with hunger, watched sorrowfully as the provisions from the steamer floated down the river. Some people put out in small boats to try to salvage what they could, and he, diving into the Cumberland, managed to swim out and rescue part of a box of crackers, which his mother whipped into batter and made into cakes.[11]

Even his earliest memories of the stars were associated with the war and tinged with the intense loneliness and sadness of that period of his life. Thus he wrote in a later memoir: 'When I was very small I saw a comet; and I have a vague remembrance that the neighbors spoke of this comet as having something to do with the terrible war that was then desolating the South.'[12] Just which comet may have captured Barnard's early attention is a matter of some interest, given his later interest in these bodies. Indeed, there were several brilliant wayfarers which swept across the sky during the late 1850s and early 1860s. One was the splendid Donati's comet of 1858, with a 50° long tail that curved like a scimitar. Barnard, of course, was too young to have seen it. The one he remembered was probably the Great Comet of 1861, discovered by the Australian farmer and amateur astronomer John Tebbutt from Windsor, New South Wales. Northern hemisphere observers did not catch sight of it until June 30, when it emerged spectacularly out of the evening twilight, its head brighter than Jupiter and its tail stretching across 105°. For the next several weeks it remained a splendid object in the evening skies.

During those years when the war was raging, Barnard used to lie out in the open air on summer nights, flat on his back in an old wagon bed, watching the stars, a pastime which he afterwards recalled 'helped to soften the sadness of my childhood.'[13] Among the stars that he viewed from that wagon bed he recalled 'a very bright one, which during the summer months shone directly overhead in the early hours of the evening.'[14] It would be many years before he learned its name – Vega.

On Barnard's seventh birthday, December 16, 1864, the fighting between the Union and Confederate armies came within only a few miles of Nashville. Barnard himself heard the cannon booming in the distance. General George Thomas's Union forces routed the Confederate troops under General John Hood. Hood's troops had already been decimated at the battle of Franklin two weeks earlier when Hood had ordered a desperate frontal assault on the Union line, 'sending 18 000 men forward through the haze of an Indian-summer afternoon in an attack as spectacular, and as hopeless, as Pickett's famous charge at Gettysburg,' in the words of Bruce Catton.[15] In a few hours' time, Hood's army lost more than 6000 men. These actions were part of a final rearguard attempt on the part of the Confederacy to regain Nashville. For all practical purposes the war was over, with Appomattox looming only a few months away.

Unfortunately, the Barnards' situation did not improve much during the early years of Reconstruction. Nashville remained under Union occupation, serving as a major supply base for the Union forces in Tennessee (the last troops were not removed until 1877). As with the entire South, the city was economically depressed, and even those who had been well off before the war found conditions difficult. Added to the economic hardships was the devastating cholera epidemic which swept the city in 1866. Eight hundred people died, and young Ed Barnard came close to being one of them.

4

Not long after the cholera epidemic, in the latter part of 1866, Elizabeth Barnard happened across 'a photograph bearing the name of a photographer in Nashville which recalled a young man she had befriended in her better days long ago in Ohio. She called upon him and found that he was really the same person.'[16] The man was John H. Van Stavoren, who since the beginning of the Civil War had owned a photograph gallery in Nashville. It was located in the second and third stories of a brick building (long since demolished) at the southeast corner of Union and Cherry Streets (now Union Street and Fourth Avenue). Van Stavoren was looking for a boy to run errands for him and to guide an immense solar camera that he had mounted on the studio roof. It says much that Elizabeth decided to bypass her older son and instead suggested Edward for the job.[17]

Van Stavoren's solar camera was one of the largest such cameras in existence at the time and had been designed for making life-sized enlargements on silvered paper from negatives. Because of the relative insensitivity of the silvered paper then in use, an intense solar beam was required, produced by concentrating the sunlight with a huge condensing lens. An artist could also paint from the image formed by the camera, and Van Stavoren often used it for portrait-painting.

In form the camera was a large rectangular box. Attached to the front and projecting just beyond the edge was fastened a piece of sheet metal, through which a small hole had been bored. As the Sun's rays passed through this hole, a bright spot was formed on a screen several feet away. By holding this bright spot in a fixed position on the screen, the image formed by the condensing lens remained steady for the painter. However, because of the Sun's steady movement across the heavens, the camera had to be panned slowly in order to keep the bright spot fixed on the screen. This was done by means of a pair of hand wheels, one turning the camera to the west, the other tilting it up or down. Because of the camera's great size, a boy could reach the hand wheels only by standing on top of a stepladder.

Barnard later recalled that this instrument had for sometime been a source of 'much apprehension and annoyance' to Van Stavoren:

> It was necessary to keep it moving precisely with the motion of the sun. If it deviated much from the solar motion the intense heat collected by the great condensing lens and brought to a focus would touch the wooden part of the instrument and set it and the house on fire. The boys that had previously attended the great camera had nearly burned the house up several times by going to sleep in the warm sunshine. 'But will your boy keep awake?' asked the photographer after explaining his difficulties. 'My son will not go to sleep!' replied the mother with

Edward Emerson Barnard as he looked when he began work at J. H. Van Stavoren's photograph gallery, 1866. Vanderbilt Photo Archives

Edward Emerson Barnard about 1867. Vanderbilt Photo Archives

confidence. And he never did go to sleep while on duty. Through summer's heat and winter's cold he stood upon the roof of that house and kept the great instrument directed to the sun. It was sleepy work and required great patience and endurance for one so young, and at this distant day he realizes that this training doubtless developed those qualities – patience, care and endurance so necessary to an astronomer's success.[18]

The giant camera was called *Jupiter*, and Barnard later recounted Van Stavoren's reasons for choosing that name which was afterwards to be so fateful in his own life:

There are coincidences in every person's life some of which are ordinary and attract no attention and others that have an important bearing upon one's future. To those who are familiar with my astronomical work, the name that had been bestowed upon this instrument by its owner will have a peculiar significance. The

This photograph, taken about 1870, shows the *Jupiter* camera on the roof of J. H. Van Stavoren's photograph gallery, Nashville. J. H. van Stavoren on left, James W. Braid in camera. Young Edward Emerson Barnard's first job was to keep this camera pointed toward the Sun by turning the large wheels on its side. Vanderbilt Photo Archives

photographer was certainly not learned in astronomy. He knew there were planets and he knew the names of some of them. He liked to name everything he had. Since this great instrument's main duty was to do obeisance to the sun, what name more appropriate than that of one of the planets? It was a mere selection at random, and so the photographer had incidently named it 'Jupiter'! Little thought the poor child when he was put in charge of 'Jupiter' that his own name would forever be linked with that of the mighty planet which it represented![19]

Barnard started working as an attendant and satellite to the great *Jupiter* when he was not quite nine. He was to remain at the Photograph Gallery until he was twenty-five. No doubt the evils of child labor were great and many – one thinks of Dickens, who was taken out of school at eleven to paste labels in a London blacking warehouse and for whom the experience, though brief, remained a painful memory for the rest of his life. Barnard's job as a human driving clock was undoubtedly the epitome of tedium, but on the positive side it gave him a chance to rise from crushing poverty by providing a measure of economic security for himself and his invalid mother, who by this time was entirely dependent upon his income. Moreover, the experience he gained over the years – at first in guiding the solar camera, later in photographic techniques – prepared him for his then unimagined future as a photographer of the heavens. Not least important, he was brought into contact with a number of stimulating characters who helped to widen his intellectual horizons.

The stepladder on which the small boy had to stand in order to reach the hand wheels of the solar camera was, indeed, symbolic. That stepladder furnished the first few steps by means of which he would begin his life's dizzying climb. However humble the job to which he was assigned, he took great pride in having done it well, especially where so many others had failed. Indeed, psychologists have suggested that children who thrive despite hardship are often set apart from their fellows by having such a success that gives them confidence and builds their self-esteem.

Barnard, like some saints, had the ability to win grace from hardship. 'The harsh experiences of his earlier life coupled with his mother's molding of his character developed in him a strong sense of purpose and conscientiousness which became apparent in his first job,'[20] wrote Vanderbilt astronomer Robert Hardie. And there is truth to Barnard's own notion that his work with the solar camera developed those qualities, 'patience, care and endurance,' which contributed to his later success as an astronomer.

5

Having somehow to keep his mind from becoming completely anesthetized by the tedious work of guiding the solar camera, young Barnard, as he cranked the hand wheels, diverted himself by analyzing the motions of the Sun which he so faithfully followed. 'The camera was mounted so that two motions were necessary to make it follow the sun,' he later wrote in reconstructing his train of thought. '. . . The process whereby the sun was followed, was to move the instrument west with the one wheel and vertically with the other':

> I knew that the sun attained its greatest altitude at noon. I amused myself by determining when noon had arrived by the fact that then I ceased to raise the instrument to follow the sun. I also established a noon mark by the aid of the shadow of a chimney. But I was surprised soon to find that neither of these signs agreed for any length of time with the noon ringing of the bell in a Catholic church [St. Mary's] near by – sometimes the noon, as indicated by the highest altitude of the sun or by his noon mark, was too soon and at other times too late according to the church bell – the difference sometimes amounting to a considerable fraction of an hour. This set

me to thinking and wondering, but it was many years afterwards before I found out the explanation of this singular phenomenon, which was due to the equation of time.[21]

The explanation for what Barnard had found has to do with the fact that the Earth travels in an elliptical orbit, and thus the apparent motion of the Sun is not constant: it is faster during the northern hemisphere's winter as the Earth is then nearest to the Sun, slower in the summer when the Earth is farthest away. As it would be impracticable to have a clock keep time with such an irregular motion, a *mean* Sun is assumed, whose motion is the average of the true Sun's motion during the whole year. It is with respect to the motion of this mean Sun that clocks keep time (hence *mean time*). The *equation of time* is the difference between the time that the true Sun crosses the meridian and the time that the mean Sun crosses it.

6

Barnard saw his first solar eclipse on August 7, 1869. The path of totality swept across North America from Alaska to North Carolina. Though not total at Nashville, Barnard noted that 'the sun was so nearly hidden that the spectacle presented some of the awe and sublimity of the total phase,' and the event increased his 'wonder at the phenomena of nature.'[22]

At about this time, the Barnard family moved yet again, to the junction of Lebanon Pike and Wharf Avenue in South Nashville – an area known as 'Varmint Town, from the toughness of some of its people.'[23] Barnard was putting in long hours at the Photograph Gallery, and it was often dark when he walked home. In a letter written in 1892, Barnard described to Joseph S. Carels something of his quiet desperation and lonely struggle, also his appreciation for the unexpected kindness shown to him by that man:

> Away back yonder . . . when I was small and ragged and sick and desolate just at the close of the war, when even those who had rolled in wealth but a few years before were struggling for subsistence, and few there were who could bestow even a kind word – so terrible had been the desolation and its effect on the people – in these times when I used to trudge home some two miles every night from work, timid and frightened, I frequently would meet a gentleman who always had a nod and a smile for me – in bad or cold weather he always wore a cloak. Sometimes he would stop me and ask how I was getting on but he never passed me without a recognition. I did not know who that man was, but his smile lighted up my heart, for years he never failed to greet me. Soon I learned to my awe that he was assistant Postmaster! Had he been President his position would not have appeared higher and more exalted to me – and that he should notice me and should stop to speak to me – I could not understand it, and I cannot understand it to this day – unless it was indeed an inborn desire in him to sympathise with the friendless and wretched for friendless and wretched I was in those days if any one was ever friendless and wretched.[24]

During this sad time Barnard had a sore on his face that refused to heal and afterwards left a prominent scar on his right jaw. Whenever Carels stopped Barnard, he would always ask how that sore was coming along. To Barnard's niece Mary Calvert, this episode always 'seemed eloquent of the loneliness of his life in those days.'[25]

In addition to the friendly greeting of Carels, Barnard derived some cheer during his lonely walks from his faithful friends the stars. 'I often noticed in my long walks homeward in the early night,' he recalled, 'an ordinary yellowish star which, to my surprise, seemed to be slowly moving eastward among the other stars':

> This attracted my attention because in all the time I had noticed the stars, though they came and went with the seasons, they seemed all to keep to their same relative positions. This one must be quite different from the others though it resembled them . . . I watched it night after night and saw that though it moved eastward with reference to the other stars, it also partook of their general drift westward and was finally lost with them in the rays of the sun. In later years when I had become more familiar with astronomy it occurred to me to look up this moving star, and I found then that what had attracted my boyish attention was the wonderful ringed world of Saturn.[26]

7

By the late 1860s, Van Stavoren was struggling financially, and he finally went out of business in 1871. To liquidate his debts, his property was sold to the People's Bank in Nashville, and subsequently the gallery and its equipment were purchased by Rodney Poole, who for some time previously had held a part interest in the property. Poole continued the business under the name 'Poole's Photograph Gallery.' The *Jupiter* camera was dismantled, but Barnard and all the other employees who had worked for Van Stavoren were retained. Not only did he take over Van Stavoren's specialties of 'life size portraits in ink or crayon, and pastel and water color work,'[27] but Poole also introduced a new department specializing in outdoor photography which was then becoming fashionable.

At about this time, the Barnard family moved to a place on the Franklin Pike near Forts Moulton and Negley. They were still there in 1873, when another cholera epidemic swept Nashville and especially, Barnard recalled, 'that particular region a little beyond which was a colored settlement known as New Bethel, where the cholera made dreadful havoc among the colored people.'[28]

Young Barnard's chief ambition during this time was to be a sign painter. 'I was constantly trying to make letters like those on the signs I saw when on my way to and from work. In this I became quite proficient, and I frequently painted signs for the gallery in which I worked.'[29] This led in time to a still greater ambition – to become an artist. He had inherited some of his mother's artistic talent, and some of his sketches from the period show considerable proficiency. He especially admired the work of the cartoonist Thomas Nast, who was then on the staff of *Harper's Weekly* and at the height of his fame. Taking Nast as his model, Barnard 'sketched until I thought I could do pretty well in this line.'[30]

Fortunately for astronomy, his ambition to become an artist lasted only until 1875, when Poole hired Peter Ross Calvert, a young immigrant from Yorkshire, to work as a retoucher and colorist at the Gallery. Calvert had studied art at South Kensington and Paris, and at once Barnard recognized his master: 'In comparing my work with that of the young Englishman, I came to the conclusion that I was not born to be an artist.'[31]

By now, Barnard's mother was suffering from such poor health that she had to give up any attempt to have a home, and Barnard moved into a room with board for $4.00 a week in the top story of the Hotel St. Charles, in downtown Nashville. He was now completely on his own. Though little is known about this period of his life, he undoubtedly spent many evenings after work buried in study of whatever second-hand books he could get his hands on.

Despite his gradual and still largely unconscious awakening to the wonders of the sky overhead, he had as yet no reliable source of information about astronomy. Finally a book came his way, entirely by chance. He would later believe that it had changed his life:

> One night, while poring over some old books on mathematics which I had purchased second-hand and perhaps not wisely, a young man came to my room . . . We had been children together, but he was a born thief. As a boy he stole, and when he got older the law often laid its hands upon him. On several occasions I had helped him out from my meager earnings, for my sympathies were easily worked upon. At one time I had paid his fine when a policeman brought him around where I was at work. On this night in particular I was in no mood to be gracious; for he had come to borrow money from me, which I knew from previous experience would never come back. As security for the return of the money he had brought a large book. This I refused to look at; and finally, to get rid of him, I gave him two dollars (which was the amount asked for). I never saw him or the money again. Shortly after he had gone I noticed he had left the book lying upon my table. I felt very angry, because the money was a large amount to me then, and it was some time before I would open the book.[32]

When Barnard finally did open the book, he discovered that the first part consisted of sermons on covetousness, with horrible accounts given of the lives of misers. One of these was the Englishman Edward Nokes, who at his death was found to be possessed of between five and six thousand pounds. 'In order to save the expense of shaving he would encourage the dirt to gather on his face to hide in some measure this defect. He never suffered his shirt to be washed in water; but, after wearing it till it became intolerably black, he used to wash it in wine to save the expense of soap. I must have groaned aloud when I saw this,' Barnard recalled.[33] But fortunately, there was more attractive matter within the book's covers. The Rev. Thomas Dick, who was the author of these sermons, was also a popular astronomy writer. The book left by the thief was the second volume of an edition of Dick's *Works*,[34] and contained, in addition to the essay on covetousness, a number of his astronomical writings: his *Celestial Scenery; Or the Wonders of the Planetary System Displayed, The Sidereal Heavens and Other Subjects Connected With Astronomy, The Practical Astronomer*, and *The Solar System With Moral and Religious Reflections in Reference to the Wonders Therein Displayed*. Of greatest immediate interest was the fact that the volume contained a set of rudimentary star charts, and as soon as he found them Barnard set out to compare them with the patch of sky visible from the open window of his small apartment:

> In less than an hour I had learned the names of a number of my old friends; for there was Vega and the stars in the Cross of Cygnus and Altair and others that I had known from childhood. This was my first intelligent glimpse into astronomy. It is to be hoped that my sins may be forgiven me for never having sought out the rightful owner of that book in all these long years.[35]

Young Barnard studied the book, as his later friend and colleague S. W. Burnham put it, 'with great avidity, and it awakened a thirst for astronomical knowledge which . . . never ceased to be controlling . . . He had never seen either a telescope or an observatory. All he knew of the literature of the subject was found in this old book.'[36]

Possibly the tendency of the 'Christian philosopher of Dundee' to blend religion and astronomy made his book more appealing to Barnard, who in Nashville was immersed in Southern fundamentalist Christianity and who even after leaving Nashville continued to resonate to the religious implications of the starry heavens. But whatever the source of its inspiration to him, the effect on him of Dick's *Works* was immediate and profound. Barnard himself dated his informed interest in astronomy to his receipt of this volume (still preserved in the Yerkes Observatory library), and afterwards pasted the following typewritten memorandum inside the front cover: 'This was the first book in the astronomical library of Edward E. Barnard into whose hands it accidentally came as an unsought pledge of a borrower in 1876. It gave him his first chart of the stars and was an important factor in determining his future scientific career.'

At the same time it must be allowed that there had been significant foreshadowings of this interest. It had begun in the wonder of childhood, when from his wagon bed he had looked up at the summer stars and viewed the Great Comet flourishing its mighty tail. Without taking any of the credit away from Dick, one can hardly avoid the conclusion that Barnard's mind had long been charged for just such a stimulus. Had he not chanced across this volume of Dick's *Works* when he did, some other book sooner or later would have come into his hands, with exactly the same result.

8

That the former aspirant to sign painting now had a new ruling passion soon became well known to his associates at Poole's, and he was fortunate to find a helpful mentor in James W. Braid. Braid was Poole's chief photographer and a man of exceptional technical abilities. His importance in Barnard's life cannot be underestimated, as

View Wagon operated in the 1870s by James W. Braid for Poole's Photographic Gallery. Barnard, pictured here, was Braid's assistant in his outdoor photography. Vanderbilt Photo Archives

Barnard himself was the first to insist. 'To Mr. J. W. Braid,' he wrote near the end of his life, '. . . I owe every obligation. . . . To those who do not know the conditions of my early life, it is not possible for me to make clear how helpful he was to me.'[37]

Braid was ten years older than Barnard. Though originally from Scotland, he had come as a boy to New York City. For a time he was employed in making telegraphic equipment, and later worked as a photographic chemist for the New York firm of Gurney & Sons, who recommended him to Van Stavoren. He arrived in Nashville in the spring of 1870, and became Van Stavoren's chief photographer. After Van Stavoren was bought out by Poole, Braid continued to be chief photographer and was placed in charge of the outdoor photography department which Poole was trying to develop. Barnard, who meanwhile had been promoted from human driving clock to printer, became his assistant. For several years the two young men went about Nashville with a portable dark room mounted on a one-horse cart known as the 'View Wagon,' and among their first subjects were the grounds and structures of the new Vanderbilt University which was then abuilding.

Braid would eventually achieve local fame for his electrical apparatus and experiments, in which his interest dated back at least to his association with the telegraph firm in New York City and only intensified after his arrival in Nashville. But at least for a time his skill was, as Barnard put it, 'turned . . . from induction coils, etc., to the making of telescope accessories.'[38]

Indeed, Braid put together for Barnard his first telescope – a very rudimentary affair, constructed from a simple tube and a one-inch spyglass lens Braid happened across in the street on a trip to the Union army barracks in north Nashville. This first telescope was later described by Barnard as 'a paper tube and lenses that looked as if they had been chipped out of a tumbler by an Indian in the days of the Mound Builders. Yet it filled my soul with enthusiasm when I detected the larger lunar mountains and craters, and caught a glimpse of one of the moons of Jupiter.'[39] He supplemented his views with this instrument with occasional looks through a somewhat better telescope owned by a street showman who regularly came to town, 'whenever nickels were sufficiently plentiful to warrant such a dissipation.'[40]

But these primitive telescopes only whetted Barnard's appetite for a better instrument, and aided by the descriptions in Dick's *Practical Astronomer*, Braid helped him to make such an instrument. Stopping one day at the shop of Charles Schott, Sr., a German-born instrument-maker, Braid, by his own account,

> . . . noticed the tube of an old ship's spyglass hanging on the wall. It had no lenses. [Schott] had bought it for old brass and intended to cut it up. I made an offer to buy it, and when I told him I wanted it for Ed. Barnard he let me have it for two dollars. It had two draws to it and was about two and one-half inches in diameter, allowing for a two and one-fourth inch optic glass of thirty-four inches focus. We sent to [Queen] & Co., of Philadelphia, and got an object glass of thirty-two inches focus which we fitted into the old brass tube. We made an eyepiece out of the wreck of an old microscope, which gave a power of about thirty-eight. A simple altazimuth mounting was made. Barnard had a good tripod which had been a surveyor's instrument stand. It answered very well as a support for the telescope.[41]

Braid's interest was always chiefly in the making of instruments, Barnard's in actual observation. Before long Braid had returned to his experiments with electrical apparatus, so that Barnard had the $2\frac{1}{4}$-inch telescope to himself, usually mounting it on the platform on the roof of Poole's where the *Jupiter* camera had formerly stood. The first object toward which he turned his telescope was the planet Venus, whose crescent phase, he later noted, 'made a more profound and pleasing impression' than his celebrated discovery of the fifth satellite of Jupiter.[42] The period in 1876 during which Venus would have shown a crescent phase in the evening sky extended between its greatest elongation east of the Sun on May 4, 1876 and inferior conjunction in early July. Thus the date at which Barnard began observing can be fixed quite exactly to this interval.

Other showcases of the sky which attracted his early attention were Jupiter and the four moons discovered by Galileo in January 1610, the craters and mountains of the Moon, and the double star Mizar. The Orion nebula fascinated him with its 'great wisps of soft light, strangely resembling in form the wings of a bat.'[43] He was beside himself with excitement. Braid put it in few words, 'This simple telescope gave Barnard more pleasure than anything else in his life,' to which Barnard's later colleague and friend S. W. Burnham added: 'Such excitement comes but once in a lifetime, although the enthusiasm and interest in the subject may never be abated.'[44]

From the most inauspicious circumstances, then, in which he could only grope for knowledge of the heavens, Barnard at eighteen had received his first real taste of the pleasures of the telescope. The career of the greatest astronomical observer since William Herschel had begun.

1   ESH to E. B. Knobel, June 16, 1893; YOA
2   O. H. Landreth, 'Barnard at Vanderbilt University Observatory,' *JTAS*, 3:1 (1928), 15
3   J. A. Parkhurst, 'Edward Emerson Barnard,' *Journal of the Royal Astronomical Society of Canada*, **17** (1923), 97
4   Curiously, the date of birth given on his tombstone in Mt. Olivet Cemetery–October 16, 1857–is incorrect.
5   EEB, information furnished to John Trotwood Moore, Tennessee State Historical Commission, April or May 1921; VUA
6   According to information from the US Federal census of 1860, which was very kindly called to my attention by Edwin W. Coles.
7   ibid.
8   EEB, autobiographical sketch in VUA, written by Barnard probably while at Lick Observatory, referred to henceforth as 'autobiographical sketch.' In the interests of clarity, I have changed the account from third to first person.
9   Henry McRaven, *Nashville: 'Athens of the South'* (Chapel Hill, Scheer and Jervis, 1949), p. 87
10 EEB, 'autobiographical sketch.' Mary R. Calvert, Barnard's niece, later recalled: 'He never talked much about those days. To one person who suggested that he should write an account of his boyhood he replied, "I couldn't do that. It was too sad." MRC to RGA, November 11, 1927; SLO
11 MRC to RGA, November 11, 1927; SLO

12 E. E. Barnard, 'A Few Unscientific Experiences of an Astronomer,' *Vanderbilt University Quarterly*, 8 (1908), 273-288:275. This was the text of an address delivered before the Alumni Association of Vanderbilt University on June 17, 1907. This text, except for a few prefatory remarks, was also published in the Nashville *Christian Advocate*, July 5, 1907.

13 Barnard, 'Experiences,' p. 275

14 EEB, 'autobiographical sketch'

15 Bruce Catton, *The Civil War* (New York, The Fairfax Press, 1980), p. 260

16 EEB, 'autobiographical sketch'

17 The two brothers must have presented the most remarkable study in contrasts. To judge from the extant correspondence in VUA, Charles remained to the end almost an illiterate, a very simple man who grasped only dimly what his younger brother was achieving. The marked difference between them must have been obvious early on—even in childhood they seem to have thought it best to go their separate ways, for there is not a shred of evidence to suggest that they were ever very close, and the responsibility for supporting their mother seems to have fallen entirely on young Edward's shoulders.

18 EEB, 'autobiographical sketch'

19 ibid.

20 Robert Hardie, 'The Early Life of E.E. Barnard,' *Leaflets of the Astronomical Society of the Pacific*, Nos. 415 and 416 (January 1964), p. 2

21 EEB, 'autobiographical sketch.' St. Mary's Catholic Church, at the corner of Fifth Street and Charlotte, was built in 1847 and is still standing.

22 EEB, 'autobiographical sketch'

23 EEB, information sent to John Trentwood Moore; VUA

24 EEB to Joseph S. Carels, September 21, 1892; VUA

25 MRC to RGA, November 11, 1927; SLO

26 EEB, information sent to John Trentwood Moore; VUA

27 William Waller, ed., *Nashville in the 1890s* (Nashville, Vanderbilt University Press, 1970), p. 52

28 ibid.

29 EEB, 'autobiographical sketch'

30 ibid.

31 ibid.

32 Barnard, 'Experiences,' pp. 275-276

33 ibid., p. 276

34 Rev. Thomas Dick, *Works*, vol. 2 (St. Louis, Missouri, Edward & Bushnell, 1854)

35 Barnard, 'Experiences,' p. 276

36 S. W. Burnham, 'Early Life of E.E. Barnard,' *PA*, 1 (1894), 193

37 E. E. Barnard, 'Letter from Mr. Barnard regarding Mr. Braid,' *JTAS*, 3:1 (1928), 10-11

38 ibid., p. 10

39 E. E. Barnard, *Artisan*, August 10, 1883

40 S. W. Burnham, 'Early Life of E.E. Barnard,' p. 194

41 J. W. Braid, 'First Employment of Barnard; his First Telescope,' *JTAS*, 3:1 (1928), 9. Braid's account had McQueen for Queen; the name of the company from which he and

Barnard ordered the lens was James W. Queen & Co., of 924 Chestnut, Philadelphia.

42   Newspaper clipping dated May 1, 1893 from *The Standard Union* of Brooklyn, New York. VUA

43   EEB, 'Mars; His Moons & His Heavens,' unpublished MS in VUA

44   Braid, op. cit., p. 9; Burnham, 'Early Life of E. E. Barnard,' p. 194

2

Ardent and faithful work with a telescope

1

Soon after the completion of the $2\frac{1}{4}$-inch refractor, Barnard learned of the existence of another telescope in Nashville, owned by the wealthy Acklen family, whose Belmont Mansion was celebrated as the most magnificent of the antebellum mansions that were built in Nashville during the 1850s. Designed by an Italian architect, it was the home of Colonel Joseph Alexander Smith Acklen and his wife Adelicia Hayes Acklen.[1] The estate consisted of two land grants of 640 acres each, of which nearly 200 acres were landscaped with extensive lawns and gardens, greenhouses, and a private zoo. According to a contemporary description, 'The mansion faced both the north and south, with its white walls and columns ornamented with lace-like iron balconies and window guards. Entrances to the estate were from the east and west, with winding drives bordered by cedars and magnolias leading to the main house.'[2] The east wing of the mansion contained bowling alleys and billiard rooms, the west wing boasted the largest private art collection in the South – hundreds of paintings, engravings, prints, and statues which Mrs. Acklen had assembled during numerous trips abroad. Of the mansion's many treasures, Barnard coveted but one – its telescope.

Barnard was eager to examine the instrument and hoped to buy it if it proved satisfactory. Only too conscious, however, of his lack of formal education and social polish, he did not believe himself capable of writing an acceptable letter of introduction. He therefore turned for help to his better educated friend Peter Calvert, who would know how to frame the sort of letter with which to approach the high and mighty Acklens.

Calvert, a devout Baptist, was willing to write the letter, but only on condition that Barnard agree to come with him to Sunday School. Calvert later explained his motive: 'Admiring his scientific love of the stars, [I] was eager that he should know something more of their creator.'[3] This condition proved acceptable to Barnard, and the letter was duly dispatched to Belmont.

In order to fulfill his side of the bargain, Barnard prepared himself for his first visit to the Sunday School, recording in an account book payment of fifteen cents for a shave – his first – and on the following day noted: 'I went to Sunday School first time on Sunday September 17th 1876.'[4]

This Sunday School was a class for young men taught by Fanny Dickinson Nelson, and it convened at a Baptist church then on Fifth Avenue just north of Union Street. Nelson was a popular teacher; both Braid and Calvert regularly attended her classes, the latter describing her as 'a refined and cultured lady, a beautiful character.'[5] Barnard too came to hold her in the highest esteem. Indeed, he once claimed that meeting her and her husband, Anson Nelson, who was for many years City Treasurer of Nashville, was the 'turning point of my life.'[6] Henceforth Barnard faithfully attended Nelson's class every Sunday except when illness prevented him, and usually contributed a nickel to the collection.

Here and there in his account book, Barnard jotted down a biblical verse or poetic passage which had struck a responsive chord in him. He wrote down the verse from Luke, 'Birds have nests and foxes have holes but the Son of Man hath not where to lay his head,' which must have spoken to his own waif-like existence at the time. The lines from the madrigal,

> How sweet the hours of early dawn
> When first the lark her matins sings,

expressed a sentiment which was undoubtedly becoming well known to him at first hand from his own long vigils of the night.

2

Eventually Calvert's letter concerning the Acklen telescope brought the awaited response. On December 15, 1876, Barnard received the telescope from Belmont, paying the $9.35 charge for expressage. He carefully examined the instrument, but to his keen disappointment he perceived that it was not as good as he had hoped, and three days later he sent it back, again paying the $9.35 charge for expressage. This was a hefty sum, as at the time Barnard was making only $12.00 a week at Poole's, of which $4.00 a week went for room and board at the Hotel St. Charles.

All of these expenditures were duly recorded in his account book. A thorough examination of its entries shows that he practiced the most rigid economy. Aside from necessary expenses, he allowed himself only an occasional entertainment – usually a play or minstrel performance. Only in astronomy was he lavish. In 1876 he recorded the purchase of several new books, Elijah Burritt's *Geography of the Heavens*, Richard A. Proctor's *Half Hours with the Telescope*, and the 1877 *American Ephemeris and Nautical Almanac*. But the main thing he was saving for was a first-rate telescope.

At this time he acquired a 3-inch equatorially mounted refractor which he found mediocre. Soon afterward, he learned from Braid, who had relatives in New York City, of John Byrne, an optician of that city, who offered first-rate refracting telescopes for sale. Byrne listed a 5-inch equatorially mounted refractor for $550, but as a favor to Braid, was willing to part with it for only $380.[7] Even this was a full two-thirds of Barnard's annual salary at the Photograph Gallery at the time. Swallowing hard, and

Pen sketch by Peter R. Calvert, Barnard's future brother-in-law, showing Barnard with the 5-inch refractor he acquired from John Byrne of New York City in 1877. Vanderbilt Photo Archives

with a generous advance in wages from Poole, Barnard went ahead and bought it, 'with the last cent I had in the world besides going heavily into debt to make up the requisite price.'[8]

Barnard began using his new telescope during the winter of 1877.[9] A single undated typewritten sheet in the archives at Vanderbilt University contains a fascinating account of his first night of observing with it:

> The first clear night after receiving my large telescope, I sat out on the roof of a three story house all night long, surrounded by ice and snow, the night being bitterly cold. After exploring the wonders of the moon until it sank from view beneath the western horizon my telescope sought the Milky Way. Here amid the splendors of that mighty zone of stars, I spent hour after hour sweeping among its marvelous fields of glittering suns, never wearying of the wonders constantly presented with each movement of the telescope, but gaining additional enthusiasm as the night drew apace. Nor did I forget the many double stars and clusters I had learned with my smaller instrument for they were each examined and I wondered at the beautiful contrasts of color in some of the binary systems and the myriads of stars revealed in the clusters that I had but dimly seen before with that small telescope. But from these lesser lights my telescope constantly swung back to the Milky Way, again to gaze on the 'broad and ample road where dust is stars.' So enraptured was I with these glimpses of the Creator's works that I heeded not the cold nor the loneliness of the night. And when the approaching dawn began to whiten the eastern skies, I sought out the great planet Jupiter, then only just emerging from the solar rays, and beheld with rapture his four bright moons and vast belt system. But when the dawn had paled each stellar fire the coldness of the night forcibly impressed itself upon me and I retired from the field of glory . . .[10]

It is important to keep in mind that in the 1870s artificial lights were nonexistent in Nashville. Thus, in reading Barnard's comments, one must keep in mind the impression the Milky Way makes on the observer in really dark skies rather than on the typical city dweller of today. Little did he realize then how much he would eventually contribute to knowledge of the Milky Way as well as to the planet Jupiter, the two objects which most captivated him on that first night.

Barnard was very justly proud of his telescope, which was superior to any that existed in Nashville at the time with the sole exception of the newly erected 6-inch Cooke refractor of the Vanderbilt University. He sometimes set it up on Capitol Hill, where it enjoyed an unobstructed horizon; more often, as on his first night of observing, on the platform of the old *Jupiter* camera on the rooftop at Poole's.

He and his telescope soon became celebrated throughout Nashville. J. T. McGill later recorded that 'at times the roof of the building at the corner of Fourth and Union was so crowded with guests that he was solicitous lest his treasure might be knocked off. After the friends departed he would begin his quiet work, and often the early morning would find him sweeping the heavens with his telescope.'[11]

It may well have been from that first winter of work with his telescope that, on showing up at Poole's one morning with a raw scratch on the side of his nose, Barnard was asked by Calvert what had happened. He explained that the night before, while working with the telescope in the intense cold, 'the rim of the eye-piece had frozen to his skin and when he moved his head away it tore off the narrow strip of skin.'[12]

3

In August 1877, the month in which Asaph Hall discovered the two moons of Mars – a discovery which electrified Barnard – the annual meeting of the American Association for the Advancement of Science was held at the State Capitol in Nashville. It was the first such meeting held in a southern city since the Civil War and was intended to 'again bring into the ranks of the Association the southern men of science who had returned but slowly'[13] after that bitter conflict, and to show off the recently founded Vanderbilt University. 'Notwithstanding the dread of the summer heat, which would normally be expected at a place so far south as Nashville, many northern and eastern members were present.'[14] Simon Newcomb, president-elect of the Association and the most famous American astronomer of his day, presided over the meeting.

Born in 1835 in Nova Scotia, though he always considered himself a New Englander by descent, Newcomb like Barnard had risen from a humble background. In his autobiography he noted that among all his ancestors, 'I do not know of any college graduate in the list.'[15] But he early discovered a talent for mathematical computations, and by the time he was six had studied arithmetic through the cube root. At eighteen, Newcomb left Nova Scotia with hardly a penny to his name and made his way on foot to Maine to join his father, an itinerant country school teacher. After a brief stint as a country school teacher himself, Newcomb obtained a position as a computer at the Nautical Almanac Office, then located in Cambridge, Massachusetts. In his spare time he earned a Bachelor of Science degree at Harvard. In 1861, Newcomb moved to Washington, DC, to become a professor of mathematics at the US Naval Observatory, and in 1877 had just been appointed Superintendent of the Nautical Almanac Office. At the Nautical Almanac Office, Newcomb distinguished himself for his investigations of the motions of the Moon and planets, which led to a great improvement in the precision of the *American Ephemeris and Nautical Almanac*. In recognition of these investigations, he was awarded in that same year, 1877, the gold medal of the Royal Astronomical Society. There is no doubting his genius, but he seems to have been a rather intimidating individual, 'more feared than liked.'[16]

Barnard's meeting with Newcomb in the State Capitol building is one of the best known events of his life. Barnard's friends Anson Nelson, who as City Treasurer 'met and knew every man of importance,' and Albert Roberts, editor of the Nashville *American*, were able to secure for him an introduction to the great man. After hearing from Nelson and Roberts of the 'ardent and faithful work' Barnard had been doing with his small telescope, Newcomb agreed to an audience, and met with the young man for about twenty minutes in a side room of the Capitol.

Barnard asked Newcomb how a young man with a small telescope might make himself useful in astronomy. Newcomb suggested searching for comets and making drawings of nebulae, and mentioned Chicago amateur S. W. Burnham's observations of double stars, made with only a 6-inch refractor, as an outstanding example of what was possible. Finally, Newcomb suggested that Barnard write to him sometime and took his leave.

Simon Newcomb. Mary Lea Shane Archives of the Lick Observatory

Years later, Newcomb recalled some of the circumstances of this meeting, and admitted that they were 'not flattering to my pride':

I was told of a young man a little over twenty years of age, a photographer by profession, who was interested in astronomy, and who desired to see me. I was, of course, very glad to make his acquaintance. I found that with his scanty earnings he had managed either to purchase or to get together the materials for making a small telescope. He was desirous of doing something with it that might be useful in astronomy, and wished to know what suggestions I could make in that line. I did not for a moment suppose that there was a reasonable probability of the young man doing anything better than amuse himself. At the same time, feeling it a duty to encourage him, I suggested that there was only one thing open to an astronomical observer situated as he was, and that was the discovery of comets. I had never even looked for a comet myself, and knew little about the methods of exploring the heavens for one, except what had been told me by H. P. Tuttle [the noted Harvard observer of comets]. But I gave him the best directions I could, and we parted.[17]

One would hardly guess from this account the effect that Newcomb's 'directions' actually had on Barnard, who had understandably approached the distinguished astronomer only with great timidity. In addition to the suggestions about observational work to which a young astronomer might devote himself, Newcomb had asked Barnard whether he was acquainted with mathematics. Barnard had answered, honestly, that he was not. Newcomb then said to him, 'Lay aside at once that telescope and master mathematics, for you will never be what you seek to become without this mastery.'[18] This, of course, was Newcomb's own bias – he was describing for young Barnard the route that he himself had traveled to success. The result was, however, as Peter Calvert reported, that 'Barnard was much depressed at the unsympathetic words of the great man and after the interview, as he confessed to Mr. Braid, he got behind one of the big columns and had a good cry.'[19] According to another account, Barnard's tears were 'because he had been commanded by one whom he reverenced to part with the thing which meant everything to him and also in pursuit of his profession, to pursue a course which he might find it impossible to successfully follow.'[20]

Barnard decided, however, to do his best to follow Newcomb's stern admonitions, and possibly the very next day hired a mathematics tutor, Russell Marling.[21] (For good measure, he also sought tutoring in French, which he received from an 'old Mrs. Coles.')[22] For a while he burned the candle late into the night in these studies, instead of what was for him the far more congenial occupation of viewing the heavens with his telescope. Eventually, however, he found Newcomb's injunction to put away his telescope too severe, and decided on a compromise – he would devote all clear and moonless nights to working with the telescope, the rest to the study of mathematics.

The story does not end here. Eventually, of course, Barnard went on to achieve some grand things with that small telescope. In 1891, after he had become a famous astronomer in his own right, he ventured to strike up a correspondence with Newcomb, though saying nothing about the circumstances of their first meeting. Instead it was Newcomb who first brought the subject up: 'When I was in Nashville in 1877 I had one or more conversations with a man who owned a small telescope and asked my advice as to what he might do with it,' he wrote to Barnard. 'I told him the only field I saw open to him, was that of a searcher for new comets and gave him some suggestions about the method of looking for them. I have totally forgotten his name but would be glad to know more about him. I have sometimes speculated on the question whether there was any possibility that it might have been you . . .'[23] One suspects that Newcomb, now regretting his coolness on that earlier occasion, hoped it hadn't been. Barnard replied: 'I am extremely glad you have mentioned it. I was of course the person who saw you in 1877 at Nashville.' Though he did not mention how discouraging Newcomb had been, one can well read between the lines:

> You then had no means of understanding the remarkable difficulties I was laboring under, the very smallest of which was the fact that my daily work for a living began at six or seven o'clock in the morning and very seldom terminated earlier than 8 or 9 p.m. I had been thrown upon my own resources when only a child and had been at work in a photograph gallery from the time I was eight years old and had not had more than one month's schooling in my life . . .
>
> I followed the part of your advice pertaining to comet seeking with very poor success . . . I struggled with mathematics at night. My non success made me hesitate in writing to you. When finally I found my first comet in 1881 I supposed you had forgotten all about me . . .[24]

Though eventually Barnard came to be on cordial terms with Newcomb, his recollection of Newcomb's stern manner on that earlier occasion did for a long while keep him aloof. At two subsequent meetings of the American Association for the Advancement of Science, he saw Newcomb but decided against approaching him. 'I saw you at Philadelphia in 1884,' he continued in the same letter to Newcomb,

> I hesitated to speak to you for I supposed you had forgotten all about seeing me at Nashville. At Buffalo in 1886 the same feeling kept me silent. I will tell you this much however; I went to Niagara Falls and there saw you in front of the International Hotel. I determined to speak to you and actually went towards you for that purpose but a look at your face stopped me in my purpose, you will pardon me, but the simple truth is, your face appeared so stern that I feared being received coldly and so did not speak to you; I was still very timid in the presence of a stranger whom I considered so high above my humble station.[25]

4

Another memorable event that occurred at the 1877 AAAS meeting in Nashville was Alexander Graham Bell's demonstration of the recently invented telephone. Bell set up a line between Polk Place and the A. G. Adams home on the adjoining corner of Seventh Avenue North and Union Street (Mrs. James K. Polk, widow of the late president, received the message). Braid, with his strong interest in electrical apparatus, was fired with enthusiasm for the new invention. Already having, through the acquiescence of the liberal Poole, a small shop for his experiments with electrical apparatus in a back room at the Photograph Gallery, he was not long in assembling his own telephones from diagrams and descriptions published in the *Scientific American*. He established a line of his own between the Photograph Gallery and Fanny Nelson's room in the nearby Hotel

St. Cloud, which was the prelude to a far more ambitious attempt to which Barnard was witness:

> [H]e and a young friend of his, James Ross, decided that they should be able to talk several hundred miles with [one of Braid's phones]. They had friends in the management of the Western Union Telegraph Company at Nashville. It was arranged with them that a trial should be made between Nashville and Bowling Green over the A. & P. lines, a distance of some seventy miles, on a Sunday morning in January of 1878. Mr. Ross went to Bowling Green for the purpose and Mr. Braid and others (I being one of the party) met at the Western Union office in Nashville. Connections were made and at 11 A.M. Ross responded, and for an hour conversations were carried on, every word being clearly heard . . . Ross played on the guitar and sang. It was a triumph in every way. Up to that time long distance conversation (if I remember correctly) had been possible over only one or two miles, so that Mr. Braid's success was far ahead of others. The Nashville papers next morning gave an account of this remarkable experiment . . . [26]

It must have been at about this time that a customer came into the Photograph Gallery and greeted Rodney Poole, 'Well, how is business?' Poole replied, 'With an operator crazy about electricity and a printer crazy about astronomy, I'm afraid my business is going to the devil.'[27]

Bell, whose father Braid had known in Scotland, was so impressed with the Nashville-to-Bowling Green experiment that he offered Braid a contract to establish the telephone service in Nashville. For some reason Braid was not interested, and the opportunity passed to Ross. However, Braid's days as a photographer were numbered. Within a year of the telephone experiment, he handed in his resignation to Poole and set up his own electrical business, which by 1891 had 'the largest business in the South in electrical instruments and electrical furnishing by contract . . . [having] fitted up some of the finest and largest buildings, not here alone, but throughout the States adjacent.'[28] Though he and Barnard henceforth saw little of one another, Braid in later years went to the trouble of obtaining his own $3\frac{1}{2}$-inch Clark refractor, and always ran the slide projector for Barnard whenever the latter lectured in Nashville. His last letter to 'dear Ed,' written shortly before Barnard's death, shows that despite being nearly deaf he was then as keenly interested in radio as he had been earlier in the telephone, describing the invention as 'the most wonderful of all things that has come in our day, in fact in the history of science.'[29]

5

Though Newcomb had failed dismally to grasp the character of the man with whom he had spoken in 1877, O. S. Fowler, a phrenologist Barnard consulted several months later, was more successful. At that time, phrenology – the pseudoscience in which mental faculties and character were supposed to be discernible from the 'bumps' on one's head – was popularly viewed as an accurate science, and Barnard himself may well have taken this character analysis quite seriously. Indeed, even so astute a psychologist as William James, in his *Principles of Psychology*, could still write of phrenology in 1890 that 'however little it satisfy our scientific curiosity of different portions of the brain, [it] may still be, in the hands of intelligent practitioners, a useful help in the art of reading character.'[30] No doubt 'intelligent practioners' relied as much on supplementary information provided by the friends of the person being analyzed as on phrenological science per se, and Fowler certainly seems to have done so. 'Astronomy, Sir,' he told Barnard, 'is your natural study, and if you could possibly shape your life so as to study and practice it do so, for your whole genius lies in this direction. This is consequent on the very great preponderance of your perceptive faculties . . . You are wonderful for eyesight.'[31] It is hard to imagine what bump on Barnard's head could have told Fowler this. Fowler also commented on Barnard's tendency to overwork, and advised him – he would not be the last to do so – to 'rest more, or break down.' Among other comments, Fowler described Barnard as 'a great deal too humble,' and told him to try to build himself up in his own eyes. At the same time he was 'intensely ambitious,' desirous only to do what would give himself 'a good name among men.' Whatever information he used, Fowler managed to sketch an accurate picture of Barnard at twenty.

6

Barnard's first published astronomical observations were of the transit of Mercury of May 6, 1878. Several weeks before the transit, he had written to Professor George Davidson of the US Coast Survey for instructions on how to observe it.[32] On the day of the transit itself, he set up his 5-inch refractor at the State Capitol in order to make timings which he submitted to the US Naval Observatory. They were later included in the official government report.

Not until 1879, however, did his telescopic work begin in earnest. There were several factors which may have contributed to his new-found sense of purpose. First, Newcomb's admonition that he 'put aside his telescope' was beginning to wear off. Then too, on May 9, 1879, he had moved. After several years of living alone in the St. Charles Hotel, he began renting, from a Mr. Davis for $9.00 a week, a 'large brick house with a mansard roof' at 1919 Patterson Avenue in west Nashville. This was near and north of the Vanderbilt University. His ailing mother, about whom he was always anxious, moved in with him, and he enjoyed a more convenient location for astronomical observations than the rooftop of Poole's.

Most important, however, was the fact that the Calverts were becoming increasingly entwined with his life. The extrovert Peter Calvert had, ever since joining the Photograph Gallery in 1875, been Barnard's close friend and mentor, and soon afterwards Peter's older brother Ebenezer, 'Ebby' as he was always called, took a job there as an artist. Through them, Barnard was introduced to their older sister, Rhoda Calvert. It was to Rhoda that Barnard was soon paying the most attention.

Rhoda had left her native Yorkshire with Ebby in 1873, the two of them drawn to

Barnard about 1878. Vanderbilt Photo Archives

Middle Tennessee by a friend's description of the beauties of its rural countryside. They settled first at Tucker's Cross Roads, near Lebanon, in Wilson County, but after Peter's success in getting on in Nashville – his ebullient personality no doubt helped him get a position, but so did the fact that by the mid-1870s the city was enjoying its first business boom since before the Civil War – the quieter and more reserved Ebby came to join him, and Rhoda with him.

When Barnard met his blue-eyed, red-tressed sweetheart, she was already in her mid-thirties – thirteen years older than himself – and at a point in her life when she may well have begun to reconcile herself to the prospect of being an old maid. It is easy to

1919 Patterson Avenue, described by Barnard as the 'house with the mansard roof.' It was from the front yard of this house that Barnard enjoyed observing Jupiter and discovered his first comet in 1881. Courtesy Professor Robert T. Lagemann

imagine that she may have seen the budding young astronomer as her last best hope and set her cap for him. If so, her attentions were not unwelcome.

Before long, Rhoda was helping Barnard with the necessary but time-consuming tasks required to keep a household running, and relieving him of much of the responsibility of caring for his invalid mother. This spared him precious time when he was not working at the Photograph Gallery for astronomy.[33] With her emotional support, he felt more secure and self-confident. It can hardly be a coincidence that the move to Patterson Avenue and her entry into his life marks the beginning of an incredible flurry of astronomical activity.

During this time he concentrated his attentions on Jupiter, chiefly because of the Great Red Spot which was just then becoming a prominent feature of the planet. As he later wrote: 'I was passionately fond of observing the planet Jupiter with my new telescope before I began seeking comets. This telescope, which was on a tripod and could easily be moved about, would be lifted out in the yard, where I could sit and watch Jupiter to my heart's content.'[34]

Barnard made his first observation of the Great Red Spot on August 3, 1879. Taking his telescope out early that morning to observe and sketch the planet, he found the spot unexpectedly 'come into view around the following limb . . . so remarkable in form and color that I was at once struck by its appearance.'[35] As he had at the time heard nothing of it, it was in fact an independent discovery – only afterwards did it come to his attention that the Great Red Spot had already been observed by Professor Carr Waller Pritchett, at the Morrison Observatory at Glasgow, Missouri, as early as July 1878. Also, Rev. William Rutter Dawes, William Lassell, and the fourth Earl of Rosse had in still earlier years recorded the spot, though it had then been a relatively inconspicuous feature and attracted little attention. By the time Pritchett identified it, it was a 'cloud . . . almost completely oval in shape, and . . . preeminently rose-tinted,'[36] well on its way to becoming so marked that, in the words of the eminent historian of astronomy Agnes M. Clerke, it 'attracted the wonder and attention of almost every possessor of a telescope.'[37]

There was, then, nothing especially meritorious in Barnard's sighting it at such a late date – the fact merely attests to how out of touch he was with developments elsewhere. But from his first observation on August 3, 1879, he began following the Great Red Spot and other features on the planet on almost every available night.

When observing Jupiter, he generally used a magnification of $\times 173$ on the 5-inch Byrne refractor – 'the instability of the mounting,' he explained, 'preventing the use of a much higher power.' A right-angle prism used with the eyepiece showed the planet's image reversed but not inverted, and in sketching,

> a small wooden box or desk was used, with a glass on top, on which the paper was laid. A faint illumination from below, through the sketching paper, was obtained from the reflected light of a candle. By this means I could sit with the desk on my knees and compare the telescopic image directly with the drawing.[38]

The drawings were very carefully made, and Barnard was confident that they 'faithfully show the markings as they appeared upon the planet at the time of observation.' He showed some of them to Peter Calvert, who did not hesitate in pronouncing them 'wonderfully good.'[39] With this encouragement from his artist-friend, Barnard sent them to the *English Mechanic*, whose editor published them with the comment that they were 'of the very best I have seen.'[40]

A pen sketch by Calvert made at about this time shows Barnard as a very sober looking young man seated beside his 'pet,' as he called his 5-inch Byrne refractor. Indeed, though Barnard had a well-developed sense of humor, it was his serious side which made the strongest impression on his friends. As Calvert saw him

> Barnard was never like other young men, bent on pleasure and frivolous pursuits; necessity and responsibility made him serious, for from the age of about 12 years he was the main support of his invalid mother who was unable to walk and with beclouded mind, her one joy in life was her devoted boy. Under such circumstances no wonder he was thoughtful.[41]

And Alfred E. Howell, an acquaintance from Fanny Nelson's Sunday School class, added independent testimony to the accuracy of the artist's sketch: 'Many the time I have noted his hollow eyes and faded cheeks and wist not that he was to be world famous from his vigils of the night, after an all-day's work at Poole's photograph gallery, his hands still stained with the chemicals.'[42]

Not only did Barnard have to battle strain and fatigue during his 'vigils of the night,' he had to contend with intense loneliness as well. At the house on Patterson Avenue, 'there was but one neighbor near,' and Barnard would recall that the loneliness of the place 'oftentimes impressed me with a kind of dread, for I was out at all hours of the night.' Once, observing Jupiter at about 2 a.m., he paused to rest his eyes for a moment from the telescope, only to be 'horrified to see two glaring, greenish-red balls of fire a few feet away in the obscurity of the bushes':

> Cold chills played up and down my backbone, and I was too frightened to move. These balls of fire came slowly toward me. A supernatural horror seemed to bind me to the spot as a nightmare holds one in its pitiless grasp. I could not have moved if I had known the foul fiend himself was behind those lights. Just at the point when I felt that I must collapse, as the hateful lights came close to me, I felt the warm touch of the tongue of some animal licking my cold hand; and in the obscurity I saw a great dog wagging his tail in a friendly manner. The relief was tremendous, and a warm flow of blood seemed to infuse my veins. It proved to be a great and fierce bulldog belonging to my neighbor. This dog had always looked so savage, with his cruel teeth, that I had not attempted to cultivate his friendship. Why he had not attacked me in the darkness and what made him seek this friendship, I do not know; but from that time on he was my good friend and made it a habit to lie down near me every night I was out observing. His friendship was a blessing, for I no longer felt the nameless dread of the night with his powerful form at my feet.[43]

It may be, then, that we owe more than we know to this friendly bulldog for the astronomical achievements of Edward Emerson Barnard.

Throughout 1879 and 1880, Barnard documented every nuance of variation in the appearance of the Great Red Spot – he was not just watching but *observing* the planet, something much more active and requiring the greatest intensity and concentration. On the first night, August 3, 1879, he had found 'the south edge of the spot . . . nearly straight and the following end blunt, the north edge tapered towards the preceding end,' while in color the spot resembled 'red-hot iron.' As he continued to observe, he noted that changes in the form of the spot occurred mainly in the ends, 'which occasionally appeared rounded, and, again, were very much pointed, or cigar-shaped.' At still other times they extended thinly in faint trails. His vivid descriptions of the changing color of the spot suggest a close familiarity with the artist's palette, presumably acquired in part through his long association with the artists at the Photograph Gallery. Thus the spot was 'clear, darkish Indian red' (1880 July 11), 'brick-dust color' (August 16), and finally 'brick red' (September 16). He estimated the spot's length to be about 22 490 miles and its breadth 6900 miles, giving it a total area equal to three-quarters of the entire surface of the Earth.

Barnard early called attention to the tendency of the Great Red Spot to repel other features, causing it to be encircled 'by a sea of light . . . for a distance of three or four thousand miles . . . which appeared as a visible barrier against the approach of any spot

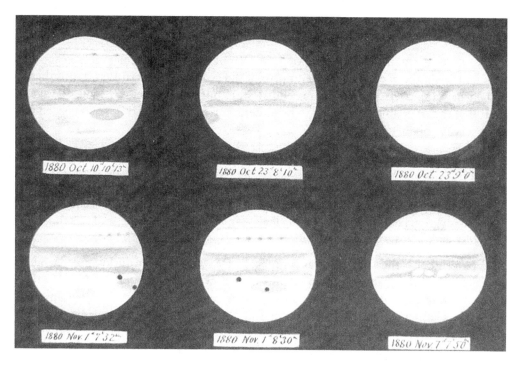

Some of Barnard's sketches of Jupiter made with the 5-inch Byrne refractor in 1880. Yerkes Observatory

or marking.' The only exception to this general rule was a 'dense, smoky shading' which during July 1880 seemed to attach itself to the south side of the Great Red Spot, eventually passing it and leaving it far behind. It was among several features he described in a paper, 'Changes on Jupiter,' which gave an account of his observations and was published in the *Scientific American* for December 4, 1880.[44]

Meanwhile, there were yet other interesting developments on the planet, including a brilliant white spot in the Equatorial Zone, first seen by Barnard in August 1880 and kept under observation by him for many months. His description of this white spot's appearance is characteristically vivid:

Among the surprising things about the spot were its great changes, both of form and brightness. At times it became so bright as to glisten like a star. When in this condition it was by far the brightest object on the planet. For a while it would appear as a rather small, inconspicuous, light, oval spot, imbedded in the dark matter of the north edge of the south equatorial band. In this state it would scarcely attract attention. It would next be seen brilliantly white, burying its head in the dusky matter of the belt, with a vast, luminous train streaming backwards along the equatorial regions, like the tail of a comet. Sometimes this train was composed of white, cloud-like balls, that streamed eastward on the planet. After continuing thus for some time, it would seem to have wasted its energies, and would then assume the quiescent state . . . When at its brightest it seemed to burrow in the south band and plow the matter before it . . .'[45]

Even more dramatic was the eruption, in October 1880, of a remarkable group of spots on the southern edge of the belt north of the North Equatorial Belt. Until then, the belt had appeared as a very thin reddish line, but on October 23, 1880, Barnard noted that the entire planet seemed to be undergoing a great upheaval. He recorded in his observing book:

> Jupiter is undergoing some remarkable changes now; there are a great many degrees of shade, somewhat like ill-defined spots and light spaces, appearing in the southern hemisphere near the Great Red Spot. The space between the north edge of the north equatorial band and the first linear belt is deepening in tint . . . [It is now] a grayish green [and] near the following limb is knobbed in appearance, as if several little dark beads were strung on it.

On November 1, the same belt was composed of 'a string of large dusky spots,' which were as black as the shadows of two satellites which were just then projected upon the planet. Each of these dusky spots had a very dark center surrounded by a penumbra, 'not unlike a sun-spot,' and showed a very rapid westward motion on the planet. Within another week the belt had become 'heavily marked all the way across the disc, and dark, with remarkably large, distinct, knotty lumps, in places quite broad with them.'[46]

What Barnard had witnessed was something that had never been seen before, and has been seen only a few times since – an eruption of dark spots at the south edge of what has since come to be known as the North Temperate Belt. He himself wrote that 'probably no more remarkable outburst has been witnessed.'[47] The motion of these dark spots was very rapid, indeed the most rapid of any spots ever observed on Jupiter.

By early December 1880, these fast-moving spots had elongated into oblong patches, and over the next few months 'diffused into a "veiling," with condensations on it,' which for a while gave the belt a beautifully scalloped appearance, its southern side consisting of 'graceful, light-rimmed curves.' Finally, by the summer of 1881, the appearance again became that of an ordinary reddish belt.

Barnard's observations, in addition to vivid descriptions and careful drawings, included eye-estimates to the nearest minute of the times that the various spots on the planet transited the central meridian of the planet's disk. From such determinations the periods of rotation of these spots – which vary with latitude because all we see of Jupiter is gaseous – could be worked out, and the periods found by Barnard compared favorably with those obtained by George Washington Hough, a professional astronomer who carefully measured them with a micrometer and a much larger telescope, the $18\frac{1}{2}$-inch Clark refractor of the Dearborn Observatory.

Lastly, Barnard paid more than passing attention to the four Galilean satellites. His Byrne refractor, optically of the finest, easily showed the satellites as 'neat little discs,' that of Ganymede being noticeably larger than the others, he found, and 'round like a small planet of a yellowish color.'[48] Barnard was especially intrigued by the unexplained 'dark transits' of satellites III (Ganymede) and IV (Callisto) across the

Jovian disk. Even when seen against the same region of the planet, these satellites would sometimes appear bright, at other times like drops of ink. Barnard carefully noted the times of each of the dark transits he observed, and published several reports about them in the then just founded popular astronomical journal *Sidereal Messenger*, edited by W. W. Payne of the Goodsell Observatory at Northfield, Minnesota.

Many an amateur has begun with Jupiter as Barnard did, hoping to contribute 'his mite in drawings and observations of that changing old planet,' as Barnard once modestly summed up his own contributions. What is remarkable is the skill and thoroughness with which he carried out this apprenticeship. The lad who had discovered for himself the equation of time was a born observer. Now he had found something worthy of his talents, and like one of the satellites of Jupiter coming out of eclipse, he would soon emerge from obscurity.

1   The mansion is now on the campus of Belmont College and remains one of the sites of interest in Nashville.

2   Henry McRaven, *Nashville: 'Athens of the South'* (Chapel Hill, Scheer and Jervis, 1949), p. 91

3   *Nashville Tennessean and the Nashville American*, April 30, 1915; newspaper clipping in VUA

4   EEB, Lagemann notebook no. 1, which consists of Barnard's day book from 1876. It is one of several of Barnard's private notebooks in possession of Robert Lagemann and was acquired by him at the estate sale of Barnard's niece, Mary R. Calvert, in 1973. I express my appreciation to Professor Lagemann for allowing me to examine and quote from these notebooks.

5   P. R. Calvert, 'Reminiscences of Barnard,' *JTAS*, 3:1 (1928), 12. Calvert, who fancied himself a poet as well as an artist, wrote an acrostic on her name which however gives little insight into her character. After her death in September 1914, the *Nashville Tennessean and The Nashville American* for April 30, 1915 ran an article, 'Mrs. Fannie Nelson Was Inspiration of Noted Astronomer – Edward Barnard Attended Sunday School Class of Beloved Woman Here,' by Robert Ewing, of which the following should give the flavor: 'Her pupils were of an age to appreciate the truth and very positive value of the lessons she taught him . . . Mrs. Nelson's individuality stood out very strongly, but her intense devotion to her husband, the well-remembered Anson Nelson, was so great that for many years before he died they were rarely seen apart. Together they went about doing good, alleviating suffering, and comforting the sorrowing. But they always kept themselves cheerful, cheerily helping to advance in life those who were ambitiously striving.'

6   EEB to Joseph S. Carels, September 21, 1892; VUA

7   Fanny D. Nelson, 'Edward E. Barnard – Fellow of Vanderbilt University,' *Our Day*, 2:2 (1883), 37

8   EEB to SN, April 27, 1891; VUA

9   Lagemann notebook no. 1 contains daily entries for September 16, 1876 to January 8, 1877, then a hiatus occurs until October 29, 1878, when Barnard resumed making occasional entries until October 14, 1880. In this notebook there are also a few random

entries which were made out of chronological sequence, including an observation of Uranus dated March 10, 1877 which was made with the 5-inch Byrne refractor; this, as far as I know, is the earliest dated observation by Barnard that survives. This same page contains a small pencil sketch of Uranus and an ink miniature showing the 5-inch equatorial. The text, in pencil, reads as follows: '1877 Saturday March 10th. Saw the planet Uranus this night in the constellation Leo about 7 degrees west and 3 degrees north of Regulus with power 52 it looked like a bright round star with 104 it had a small disc, and appeared about this size with 173 – the highest power I could use on the house for shaking.'

10 The text ends with a note in Barnard's handwriting, 'MS torn here.' The line of poetry slightly misquoted is from John Milton, *Paradise Lost*, VII, 577–8: 'A broad and ample road, whose dust is Gold/ And pavement Stars . . .'

11 J. T. McGill, 'Edward Emerson Barnard,' *JTAS*, 3:1 (1928), 34. A Mr. Charles S. Mitchell, who was afterward a confectioner in Nashville, remembered in old age: 'Part of our education came from visiting the roof of Poole's Gallery where E. E. Barnard with his small telescope showed us the wonders of the sky.' Quoted in William Waller, ed., *Nashville in the 1890s* (Nashville, Vanderbilt University Press, 1970), pp. 168–9.

12 P. R. Calvert, 'Reminiscences,' p. 13

13 Remarks by Simon Newcomb in *Proceedings of the AAAS*, **26** (1878),367.

14 ibid.

15 Simon Newcomb, *The Reminiscences of an Astronomer* (Boston and New York, Houghton, Mifflin & Co., 1903), p. 3

16 Joseph S. Tenn, 'Simon Newcomb: The First Bruce Medalist,' *Mercury* Jan./Feb. 1990, p. 30

17 Newcomb, *Reminiscences*, pp. 191–2

18 *Nashville Tennessean and the Nashville American*, April 30, 1915

19 P. R. Calvert, 'Reminiscences,' p. 12

20 *Nashville Tennessean and the Nashville American*, April 30, 1915

21 The 'very next day' is according to ibid.

22 J. W. Braid to EBF, February 9, 1923; YOA

23 SN to EEB, April 17, 1891; VUA

24 EEB to SN, April 27, 1891; VUA

25 ibid.

26 E. E. Barnard, 'Letter from Mr. Barnard Regarding Mr. Braid,' *JTAS*, 3:1 (1928), 10–11

27 J. T. McGill, op. cit., pp. 34–35

28 Waller, *Nashville*, p. 61. The Braid Electrical Company was still in business in Nashville a hundred years later, in 1991.

29 J. W. Braid to EEB, January 18, 1923; VUA

30 William James, *The Principles of Psychology* (New York, Dover, 1950 reprint of 1890 ed.), vol. 1, p. 28

31 'Phrenological Character of Edward Barnard,' given by Prof. O. S. Fowler, phrenological author, editor and lecturer, March 16, 1878; VUA

32 EEB to GD, April 19, 1878; BL

33 The moving expenses, carefully recorded in Lagemann notebook no. 1, were as follows:

Towels pillows slips etc.	$3.60
Dishes etc	8.20
Carpet	22.45
Chairs	5.70
Bedstead bed washstand etc.	19.25
Pillows	3.00
Moving	2.25
groceries basket etc.	1.40
gave Mrs. Underhill to move	1.00
" " " to send away boy	2.50
gave Miss Calvert to buy groceries	6.00
wood & straw for carpet	1.50
Hauling straw	0.35
Paid Miss Calvert for safe	1.00
Pd Miss Calvert on more groceries	1.00
Hauling safe & coal box	0.50

34 E. E. Barnard, 'A Few Unscientific Experiences of an Astronomer,' *Vanderbilt University Quarterly*,' **8** (1908), 273–88:281

35 E. E. Barnard, 'Observations of *Jupiter* with a Five-inch Refractor, during the years 1879–1886,' *PASP*, **1** (1889), 89–111:90

36 C. W. Pritchett, 'Markings on Jupiter,' *Observatory*, **2** (1879), 307–9

37 A. M. Clerke, *History of Astronomy During the Nineteenth Century* (London, Adam & Charles Black, 3rd ed., 1893), p. 358

38 Barnard, 'Observations of Jupiter,' p. 89

39 P. R. Calvert, 'Reminiscences,' p. 11

40 Quoted in Fanny D. Nelson, op. cit., p. 39. See *English Mechanic*, **30** (1879), 165–6

41 ibid.

42 A. E. Howell, *JTAS*, 3:1 (1928), 8

43 Barnard, 'Experiences,' pp. 281–2

44 E. E. Barnard, 'Changes on Jupiter,' *Scientific American*, **43** (1880), 356

45 Barnard, 'Observations of Jupiter,' pp. 96–7

46 ibid., pp. 100–2

47 ibid., p. 102

48 EEB, Lagemann notebook no. 2, which bears the inscription 'Telescopic Observations by Ed. E. Barnard,' and contains observations, mostly of Jupiter, from 1880

3

Mars; His Moons and His Heavens

After Jupiter, Mars was the favorite object of Barnard's early telescopic attentions. Somewhat surprisingly, he does not seem to have made many observations at the unusually favorable opposition of 1877 – opposition date, September 5, occurred during the very week of the AAAS meeting in Nashville. Approaching within 35 000 000 miles of the Earth, the planet then came closer than it had for fifteen years and outshone even mighty Jupiter in the night sky. Under these favorable circumstances, Asaph Hall was able to discover the satellites of Mars, and from Milan came Giovanni Schiaparelli's first reports of the strange surface markings he called *canali*, of which more will be said later.

Possibly Barnard's meeting with Newcomb at the AAAS meeting, which led him to temporarily put aside his telescope, suffices to explain his failure to take better advantage of this golden opportunity. However this may be, by the next opposition, in November 1879, he was elated by his previous summer's success as an observer of Jupiter and well prepared to take up the challenge of the more elusive red planet.

With the 5-inch Byrne refractor, he duly noted the 'lucid spots at, or near [the planet's] poles,' and followed their changes, which suggested to him that these spots were 'in all probability snow and ice, or something analogous.' He also made out the dark markings, about which he wrote that 'if everything is in good condition for observation . . . these dusky particles or marks have, in most instances, well defined outlines, and are of a slight greenish tinge, while the other portions of his disk . . . are of a brick-dust color. The patches of a greenish cast . . . may be oceans and seas. The ruddy part of his surface is doubtless land.' Barnard perceived that at times parts of the surface were temporarily obscured, and the effect of these obscurations could be traced 'as they drift along over the lands and seas, blurring the definition of the markings more or less whenever they pass.'[1] He naturally thought that these were 'masses of vapor floating in the Martian atmosphere'; we now know that what he had witnessed were, in reality, Martian dust clouds.

By the early spring of 1880, Mars was fast receding from the Earth and growing steadily dimmer in the night sky. Nevertheless, though its dwindling disk no longer was a rewarding object in the telescope, the planet remained very much in Barnard's thoughts, for he was by then writing an ambitious monograph – *Mars; His Moons and His Heavens*.[2]

Barnard was, of course, very young, and completely without formal training in either literature or astronomy. Under the circumstances, the odds against success in such a venture were very long. Though he began in the flush optimism of youthful inexperience and was helped by the growing consciousness of his own powers, it was probably not false modesty that led him to claim in the preface that the book was 'prepared under many unfavorable conditions,' or to beg the indulgence of his readers 'should they be disposed to criticise any deficiency of description.' He planned to publish by subscription, and he went so far as to enroll his friends and some of the leading citizens of Nashville as subscribers.

The subject, at least, was well chosen. Popular interest in Mars, a world which seemed in many ways like the Earth, had been growing steadily as the century advanced, and by 1880 there were definite signs of the 'Mars furor' soon to come. But as yet the field was still far from overworked. Though books about Mars would soon outnumber those dealing with all the rest of the planets combined, not a single book devoted exclusively to the red planet had so far appeared in English.

In addition to general information about Mars and descriptions of its telescopic appearance based largely on his own impressions during the opposition of 1879, Barnard describes at length the recently discovered moons of Mars – their eclipses, transits, and mutual occultations. He conjures up the spectacular view of Mars itself as seen from the satellites, and works out in detail everything that an observer would see in the heavens of Mars were he to watch through a single night (to be specific, the night of the last day of winter of the Martian northern hemisphere) while standing on the equator of the surface of the planet. All of his descriptions were based on his own careful calculations. 'To be sure,' he noted, 'we cannot describe the scenery of the surface of the planet, but we can describe the scenery of his heavens, as accurately and minutely as if we were "dwellers of his lands."' The following portrayal of Mars's appearance from Phobos will give an idea of the kind of effects he was aiming at:

> In 7h 39m 15s [the period of revolution of Phobos] the planet would pass through all the phases our moon goes through – from new moon to new moon. What a magnificent spectacle! A huge crescent over five thousand, five hundred times as large as our crescent moon – stretching from the horizon nearly half way to the zenith, behind which the stars are swiftly passing to reappear again at the west; whole constellations of these bright fires melting away before its rapid approach! What an awe-inspiring sight to witness from this satellite the rapid sweep of sunlight over this stupendous globe! Its waxing and waning! See the sun sink behind it and darkness reign for nearly one hour at the time of the martian equinoxes; then the swift emergences of the brilliant orb from the planet's edge quickly followed by a thin rim of light glinting along the edges of the great globe.

Though for the most part intent on such descriptions, here and there Barnard does not shy away from more controversial fare. As yet he had not heard about Giovanni Schiaparelli's *canali*, but he was well aware of the speculations of the ebullient French astronomer Camille Flammarion, whose comments were widely reported in the American newspapers in 1879: 'The inhabitants of Mars,' Flammarion was quoted as saying, 'are of a different form from us and fly in its atmosphere.' Barnard regarded such conclusions as premature, to say the least:

> We have heard that the people on the moon walked on their hands. Why should not the people on Mars go a step further and not walk at all. If the good astronomer had said these Martialites might, if they were so constructed, fly in the atmosphere, we might have admired his ingenuity . . . But when the worthy man positively and emphatically states that they *do* fly, we think it even possible that his astronomical reason is becoming a trifle volatile. However, we would not say these creatures do *not* fly in the air; that would be taking issue with a great astronomer, and we are hardly inclined to do that. For a similar reason we would scarcely feel like denying that there live in the oceans of that planet – nations and tribes of mermen and mermaids. We would not feel like refusing them the luxury of four legs, instead of two; and we might even grant them a caudal appendage to lash the Martial flies from their august backs. Indeed if the occasion demanded we might do better still and concede them large pointed ears to look wise in.

The twenty-two year old amateur astronomer had a decided aversion to what he referred to as 'Rashly Drawn Conclusions,' and judiciously warned against letting the imagination run riot:

> . . . it is well to fetter the wings of our fancy and restrain its flights. It is quite possible we may have formed entirely erroneous ideas of what we actually see. The greenish gray patches may not be seas at all, nor the ruddy continents, solid land. Neither may the obscuring patches be clouds of vapor. Man is too quick at forming conclusions. Let him but indistinctly see a thing, or even be undecided as to whether he does actually see it and he will then and there set himself to theorizing, and build immense castles of conjecture on a foundation, of whose existence he is by no means certain. When the moon was first examined with a telescope great 'seas' were discovered on its surface and duly named 'seas and oceans.' Afterward, on applying better telescopes, it was ascertained that these seas and oceans were not bodies of water at all, but large plains, or level tracts of dry sterile country. Now we see Mars, in the best telescopes, scarcely better than we see the moon with the unaided eye . . . From this it is manifest how foolish is that habit of theorizing about the surface of a celestial object, which is never nearer to us than between 34,000,000 and 35,000,000 miles.

Such comments prove that Barnard was after the truth and not prepared to accept many of the unqualified statements about Mars that were floating around at the time he wrote it. Unfortunately, his ideas were never circulated. Though he produced a substantial manuscript which appears to have been nearly finished aside from a conclusionary section, for some reason – most likely his growing interest in comet-seeking after 1881 – he broke off work and never returned to it. The manuscript was to remain as inaccessible as if it had been locked away in a desk drawer. Indeed, to this day knowledge of this important trace of Barnard's youthful aspirations has scarcely extended beyond the mere rumor of its existence in the Vanderbilt University archives, where it was deposited after Barnard's death by his niece, Mary R. Calvert.

Thus one can only guess whether, if he had gone ahead with his plans to publish, it would have made a difference in the debate about Mars at a time when sound sense about the planet was fast going out of fashion and when most of what would appear in print for many years to come would be decidedly of the 'castles of conjecture' variety.

1   EEB, 'Mars; His Moons and His Heavens'; unpublished MS in VUA
2   From internal evidence, it is possible to date the composition of the MS to spring 1880. Noting that during much of 1879 Mars had been situated not far from the Pleiades, Barnard writes: 'Now in the latter part of March [1880] he has advanced some twenty degrees eastward of Aldebaran and forms a regular isoceles triangle with that star and Betelgeux, a first magnitude golden star in Orion.' Barnard added that it took a 'sharp eye' to detect any difference in color between the three objects. All quotations in this chapter are from the holograph manuscript in VUA.

4

A seeker of comets

1

Barnard's painful childhood was to haunt him throughout his life. A constant reminder of what he had endured was the scar on his right jaw marking the site of the wound that had so long failed to heal – Barnard thought because he had been so malnourished at the time. But the psychic scars were if anything more unhealing. The unspeakable hardships of his childhood and the insecurities he had known – and it is impossible to know fully what they were, for there were features of his early life that were so sad that he would not speak of them – were such that any situation that recalled those deprivations, even decades later, was apt to reawaken massive anxieties. Barnard thus reminds one of Abraham Lincoln, who was also sensitive about his humble origins and, as his law partner William Herndon pointed out, would discuss his boyhood only 'with great reluctance and significant reserve.'[1]

Psychologists have suggested that creative effort begins as an attempt at restitution for a traumatic loss, either of a physical or a psychic kind. Lincoln lost his mother, Nancy Hanks, when he was nine, and suffered throughout his youth from terrible depressions, more than once coming close to making an attempt on his own life. Howard I. Kushner has suggested that in later coming to grips with his guilt and rage, Lincoln triumphed over self-willed death, and that such reconciliation became entwined with the quest for immortality – the desire to vanquish death through work honored by posterity.[2] Barnard did not lose his mother to death as Lincoln did, but the loss he suffered was hardly less catastrophic. As her mind began to fail, she became as unavailable to him, emotionally, as if she were dead. As with Lincoln, Barnard's desire to compensate for his loss seems to have become associated with a need to achieve restitution through some immortal work, and in the nineteenth century, when there were still strong notions of the 'eternal heavens above,' what more immortal work could there be than to add to knowledge of the stars? Thus in part the secret behind Barnard's enormous drive, which made him, in a phrase Herndon used to describe Lincoln, 'a little engine that never knew rest.'[3]

2

In his efforts to raise himself from the disadvantaged circumstances of his youth, Barnard had already by 1880 climbed a very long way. Already he was registering a favorable impresssion on leading astronomers of the day. He was well-known to Vanderbilt's Olin H. Landreth, to whom he had been introduced either by W. Leroy Broun, professor of mathematics at Vanderbilt, or Braid. Just arrived from the Dudley Observatory in Albany, New York, to take charge of Vanderbilt's new School of Engineering, Landreth was able to show Barnard how to make a ring micrometer, a simple device inserted in the focal plane of a telescope which could be used to measure the position of an unknown object, say a comet, relative to a reference star in the same field whose coordinates were known. Landreth later recalled how impressed he was by Barnard's 'boundless enthusiasm, by the remarkable store of astronomical information he had already accumulated, and by his wonderfully keen mind, and his sharp eye.'[4]

Landreth in turn may have suggested that Barnard correspond with Lewis Boss, the director of the Dudley Observatory, and with Seth Carlo Chandler, Jr. of Cambridge, Massachusetts, an actuary-turned-astronomer who was a leading computer of cometary orbits and who in collaboration with John Ritchie, Jr., the president of the Boston Scientific Society and editor of the *Science Observer*, was an important disseminator of intelligence about planetary and cometary discoveries of his day. By 1880, Barnard was writing to both of them. Still more important in the influence he was to have on Barnard's future career was Lewis Swift, with whom Barnard began corresponding in August 1880. Swift, the aging doyen of American cometary discovery and an expert visual observer of nebulae, would become the young man's most valued mentor during the years before he went to Vanderbilt. Unlike the stern and forbidding Newcomb, Swift was 'of a genial and happy disposition,' and Barnard would always remember his help with gratitude: 'He was generous to the struggling amateur, and was always liberal with help and advice to those who were striving to gain a foothold in astronomy.'[5]

One of several older men to whom Barnard would look for the father-figure he himself had never had, Swift had been born on a farm in upstate New York in 1820. When he was thirteen, he fractured his hip, and was lamed for the rest of his life. 'It was through this accident,' Barnard later recounted in a brief biographical essay of his mentor, 'that he was enabled to attend school and receive the simple education of his day.' And speaking from his own hard experience, Barnard added: 'The stern necessities of the time demanded constant labor from the young as well as the mature, and for the robust child there was little or no time to be spared for the school.'[6] No longer able to labor from dawn to dusk on the family farm, Swift trudged the two miles to school everyday on crutches.

Swift's tastes ran strongly to the sciences, and during his mid-twenties, he traveled extensively, giving popular lectures on electricity and magnetism. In 1851, he became a country storekeeper at Hunt's Corner, New York, where he purchased from a pedlar a copy of the *Works* of Thomas Dick. The book had the same electrifying effect on him that it was later to have on Barnard, and it inspired Swift to acquire a small telescope of his own. In 1860 he got his hands on a much better one, a fine $4\frac{1}{2}$-inch Fitz refractor, with which in July 1862 he picked up a comet in the constellation Camelopardalis. Since he did not at that time have any better source of astronomical information than the local newspapers, he at first assumed that he was only seeing another comet that had

Lewis Swift. Mary Lea Shane Archives of the Lick Observatory

been reported two weeks earlier by Julius Schmidt at Athens. In fact Schmidt's comet was nowhere near Camelopardalis (it was far to the south in Virgo), and when three nights later Horace P. Tuttle of Harvard independently picked up the new comet – at once recognizing that it was new – Swift announced his earlier sighting in order to claim a share of the credit for his first cometary discovery. Comet Swift-Tuttle (1862 III) proved to be a splendid object, and it acquired special significance when the Italian astronomer Giovanni Schiaparelli recognized that its orbit was virtually identical to that followed by the debris causing the Perseid meteor shower each August, thereby forging for the first time a link between comets and meteors. (The comet's period is now known to be 134 years. It was recovered at its first return since the Civil War in September 1992 by a Japanese amateur astronomer, Tsuruhiko Kiuchi; it is not due back again until 2126.)

In 1872, Swift moved to Rochester, New York, to open a hardware store. At first he had no other place to mount his telescope than in the back alley behind the store, where he could view only small sections of the sky. Eventually, however, an owner of a nearby cider mill came to Swift's rescue by inviting him to set up his telescope on the mill's flat roof, half a mile from Swift's home. There Swift enjoyed an unobstructed view of the horizon, provided he was willing to make the not inconsiderable effort of getting there. As Barnard again relates the story,

> Access to the bleak place . . . was had only by climbing over several lower roofs and up a rickety wooden ladder – a process hazardous enough for one with sound limbs, but doubly so to him because of his lameness. The tube and stand were left on the roof permanently, but the object glass and eyepieces were carried back and forth in a basket, and it was by this means that a fall broke the flint lens of the $4\frac{1}{2}$-inch.[7]

Swift subsequently replaced the damaged Fitz lens with one made by Alvan Clark, and with this refurbished telescope the budding astronomer discovered a comet a year in 1877, 1878, and 1879.

Rochester grew justifiably proud of the achievements of the local hero, and the interest generated by his work led to the establishment of a first-rate public observatory, of which Swift became director. A local entrepreneur, H. H. Warner, purveyor of a popular patent medicine known as 'Warner's Safe Remedy' (a preparation promoting good bowel hygiene, a late-Victorian obsession) put up $100 000 for a fine building at the corner of East Avenue and Arnold Park, while the citizens of Rochester raised the funds for the observatory's 16-inch Clark refractor, then the fourth largest refracting telescope in the country.

The splendid observatory was not actually ready for its occupant until 1882, and meanwhile Swift continued to sweep for comets from the roof of the cider mill. On August 11, 1880, he found a suspicious object, but he was able to observe it for only an hour when the sky suddenly clouded up. Returning to the same sector of the sky several nights later, Swift found that the object had indeed vanished from the field, but by then the Moon had begun to interfere with his observations. Before it would be out of the way, Swift was scheduled to leave on a trip, and was concerned lest the suspected comet be lost. Thus he appealed for help to a number of other ardent observers, which by then included Barnard, to whom he wrote on August 18:

> I wish you would take part in the search. It [the comet] was faint but not very faint and was elongated in the direction of the sun. It must have been moving very slow. Searched 2 [hours] yesterday a.m. for it but in the a.m. it is too low.[8]

Presumably Barnard did 'take part in the search,' but the comet was never recovered.

Swift found another comet, his sixth, in October 1880, and wrote to Barnard asking him to try to measure its position.[9] Barnard was already skillful enough with the ring micrometer to do so, and Chandler used his measurement along with several obtained elsewhere to work out a preliminary orbit. Barnard was duly credited in Chandler's report to the *Science Observer*, though his address was incorrectly given as Jersey City.[10]

3

Meanwhile, Barnard was taking time out from his heavy schedule to go from bachelor to benedict, marrying Rhoda Calvert on January 27, 1881.[11] The ceremony took place on a Thursday – presumably Poole gave Barnard the day off – and was performed by a colorful Baptist preacher of that era, Tiberius Gracchus Jones.[12] Anson Nelson stood up for Barnard as best man.

At least by the standards of the day, it was an unusual match. Like Swithin St. Cleve, the young astronomer in Thomas Hardy's *Two on a Tower*,[13] though with a happier outcome, Barnard had fallen in love with a woman much older than himself – Rhoda, thirty-seven years old to Barnard's twenty-three, was almost old enough to be his mother. The fact may become more understandable if, psychologically, a mother figure was indeed what he was looking for. It is possible that in marrying Rhoda he was unconsciously trying to recreate the situation with his mother, who had long been ill and could not be expected to live much longer.

Though one suspects that their romance may not exactly have been high passion, the devotion of Barnard and Rhoda to one another is beyond question and seems only to have increased with the years. Barnard later wrote movingly of 'the blessing of taking a partner for life and the renewed struggle to save something with the addition of a cheerful helper in the matter,'[14] something that only too eloquently testifies to his chronic scarcity of funds and the loneliness of the struggle he had faced until then.

With a wife and an invalid mother to support on his small salary from the Photograph Gallery, Barnard became a dedicated seeker for comets. In this new quest, pecuniary considerations must have motivated him as much as the piquancy of fame, for just then Warner, following up on a suggestion by Swift, established a $200 prize to be awarded for every new comet discovered in 1881 by an observer in the United States or Canada. The amount was almost half Barnard's annual salary at the Photograph Gallery at the time, so it was no slight incentive. The first Warner prize was won by Swift himself, who discovered his seventh comet on April 30, 1881.

4

In quest of the Warner prize, Barnard was out sweeping the skies at 3 a.m. on May 12, 1881. In the course of his sweeping he came to the star α Pegasi. He noticed an object in the same field with that star where there was no known nebula. In the hour he kept it under observation, no motion was detected, but by the following morning the object had disappeared from the place. Searching frantically for it, Barnard finally recovered it just 'as daylight was whitening the sky, very closely north following α Pegasi, and only visible when the bright star was hidden by the ring micrometer.'[15]

Warner's terms had stipulated specifically that anyone who believed himself to have captured a new comet must announce the discovery to Swift alone, who was charged with responsibility for cabling the news to astronomers elsewhere. Now sure that he had a comet in his grasp, Barnard cabled Swift early on the morning of May 13. Swift in turn communicated the information to half a dozen other astronomers, including Chandler, William Robert Brooks of Phelps, New York, and Pittsburgh telescope-maker John A. Brashear. The morning papers in Nashville also picked up the news and heralded the discovery by the local astronomer – now, it seemed, on the verge of a much wider celebrity.

Unfortunately, and to his subsequent mortification, Barnard was unable to recover the comet on the following morning, and thereafter cloudy weather and moonlight interfered. Chandler searched from Cambridge, Massachusetts, without success, and on May 15 wrote to Barnard:

> Friday night (Sat. A.M.) from 1 A.M. to 4 A.M. [I] was unsuccessful after careful search, in finding your comet using 3-inch aperture. Telegraphed Swift, thinking he would have better luck, but he telegraphed 'couldn't find it.'
>
> Last night was cloudy, and I think tonight will also be, and the moon is getting so near that I am afraid it is all up with my seeing it . . . We are very anxious to get all obs[ervatio]ns of position as soon as possible for orbit purposes . . . I cordially trust that you may pick this up again, or failing that, that fortune will soon favor you with another find. Do not be discouraged by bad luck. Success to your efforts!![16]

Though the sky was cloudy at Cambridge that night, it was clear at Rochester, and Swift tried again. He telegraphed Chandler: 'Comet not found. Hazy. Fear it was a ghost,' i.e., an optical reflection of the nearby star α Pegasi. Swift wired the same message to astronomers elsewhere, and in a long letter explained to Barnard:

> . . . the statement that [the comet] was bright and that you saw it as late as 4 o'clock raises the suspicion in my mind that you was treated to a ghost of α Pegasi. The fact that its motion was so very slow is another strong evidence of its ghostliness. Still as late as 4 o'clock it does not seem as though the star would be bright enough to produce much of a ghost. If my suspicions are true, and they are only suspicions, then you have the consoling reflection that you are not the only man that has been thus deceived . . .[17]

Though obviously dubious of Barnard's claim, Swift tried to soften the blow somewhat with a story from his own early career. He admitted that he himself had once announced the discovery of a 'comet,' revolving once in twenty-four hours around Jupiter, which further study had shown to be an optical ghost. As a future aid to Barnard in comet-sweeping, Swift generously offered to mark in Barnard's copy of

Burritt's *Geography of the Heavens* the positions of 400 nebulae which he had carefully entered into his own copy – no mean task. These nebulae were easily confused for comets, and Swift thought that Barnard would profit by having a handy reference of their positions. Swift closed on a note of commiseration:

> Your disappointment that you cannot find [the comet] must be great for the jingle of those yellow boys in your pocket would have made captivating music not only to yourself but to your invalid mother and the honor it would have brought you would be worth more than the prize.

Barnard himself strenuously objected to the 'ghost' theory. Despite his youth, he was already a seasoned observer, and he insisted that he had taken every precaution to guard against such an innocent deception. Chandler, meanwhile, had reached the conclusion that Swift was really to blame for the whole affair. As he wrote confidentially to Barnard on May 25:

> I am very much provoked with Swift for cabling that the comet was probably a mistake. This he had no manner or right to do, at least in that shape. The only reasons he had were insufficient ones to justify his action . . . All the evidence he had was, 1st, that he with only $4\frac{1}{2}$ in., on a hazy morning, moon a day older, could not see it. 2d that I with only 3 ins. could not see it. On *this*, he took the responsibility of butting foreign astronomers off the track of searching for it. If he had simply cabled the fact that he could not find it, it would have been very well, but to say it was a *mistake* conveyed a very false impression.

Alluding to ill feeling between Swift and C. H. F. Peters of Hamilton College (Clinton, New York) after the latter had slashingly dealt with Swift's supposed observation of an intra-Mercurial planet at a total solar eclipse in 1878, Chandler added:

> I think your claim will be endorsed by fairminded men. Prof. Peters has written us one or two letters. You know there is no love lost between him and Swift. Peters called Swift publicly a 'bungler'; and criticized pretty severely his claims to seeing the Intra-mercurial planet. If he had the facts in the case of your comet, he would go for Swift with a sharp stick.[18]

Chandler and Ritchie encouraged Barnard to go ahead and claim the Warner prize anyway. However, they must have known that it was a lost cause. If Barnard ever entertained a thought in that direction, he dismissed it promptly on receiving a letter several days later from Warner himself. Concerned about the integrity of his prize (and his own reputation), Warner was miffed that news of a 'discovery' had been leaked to the press before it had been fully confirmed. He went so far as to suggest that Barnard publish an open letter of retraction: 'Many of them who have published the accounts of its discovery may feel provoked at us for notifying them of something which they may feel has not occurred,' he wrote cooly, 'and when we cry "wolf" again, close their columns entirely.'[19]

Needless to say, Barnard's pride was injured by the whole affair. Years later he wrote from bitter experience of comets found only to vanish seemingly without a trace from the sky:

> [T]he observer hastily seeks the suspicious object, only to find that it . . . has vanished . . . Experiences such as these have time and time again been the lot of those who make comet work their especial study . . . The student must not let such an experience embitter him for the work. Ambition should be made of sterner stuff.[20]

5

After this near brush with fame and $200 of Warner prize money, Barnard dedicated himself even more wholeheartedly to comet-seeking. Presumably the spectacular display put on in June 1881 by Tebbutt's Comet – it became almost as bright as the Great Comet of 1861 – also did much to awaken the 'sterner stuff' of his ambition.

He was certainly familiar with the careers of such great comet discoverers as Charles Messier of France, who found more than a dozen at the end of the eighteenth century from his small observatory atop the Hôtel de Cluny in Paris, and the remarkable Jean-Louis Pons, who began his career as a humble doorman at the Marseilles Observatory but went on to become one of the world's greatest astronomers. Of the forty-three comets reported between 1801 and 1828, he discovered, or shared in the discovery of, thirty-four (the often quoted figure of thirty-seven, which was the number Pons actually claimed, includes several which were not well enough observed to have their orbits calculated).

Barnard's dream was now to emulate their discoveries and to write his name across the sky. Indeed, as the custom was, and is, to name comets after their discoverers, there was a certain immortality to be won thereby. Here at last was a way for the poor boy to make good.

The work required no advanced equipment, and Barnard's 5-inch Byrne refractor was excellent for the purpose. Barnard's method was to sweep the telescope sequentially across horizontal zones of sky. With the completion of each sweep, he raised the telescope slightly and began a new sweep of a zone partially overlapping the first. As he aptly described it, the sweeps, 'if plotted on a piece of paper, would resemble in length, breadth, and overlapping the weather-boarded side of a house.'[21]

Though many searchers for comets confined themselves to the regions near the Sun after twilight or before dawn, hoping to take advantage of the fact that comets grow brighter as they approach the Sun, Barnard distrusted this practice. 'It is doubtful,' he later wrote, 'whether the brightness of the average telescopic comet under these conditions is not more than offset by the illumination of the sky and the comet's nearness to the horizon.'[22] He himself took no chances, making use of every hour of the night in which the Moon did not interfere. Indeed, in describing what soon became his own habitual practice, he noted that 'frequently, starting in the west at nightfall, the comet-seeker finds the first gray streaks of dawn illuminating his telescope as he completes the last few sweeps low down in the eastern sky. Throughout the long dreary night he has searched the heavens without success – repaid only by the knowledge of a duty carried out . . . A clear sky, no favor, and *no moon* is his watchword.'[23]

Needless to say, such work could be tiring and extremely discouraging. Nevertheless, even unsuccessful sweeping was not without its consolations. Above all, Barnard later noted, the comet-seeker was 'the *only* astronomer who thoroughly knows the heavens; for he must examine every portion of it time and time again . . . Everything is interesting and numberless objects are beautiful in the extreme. There is nothing commonplace in the sky.'[24]

For Barnard, the wonder of wonders was the Milky Way. In an 1883 article for the Nashville *Artisan*, a bimonthly publication whose astronomical department was placed 'under the exclusive control of E. E. Barnard,' he reflected on some of his experiences as a comet sweeper:

> Sweep on through glittering star fields and long for endless night! More nebulae, more stars. Here a bright and beautiful star overpowering in its brilliancy, and there close to it a tiny point of light seen with the greatest difficulty, a large star and its companion. How plentiful the stars now appear. Each sweep increases their number. The field is sprinkled with them, and now we suddenly sweep into myriads and swarms of glittering, sparkling points of brilliancy – we have entered the milky way. We are in the midst of millions on millions of suns – we are in the jewel house of the Maker, and our soul mounts up, up to that wonderful Creator, and we adore the hand that scattered the jewels of heaven so lavishly in this one vast region. No pen can describe the wonderful scene that the swinging tube reveals as it sweeps among that vast array of suns.[25]

The nebulae, those mystifying objects which passed in great numbers through the field of the eyepiece and could look so deceptively like comets, also intrigued him. Messier, indeed, had deemed them mere distractions in his single-minded quest for comets – his main motive in drawing up his famous catalogue of 'nebulosities,' published in the *Connaissance des Temps* in 1784, having been just that of helping him to distinguish these objects from actual comets so as to waste as little time as possible on them. Others, however, wondered what they were. By the end of the eighteenth century, William Herschel of England had made the first careful study of them, cataloguing them by the thousand and describing their beauty and variety.[26] Some, as we now know, are indeed true nebulae – vast clouds of gas and dust; others are galaxies, enormous star-systems far beyond the Milky Way. In Barnard's day, their nature was one of the great unsolved problems of astronomy.

While some nebulae could easily be recognized by their fantastic forms, the majority were, at least in the 5-inch telescope Barnard used, 'roundish patches of foggy matter, extremely like comets in appearance.'[27] Thus, the beginner had to tread carefully, avoiding at all costs the embarrassment of announcing as a new comet what was actually a well-known nebula. With constant application to his work, the comet-seeker became familiar with most of the nebulae visible in his telescope.

6

In July 1881, Barnard's heart must have skipped a beat when he came across a nebulosity near the star ρ Virginis which was not in any of the catalogues. Barnard wrote to Swift, who confirmed that there were no known nebulae in this position. Unable to search for it himself because of rain, Swift urged his young colleague, 'If you have made no mistake about the position & have not been deceived by 2 or 3 faint stars close together it is a comet *sure*. Hunt it down.'[28] Further observations showed no motion, however, and in the meantime Chandler, notified by a postcard from Swift of Barnard's observations, proposed to have the object searched for by his colleague O. C. Wendell using the 15-inch Harvard refractor. Unfortunately, by the time Wendell got around to it, twilight was drowning out that part of the sky, and it was another year before the object was finally confirmed as a nebula new to astronomy. Chandler sent Barnard his belated congratulations: 'It will prove a very creditable thing for you to have found, with five inches, an object too faint to be in any known list.'[29] The nebula was included in J. L. E. Dreyer's *New General Catalogue* as NGC 5584 – the first of a score of these objects which Barnard discovered in Nashville while comet-seeking.

There was, alas, no Warner prize awarded for new nebulae, and creditable though such a discovery might be, only a comet would fully redeem Barnard for the discouraging affair of the preceding May. Fortunately, Barnard did not have to wait long to reap the reward of his diligent work. At 7:15 p.m. on a Saturday evening, September 17, 1881, he captured a new comet, riding low in the west after sunset, some 2° west of a naked-eye star which he identified with difficulty as ξ Virginis. Barnard estimated the comet's magnitude at $8\frac{1}{2}$ and described it as round in shape, two minutes of arc across, with a definite central condensation. Acting on advice received previously from Chandler, Barnard telegraphed Swift as required by the Warner prize stipulations but as a hedge also sent its position in confidence to Ritchie at the *Science Observer*, who in turn communicated it to the director of Harvard Observatory, E. C. Pickering.[30] Sunday night, the 18th, Barnard saw the comet briefly through clouds and secured another position, but by the 19th it had moved too far north of the star ξ Virginis for him to determine its position with his ring micrometer. The next day he paid a visit to Landreth asking him to try for the comet that evening with the 6-inch Cooke equatorial at Vanderbilt Observatory; Barnard joined him in the dome as Landreth succeeded in obtaining another position.

Though Swift searched for the comet in vain and Chandler obtained only a single position because of a run of cloudy weather at Cambridge, enough positions were obtained at Nashville and elsewhere to allow Chandler to calculate the orbit of the new comet. It received the provisional designation 1881e (indicating that it was the fifth comet discovered in 1881). Barnard must have heaved a sigh of relief on receiving Chandler's postcard of September 26, which contained his rough set of orbital elements, for there was now little chance of the comet's being lost.[31] Two days later Chandler wrote to Barnard: 'I am most sincerely glad on your account for this success. I think it will give you a great deal of credit, among astronomers, to have found it, so near the sun, and when so short a time was available for seeing it after himself. It more than atones for the loss of the object in May last.'[32] Brashear, who had received notice of the discovery by telegram from Swift (and at the time of Barnard's announcement of the lost comet in May had privately inclined to Swift's 'ghost' theory) added his warm congratulations:

> I was more than delighted with your success in picking up comet e and thus effectively closing the mouth of fault finders at your last announcement, I always did believe you saw something before and altho in my letter to you I conveyed the idea that you . . . had been deceived by a false image, yet after receiving your letter I was perfectly satisfied that you knew what you were doing.[33]

Comet Barnard was eventually assigned the official designation 1881 VI – it was, of all the comets that came to perihelion that year, the sixth to do so. It proved to be a rather modest affair, remaining difficult of observation on account of its proximity to the horizon, and was not seen at all by observers in Europe until October 1. The last observation of it was made by Wendell with the 15-inch Harvard refractor on October 27. In his report on the comet in *Astronomische Nachrichten*, the distinguished German journal published at Kiel, Barnard took pleasure in noting that its discovery 'was not due to accident, being the result of a careful search begun in May 1881.'[34]

For this discovery, Barnard received his first $200 Warner prize. In point of strict technicality he should not, perhaps, have been eligible, as he had violated Warner's stipulation that Swift at the Warner Observatory was to have complete priority in receiving news of all such discoveries. But though Barnard's conscience was much more acutely developed than most, he seems to have felt little if any compunction on this account. Swift later urged him to follow procedure more strictly in any future discoveries, in order to placate the publicity-hungry Warner. 'All Warner asks,' Swift reminded him, 'is the privilege of sending the Associated Press dispatches, and as he spends his money on it, it seems not so very unreasonable that he should demand that much.'[35] Swift promised henceforth to cable astronomers at Harvard and elsewhere immediately upon receiving any telegraph, assuring Barnard that by cabling him first 'it will not delay the announcement over an hour.' Apparently Barnard found these assurances satisfactory, as he henceforth adhered strictly to procedure.

With Warner's check in hand, Barnard and his wife, so long used to sailing close to the wind financially, were faced with the pleasant question of what to do with the money:

> After due deliberation it was decided that we would try to get a home of our own with it. I had always longed for such a home where one could plant trees and watch them grow up and call them our own. So we bought a lot with part of the money, which was on what was afterwards called Belmont Avenue [now 16th Avenue South], but which was not then even a road. It was hard to find the lot after it was bought, for it was out in the open common. The place was in the midst of a scattered settlement of negro shanties, where the negroes had 'squatted' after the war, though on a beautiful rising ground which I had selected in part because it gave me a clear

horizon with my telescope.

After some saving and some borrowing, and mainly a mortgage on the lot, we built a little frame cottage where my mother, my wife, and I went to live. Those were happy days, though the struggle for a livelihood was a hard one, with working from early to late for a bare sustenance (and the hope of paying off the mortgage), and sitting up all the rest of the twenty-four hours, hunting for comets.[36]

Edwin B. Frost later quipped: 'Few, indeed, are the astronomers whose keen eyesight and extraordinary diligence in the quest for celestial discovery have literally provided them with a roof to sleep under. It was very little, however, that he slept under that roof when the sky was clear.'[37]

7

As if he did not have enough to occupy him between his work at the Photograph Gallery and his sweeping for comets, Barnard decided to try his hand at raising chickens. He purchased several of them, despite the skepticism of friends who warned him that

... the temptation would be too great for our dusky neighbors to withstand. However, we went ahead, and the chickens thrived and prospered; and not one was stolen. There were various theories to account for this extraordinary state of affairs. One was that I was up all night watching the chickens. Another was that the superstitious negroes had seen me out at night, 'communing with the stars,' and had concluded that I was some sort of astrologer, and that my intimate knowledge of the stars might make it a dangerous thing to look with envious eyes upon my chickens ... Still another – and one that appealed to me as much as any other – was that our kindness to our colored neighbors had gained their respect and consideration. However this may be, it is a fact that our chickens crowed and cackled and 'flourished as the green bay tree,' with never a thought of the danger that lurks in the night.[38]

Unfortunately, the episode did not end here. After the Barnards moved onto the campus of Vanderbilt University, their house was rented to Barnard's brother-in-law, Ebenezer Calvert. In addition to renting the house, Calvert bought Barnard's chickens and added others to his stock. Though Barnard regarded the mild-mannered Calvert as 'if anything, a more kind and considerate friend to the colored man than we had been,'

The result was disastrous; the chickens disappeared mysteriously, and the dawn would see a decimation of stock that was frightful. It got so bad that nothing that had feathers was safe overnight, and it is related, though it is doubtful if it be true, that one night the feather bed was magically transferred from beneath him even while he slept upon it.[39]

Barnard reluctantly concluded that it must have been the superstitious beliefs of his neighbors after all that had protected his chickens from destruction.

8

Barnard lived in the little cottage 'on the beautiful rising ground' for just over a year, sweeping diligently on clear and moonless nights for comets. On March 16, 1882, he discovered another comet, but he was unable to confirm it because of a streak of cloudy weather.[40] As a result, he never received credit for it.

That same month, he joined forces with fellow comet-seekers Swift and Brooks in a plan proposed by Ritchie, whereby each observer was assigned responsibility for scanning a given part of the sky for comets. Swift claimed the circumpolar region, at the same time suggesting that Barnard, because of his more southerly latitude, ought to sweep the southernmost parts of the sky.[41] The division of labor was most advantageous to Swift, who in his own comet-sweeping had already made a point of avoiding Sagittarius, Virgo and Coma Berenices – as he admitted, 'the neb[ulae] there bother me too much and as there is plenty of sky without I search elsewhere.'[42] It was outrageously unfair to Barnard, whose zone included both Sagittarius and Virgo – the latter, in particular, has a record number of what David Levy has called 'comet masqueraders.'[43] Nevertheless, Barnard acquiesced to the arrangement, and for that matter there was nothing binding in it – once each of the collaborators had swept his zone for the month, he was free to move on to 'pastures new.'

The group was forestalled, insofar as the next comet discovery fell not to one of the three but to C. S. Wells of the Dudley Observatory, who found Comet 1882 I on March 17 (Brooks, to his credit, did find it independently). Wells's Comet became an impressive naked-eye object in May and June, with a splendid tail.

In a bizarre twist, Wells's claim was disputed by a 'Professor' John M. Klein of Hartford, Kentucky, who had made a 'prediction,' published in the Louisville Courier-Journal of February 7, 1882 (and later reprinted in the New York Sun) that a bright comet would appear 'about the middle of March.' Subsequently the Courier-Journal published a grandiose letter from Klein in which he proclaimed himself 'once more victorious over the old-school astronomers,' whom he hoped to teach 'that a modern Kepler can live in an interior village of Kentucky and yet understand some of the principles of the great science of astronomy.' And he added: 'They will not cheat me out of the honor of discovery in this instance. I thank the Courier-Journal and the press generally for defending me against a hungry hord of scientists, who would claim the honor of the discovery of my comets.' With evident irritation, Swift, who sent Barnard these newspaper clippings, urged Barnard that 'If their cock and bull story begins to get a wide circulation I wish you would set the matter right before the people of the south.'[44]

Barnard's next discovery – not, apparently, disputed by the Kentucky 'Professor' – was made at 3 a.m. on September 14, 1882, while he was sweeping outside of his assigned search zone in the region near the star λ Geminorum. As usual Barnard promptly communicated news of the discovery to Swift, whose cable to the Harvard Observatory spoke of the discovery as 'not yet certain.' Wendell, however, put it beyond doubt the following night, capturing it with the 15-inch refractor at Harvard.

On the evening after he had reported this comet, Barnard paid a visit to friends, Mr. and Mrs. J. R. Brown, but said nothing about what he believed he had found. Subsequently he explained to Mrs. Brown,

> Mr. B. will likely think it rather strange that I said nothing of my discovery to him that evening at your House, but I was a little uncertain having only seen the [comet] that morning, and I do not like to raise false hopes [in] anyone.[45]

Barnard's latest comet (1882 III) was very faint when he first detected it. He described it as a round nebulous object of about the tenth magnitude, but it brightened over the next month, until by October 16 he was able to make it out easily in his $1\frac{1}{4}$-inch finder telescope. By then he also noted 'a smudge of a tail pointing away from the Sun.'[46] The comet came to perihelion on November 13, and astronomers followed it into the skies of the southern hemisphere, where it was last seen as a ninth magnitude object by Tebbutt at Windsor, New South Wales, in Australia, on December 8. As before, Barnard received a $200 Warner prize for the discovery.[47]

9

Barnard's small comet was completely overshadowed by the comet known officially as 1882 II, which appeared at almost the same time and became, in the words of astronomical historian Agnes M. Clerke, 'one of the most remarkable cometary apparitions on record.'[48] The years 1880–81–82 were indeed wonderful for comets – during that brief period, there were no less than nine which became bright enough to be seen with the naked-eye – but this was the last and the greatest. Indeed, nothing like it had been seen since the Great Comet of 1843, which had followed a similar orbit.

The Great Comet of 1882, or the 'September Comet' as it was popularly known, was first seen by a number of observers in the southern hemisphere in early September, when it was already brilliant. So poor was communication between astronomers, however, that it was independently discovered at least five times in two weeks, and as late as September 17, when the comet reached perihelion, some observatories had still not received information about it – indeed, on that date it was independently discovered yet again, by A. A. Common in England.

On that date, the comet actually passed in front of the Sun, though despite careful searches it could not be seen in transit, the bulk of the comet being, as Clerke explained, 'of too flimsy a texture, and its presumably solid nucleus too small, to intercept any noticeable part of the solar rays.'[49] Over the next few hours it emerged briefly from the solar disk then swept behind the Sun. The morning after its spectacular perihelion passage, Sir David Gill, observing at the Cape of Good Hope, reported that the comet appeared of 'an astonishing brilliancy as it rose behind the mountains on the east of Table Bay,' and when the Sun rose minutes later, he found it 'in no way dimmed in brightness, but becoming instead whiter and sharper in form as it rose above the mists of the horizon.' Later in the morning, Gill pointed it out in broad daylight to friends as he passed them on the way back to the observatory, noting: 'It was only necessary to shade the eye from direct sunlight with the hand at arm's-length to see the comet with its brilliant white nucleus, and dense white, sharply bordered tail of quite $\frac{1}{2}°$ in length.'[50]

The comet was now sweeping rapidly into the predawn skies of the northern hemisphere. Word of it first reached the United States through the Smithsonian Institution on September 17, and on that date it was observed by Chandler and Wendell at Harvard. Because of cloudy weather at Nashville, Barnard did not see it until the morning of September 22. That morning, as seen with the naked eye, the tail was 'very bright, straight and slender, about 12° long; the head and nucleus . . . rather foggy.' In the 5-inch refractor the nucleus looked to him like 'a ball of diffuse light seen through haze,' which remained visible as a speck of light even after daybreak.[51] Barnard continued to observe the comet carefully through succeeding weeks and later published detailed and highly readable accounts in the *Sidereal Messenger* and the German journal *Astronomische Nachrichten*.

On the morning of October 4, Barnard noted that the comet's nucleus appeared much elongated. The following morning it seemed actually to have split into three unequal parts. On October 13, he recorded the following impressions in his observing book:

> High south-west wind, instrument unsteady. The comet is seen to have two short filmy tails pointing *towards* the Sun and opposite to the main train . . . The head is enormous, filling the whole field of view with a mass of bright pearly vapors, which, in the unsteady view, seems to shine from separate centers, the nucleus certainly being *very* elongated. No description can do justice to the wonderful appearance of the head and its appendages on this occasion.[52]

What he saw the next morning, however, was even more remarkable. 'My thoughts must have run strongly on comets during that time,' he later wrote of that memorable morning, 'for . . . thoroughly worn out I set my alarm clock and lay down for a short sleep':

> Possibly it was the noise of the clock that set my wits to work, for they say that in a dream a whole lifetime may pass in review in the single moment of waking; or perhaps it was the presence of that wonderful comet which was then . . . gracing the morning skies; or perhaps it was the worry of the mortgage and the hopes of finding another comet or two to wipe it out. Whatever the cause, I had a most wonderful dream. I thought I was looking at the sky, which was filled with comets, long-tailed and short-tailed and with no tails at all. It was a marvelous sight, and I had just begun to gather in the crop when the alarm clock went off and the blessed vision of comets vanished.[53]

On being jarred awake from this pleasant dream, Barnard took his telescope out into the yard in order to observe the comet. He began sweeping south of it. Suddenly, some 6° southwest of its head,

I 'picked up' a large distinct cometary mass, fully 15′ in diameter. A similar object but less bright was seen close beside this, their edges touching – apparently a double comet – and on the opposite side of the first object was a third fainter mass, the three almost in a line, east and west . . . A slight displacement of the telescope toward the south-east revealed a number more of these wonderful objects; one was very elongated in form. There were at least, six or eight of those objects near one another within about 6° south by west of the large comet's head. Their appearance was that of distant telescopic comets with very slightly brighter centers and so near that several were in the field at once. Time was used in sweeping for more, so that no watch was made for motion; but from a rough comparison of a number of small stars not in the field it was evident that the brightest object was not moving rapidly.[54]

He found these small comets, which in his drawing somewhat resemble a flock of geese flying in wedge formation, at about 4:30 a.m. on October 14, and they remained visible until daylight 'killed them out.' Whether out of excitement or in order to have an independent confirmation, Barnard took the, for him, unusual step of rousing his wife, who also saw them 'easily.'

Later that morning, Barnard sent a telegram to Swift. In Barnard's later recollection of the event, Swift failed to forward the message to anyone else. 'Whether he thought that I was trying to form a comet trust or had suddenly gone demented has never been clear to me. He unfortunately did not forward the telegram,' Barnard claimed.[55] Here, however, his memory failed him; Swift did indeed cable the news of Barnard's discovery.[56] The morning of October 15, a slight haze prevented Barnard from searching for the strange objects, nor did he recover them on later occasions. They had vanished, seemingly without a trace.

After himself searching in vain for the 'multiple comets' on October 15 and 16, Swift vacillated between accepting what Barnard had seen and wondering whether the apparition had been caused, as he suggested to Barnard, 'by moisture on the obj[ect] glass probably on the inside.'[57] Barnard, however, was most emphatic: 'I do not think that doubts of the observations should be entertained . . . I am positive of the reality of my objects.'[58] And this time he was not alone. On October 8, even before Barnard's unusual observation, the seasoned German astronomer Julius Schmidt, at Athens, Greece, had made out a nebulous object 4° southwest of the main comet and travelling along with it in the same direction. Another object of the same kind was found by Brooks on October 21, this time travelling in the opposite direction. Swift later overcame his doubts and concluded that what all three observers had seen were indeed 'fragments of the great comet.'[59] It had been broken up by tidal forces as it had made its hairpin passage around the Sun.

To Barnard, quite as remarkable as the discovery of these unexpected objects was his dream about them. 'The association of this dream with the reality,' he later reflected, 'has always seemed a strange thing to me.'[60] To be sure, Barnard's financial predicament gave him reason enough to dream of comets, as at that very moment he was awaiting the arrival of the Warner prize money that was due him for the discovery of his latest comet, with the note demanding payment on his mortgage in hand.

10

The Great Comet of 1882 remained a striking naked-eye object throughout October and November, as it withdrew from the Earth on its way out to the frigid regions from whence it had come. In the mid-twenty-sixth to mid-twenty-ninth centuries AD, at intervals of about a hundred years, each of the fragments of its nucleus may return to the vicinity of the Sun as a separate comet in its own right.

After the mysterious visitor had well retreated, a friend approached Barnard asking him to make an entry in an autograph book. Instead of the usual sentimental verses which were customary on such occasions and which filled the pages of Barnard's own autograph book, the young astronomer made a freehand sketch of the Great Comet of 1882, beside which he wrote:

> I had intended writing a short tale for your album but feeling my inability to do so, I have drawn one belonging to a friend of mine, which stretched halfway from the sun to the earth. May your happiness be as long as the tail herewith presented.
>
> I can not wish that you may live to again see this wanderer, for it will not come back to us until many thousand years have passed, but I can and do wish that your life may always be as beautiful as that stranger, now gone. Sincerely yours, E. E. Barnard.[61]

1   Stephen B. Oates, *With Malice Toward None: The Life of Abraham Lincoln* (New York, Mentor, 1977), p. 4

2   Howard I. Kushner, *Self-Destruction in the Promised Land* (New Brunswick, New Jersey, Rutgers University Press, 1989)

3   Quoted in Oates, op. cit., p. 106

4   O. H. Landreth, 'Barnard at the Vanderbilt University Observatory,' *JTAS*, 3:1 (1928), 16. Landreth later recalled that he had shown Barnard how to make a ring micrometer shortly after Barnard had discovered his first comet. 'At this time,' he writes, 'Barnard's astronomical equipment did not include the mental nor the instrumental means of measuring and recording the new comet's position from night to night in terms of the accepted astronomical co-ordinates of right ascension and declination, so that, although the discoverer of the comet, he was at first unable to follow up his discovery with the quantitative observations essential to the determination of the comet's orbit.' This recollection cannot be valid, however, in that Barnard was already using such a device by October 1880.

5   E. E. Barnard, 'Lewis Swift,' *AN*, **194** (1913), 133–6.

6   ibid.

7   ibid.

8   LS to EEB, August 18, 1880; VUA

9   LS to EEB, October 11, 1880; VUA

10   As Chandler later explained to Barnard, 'from the wording of Mr. Swift's letter to me, giving among other positions of his comet yours of the 21st, I certainly had grounds to infer the place of obsn to be Jersey City. The mistake is somewhat unfortunate, as the

diff. of longitude, about 51m, affected directly the interval between that obsn and the other observations used in computing my elements. However, the orbit represents the subsequent observations, after all, better than most first elements do . . .' Chandler also noted that 'The possession of your position (altho' it was only approximately given) enabled me to get the precedence in publishing elements and ephemeris.' SCC to EEB, November 4, 1880; VUA. The comet was later shown to be an unexpected return of Comet Tempel of 1869.

11 Davidson County (Tennessee) Archives

12 He made a salary of $4000 a year, which at the time was extraordinary. R. Lagemann to W. Sheehan, personal communication, December 4, 1993

13 Thomas Hardy, *Two on a Tower* (New York, Macmillan, 1960)

14 E. E. Barnard, 'A Few Unscientific Experiences of an Astronomer,' *Vanderbilt University Quarterly*, 8 (1908), 273–88:277

15 E. E. Barnard, 'A Lost Comet,' note in *AN*, **100** (1881), 127

16 SCC to EEB, May 15, 1881; VUA

17 LS to EEB, May 16, 1881; VUA

18 SCC to EEB, May 25, 1881; VUA. An excellent account of the career of C. H. F. Peters and his bitter controversy with Swift is given in Joseph Ashbrook, *The Astronomical Scrapbook: Skywatchers, Pioneers, and Seekers in Astronomy* (Cambridge, Mass., Sky Publishing Corp., 1984), pp. 56–66.

19 H. H. Warner to EEB, May 28, 1881; VUA

20 E. E. Barnard, 'How to Find Comets,' *San Francisco Examiner*, February 5, 1893

21 ibid.

22 ibid.

23 ibid.

24 ibid.

25 Quoted in Fanny D. Nelson, 'Edward E. Barnard,' *Our Day*, 2:2 (1883), 40. In 'How to Find Comets,' op. cit., Barnard had made the same point more succinctly: 'Of all the objects . . . encountered by the seeker of comets the most magnificent is the "Milky Way;" whose wonders are truly seen in all their sublimity only by him. I have never swept over that "broad and ample road whose dust is stars" without the keenest appreciation of the grandeur of creation and the infinite glory of the Creator.'

26 William Herschel, 'Catalogue of one thousand new nebulae and clusters of stars,' *Philosophical Transactions of the Royal Society*, lxxvi (1786), 486

27 Barnard, 'How to Find Comets'

28 LS to EEB, August 1, 1881; VUA

29 SCC to EEB, June 29, 1882; VUA. In the 15-inch refractor, Wendell found it 'rather diffuse and faint, but gradually a little brighter in the middle.'

30 Chandler had written to Barnard on August 8, 1881: 'Would it not be as well for your own advantage in case you find any suspicious cometary looking objects, wh. *your* means dont permit your identifying, to write to us at the same time as to Swift. We would cheerfully look them up, and treat the query as confidential. Writing to Swift first, and his then writing to us makes delay, which might lose you a [comet] some time.' VUA. Chandler, of course, had his own motives – he wanted to secure positions as early as possible in his quest for priority in working out an orbit; but needless to say, after his experience with his first comet, Barnard required very little convincing.

31 Chandler's final orbit, which was based on his observations of September 21, October

3, and one by Wendell on October 21, gave the following parameters: date of perihelion passage September 14 785, longitude of perihelion 280°32′40″, longitude of the ascending node 274°11′39″, inclination 112°47′50″, perihelion distance 0.9652478 AU. The orbital eccentricity was not given but was presumably assumed to be 1.0000. SCC to EEB, October 27, 1881; VUA

32 SCC to EEB, September 28, 1881; VUA

33 JAB to EEB, October 6, 1881; VUA

34 E. E. Barnard, 'Comet 1881 VI,' *AN*, **102** (1882), 155–6

35 LS to EEB, May 28, 1882; VUA

36 Barnard, 'Experiences,' pp. 277–8. The house no longer stands.

37 E. B. Frost, 'Edward Emerson Barnard,' *BMNAS*, **21** (1926), 2

38 Barnard, 'Experiences,' p. 279

39 ibid.

40 EEB, memorandum; SLO

41 LS to EEB, February 27, 1882; VUA

42 LS to EEB, September 14, 1881; VUA

43 D. Levy to W. Sheehan, personal communication, January 22, 1992.

44 LS to EEB, April 31 [sic] 1882; VUA

45 EEB to Mrs. J. R. Brown, September 25, 1882; VUA

46 E. E. Barnard, 'Comet 1882 III,' *AN*, **103** (1882), 287–8

47 H. H. Warner to EEB, October 12, 1882; VUA

48 A. M. Clerke, *A Popular History of Astronomy During the Nineteenth Century* (London, Adam & Charles Black, 3rd ed., 1893), p. 433

49 ibid., p. 434

50 Gill's comments are quoted from Clerke, p. 434, and Joseph Ashbrook, 'Early Photography and the Great Comet of 1882,' in *The Astronomical Scrapbook*, op. cit., p. 166

51 E. E. Barnard, 'Notes on the Great Comet of 1882,' *SM*, 1 (1882–3), 221–3 and 255–8

52 ibid., pp. 222–3

53 Barnard, 'Experiences,' p. 280

54 Barnard, 'Notes on the Great Comet,' p. 223

55 Barnard, 'Experiences,' p. 280

56 LS to EEB, October 14, 1882, telegram: 'Probably same as Schmidts of Athens. Have notified both objects. Swift.' VUA

57 LS to EEB, October 16, 1882; VUA

58 Barnard, 'Notes on the Great Comet,' p. 257

59 LS to EEB, November 13, 1882; VUA. This conclusion was generally accepted. Thus A. M. Clerke wrote in 1893 that the multiple comets were undoubtedly 'the offspring by fission' of the body they accompanied, and added: 'Thus, space appeared to be strewn with the filmy debris of this extraordinary body all along the track of its retreat from the sun.' Clerke, *History of Astronomy*, pp. 439–40.

60 Barnard, 'Experiences,' pp. 280–1

61 Barnard made a photographic copy of this which is preserved in VUA. One can also find there Barnard's own autograph book, which contains entries by most of his early associates in Nashville, including Anson and Fanny D. Nelson, Braid, Calvert, and Russell Marling, Barnard's onetime mathematics tutor. Regrettably, few of these entries rise above the conventional to reveal much about those who wrote them.

5

Vanderbilt astronomer

1

At about the time Barnard discovered his second comet, several prominent citizens of Nashville, led by Judge John M. Lea, resolved to build an observatory for the young astronomer of their city, 'where he might carry on his nightly researches with more convenience, and more safety to his health.'[1] Some of them went to his house and looked over the grounds to decide where the new observatory building should be, and hired an architect to draw up the plans. Rhoda told Barnard when he came home that night what was afoot, but he would not take it seriously. Only when the stones for the foundations began being hauled to the front yard did he realize that their plan was in earnest. The very next day he went to the would-be donors and told them that he could not accept their proposed gift. According to Fanny Nelson, his reasoning was that 'should he accept a gift so costly and elaborate from these kind friends, it would be a strong tie, binding him to his present situation, and he wished to avail himself of whatever better and more advantageous position might be offered him, even though it should call him away from his native city.'[2]

Soon afterwards, Barnard received the better opportunity he hoped for. In late 1882, Olin Landreth invited him to observe the December 6 transit of Venus from the Vanderbilt campus. Nineteenth century astronomers viewed these transits as of the utmost importance because they were tied to a scheme – better, alas, in theory than in practice – for measuring the distance from the Earth to the Sun, which in turn provided the basis for the scale of the rest of the Solar System. The scheme depended on making exact timings from various points on the Earth's surface of the so-called contact points (first contact occurs when the limb of Venus first touches the solar disk, second contact when the disk of Venus gets just completely inside the solar disk; third and fourth contacts mark the corresponding points as Venus moves off the Sun). In order to obtain such observations, expeditions had been mounted to remote corners of the globe at the previous transits of 1874 and 1769. Conditions at the 1882 transit greatly favored observers in the United States, where the entire event would be visible with the Sun at a convenient altitude above the horizon.

On the great day, Barnard set up his 5-inch Byrne refractor 200 feet west of the Vanderbilt University Observatory, where he used a telegraphic sounder on a circuit with the observatory's sidereal clock to make accurate timings, while Landreth and J. T. McGill, a fellow in the Vanderbilt chemistry department, observed the transit with the 6-inch Cooke refractor in the observatory. Clouds over Nashville interfered with

Barnard about 1879. Vanderbilt Photo Archives

first contact, and Barnard, though not Landreth and McGill, obtained only a peep at second contact twenty minutes later through a brief parting of the clouds. Fortunately, by early afternoon the skies had begun to clear. At third contact definition was poor because of heated air from a chimney, which led to the appearance of the infamous 'black drop' – an optical effect consisting of a black ligament forming between the almost touching limbs of the planet and the Sun as a result of turbulent atmospheric conditions. The 'black drop' had often been observed at earlier transits, foiling

Vanderbilt University Observatory. Vanderbilt Photo Archives

attempts to determine the precise moment when contact took place. By fourth contact late that afternoon, conditions had improved, and a faint line of light became observable around the protruding limb of Venus – it was, Barnard recognized, a halo of sunlight being refracted by Venus's dense atmosphere. The startling effect became steadily more noticeable as the planet advanced off the Sun's disk. 'After being visible for several minutes,' Barnard wrote, 'nearly two-thirds of it disappeared and left a soft, feathery horn of light projecting from the edge of the Sun.'[3]

Barnard's timings of the contacts compared favorably with those of Landreth and McGill. Moreover, the vivid descriptions of his observations which he published in the *Sidereal Messenger* and *Astronomische Nachrichten* were among the best anywhere. With this evidence of Barnard's knowledge and ability fresh to hand, Landreth resolved to see what could be done at Vanderbilt to help him develop his astronomical potential.

He hoped for nothing less than to hire Barnard as an astronomer for the Vanderbilt University Observatory. Though one of the first priorities when the university was founded, the observatory had languished in comparative idleness during its first few years, and as the year 1883 opened this situation promised no improvement. 'No special or professional astronomer had been appointed,' Landreth lamented, 'and the prospect of the early appointment of an educated, trained astronomer was not bright.'⁴ However, Landreth saw Barnard as the perfect person to fill the gap: 'Here was offered the unusual opportunity of educating and training such an astronomer, from the most promising material imaginable, and moreover, during the educating and training period, [was offered] the opportunity of utilizing – during part time at least – the professional services of a young astronomer of far more than average promise.'⁵

At first the University chancellor, Landon Cabell Garland, opposed the idea on the grounds that Barnard had insufficient formal education. But the president, Bishop Holland N. McTyeire, 'proved to be sufficiently radical, or at least progressive, to approve the plan and convert the Chancellor to its adoption,' noted Landreth.⁶ McTyeire thus wrote to Barnard on March 1, 1883, notifying him of his election to a fellowship in the University 'connected with Astronomy.' If he accepted the position, Barnard was expected to take charge of the observatory and assist Garland, who in addition to being chancellor was chair of the physics and astronomy department. In return for these services, Barnard was to receive 'instruction in any of the Schools (nonprofessional) of the University free of charge, a salary of five hundred dollars, *per annum*, to be paid quarterly, also the use of a house, convenient to the observatory, free of rent.'⁷ He would be enrolled as a special student in the School of Mathematics, a status which exempted him from the requirements of working toward a degree.

Though the salary offered was significantly less than he was receiving at the Photograph Gallery at the time, with Rhoda's encouragement Barnard enthusiastically accepted McTyeire's offer. After seventeen years he severed his tie with the Photograph Gallery, and he, his wife, and his mother moved from the little cottage on Belmont Avenue into a frame house on campus situated just to the west of the observatory.

2

Just when Barnard was contemplating the move to Vanderbilt, his brother-in-law Ebby Calvert decided to marry. Needing a place to live and the cottage on Belmont Avenue conveniently becoming available, Calvert decided to rent it from Barnard, who as the landlord remained responsible for the mortgage notes on the property.

The Warner prize money received for Barnard's first comet had paid for the lot itself, and later, when Barnard took out a mortgage note on the lot to build the cottage, another Warner prize had paid off the first mortgage note. 'And this,' he afterward recalled, 'continued after we had gone to other scenes':

> The faithful comet, like the goose that laid the golden egg, conveniently timed its
> appearance to coincide with the advent of those dreaded notes. And thus it finally
> came about that the house was built entirely of comets. This fact goes to prove the
> great error of those scientific men who figure out that a comet is but a flimsy affair

after all, infinitely more rare than the breath of morning air, for here was a strong compact house, albeit a small one, built entirely out of them. True, it took several good-sized comets to do it, but it was done, nevertheless.[8]

For many years afterward the house, at 807 Sixteenth Avenue South, was known as 'comet house,' and it remained in the family for many years. Ebby Calvert eventually bought it from Barnard, and with the birth of his four daughters – Mary, Bertha, Zillah, and Alice – he built an addition to the modest dwelling. In the 1890s, when Ebby and his brother Peter Calvert bought out Poole and the Photograph Gallery became Calvert Bros., Ebby moved to another house on the Buena Vista Pike, and the house was sold to Peter, who still lived in it in the 1920s. At that time the building stones which had been intended for Barnard's observatory still remained in the front yard, and Peter wrote of how honeysuckle had come to trail over them, 'as if in the effort to mollify the wounded feelings of the rocks vainly protesting against such an inglorious end, after being designed for so noble a purpose.'[9] The house remained standing until 1964, when it fell victim to Music Row; it was razed to make way for a parking lot.

3

Vanderbilt University, to which Barnard was now to be attached as a fellow, had originally been known as 'Central University.' It had begun as a dream of Bishop McTyeire, Garland, and other leaders of the Southern Methodist Church who had wanted to establish 'an institution of learning of the highest order' in the South. However, such was the exhausted condition of the South after the Civil War, when even long established universities were languishing for lack of funds – some, such as the University of Alabama, had been pillaged and burned – that it did not appear likely that the plan would soon be realized.

Enter at this point Cornelius ('Commodore') Vanderbilt, the New York railroad and shipping tycoon. Not hitherto known for his philanthropic largesse, the Commodore, marrying in the twilight of life a fervent Southern Methodist and a cousin of Mrs. McTyeire, underwent a sudden change of heart. He soon afterwards assumed various philanthropic obligations, of which his most important were in behalf of the proposed university. Expressing the hope that such an institution would help 'unite this country, and all sections of it, so all the people shall be one, and a common country as they were before,'[10] he turned over to Bishop McTyeire a check worth \$500 000 in March 1873. Eventually the Commodore's donation to the University would amount to \$1 000 000. In appreciation for this generosity, the Board of Trust of the University voted that Central University, an institution that hitherto had existed only in the dreams of the elders of the Southern Methodist Church, should be renamed Vanderbilt University, and McTyeire, who commanded the Commodore's absolute trust, became the first president.

The University opened its doors in September 1875, occupying a seventy-five-acre site 'west of the city, beautiful for situation, easy of approach, and of the same elevation as Capitol Hill, which is in full view.'[11] Initially there were but two buildings on the campus, one used by the Biblical Department and the other, University Hall (later renamed Kirkland Hall), for everything else. The next building to be constructed on campus was the observatory. Ground was broken in fall 1875, and the structure was completed by the following spring.[12] Its main instrument was a 6-inch equatorially mounted refractor built by Thomas Cooke and Sons of York, England. Other astronomical instruments included a meridian circle, a sidereal clock, and solar and stellar spectroscopes, of which all except the meridian circle were on hand by September 1876.[13]

From the first, Vanderbilt University aspired to the very highest standard of scholarship – 'start nothing,' Chancellor Garland once said, 'in a crippled condition.'[14] However, it soon became apparent that this standard would have to be somewhat relaxed, at least initially. Despite an understandable desire to admit only well-prepared students, there were simply not enough of them to be found. Of those entering the University in its first year, Garland observed, 'Few had any power of fixed and prolonged attention – or any practical knowledge of the modes of successful study . . . Hence sub-collegiate classes [had to be instituted], which introduced a very large element of a boyish character which invariably tends to the deterioration of manners and scholarship.'[15] By the time Barnard's association with the University began, the standard of preparation of matriculating students had improved so far that Garland could not unreasonably oppose Barnard's appointment on the grounds that, with only one or two months attendance in the Nashville public school system, he lacked the necessary preparation. For that matter, Barnard himself always remained 'painfully conscious of his lack of educational training.'[16]

4

Known to Vanderbilt students as 'Old Grey' or 'Old Horsehead,' Garland himself was one of the faculty members who exerted the strongest influence on Barnard during his four years at Vanderbilt. Born in 1810 in Nelson County, Virginia, near the village of Lovingston, some thirty miles south of Charlottesville, Garland had received his own education at Virginia's Hampden-Sydney College, where his older brother was a professor of Greek. After receiving his B.A., Garland briefly considered becoming a lawyer. However, in 1829 he was offered and accepted a position as a teacher of mathematics and natural science at Washington College (now Washington and Lee University), in Lexington, Virginia. Thus began a distinguished teaching career that spanned more than six decades, in which he served successively at Washington College, Randolph-Macon College in Ashland, Virginia (where McTyeire had been one of his students), the University of Alabama, the University of Mississippi, and finally Vanderbilt. He once summed up his own philosophy of education:

All true education is from within – the energizing of the mind itself. Your teachers cannot study for you. We may remove some obstacles from the way; we may save you from wandering into harmful bypaths, but you will have to do the hard and steady and systematic work . . . [while] all the learning in the world cannot substitute for a moral principle in shaping the conduct of life.[17]

Many of those who knew Garland spoke of his 'impressive moral force' and the 'sheer magnitude of his character.' To Barnard, he was always 'the noble Dr. Garland.'[18]

When Barnard's fellowship began, Garland was still remarkably spry at seventy-three, and in addition to his heavy administrative responsibilities still taught the first two years of coursework in physics and astronomy. Mechanics, waves, heat, electricity, and magnetism were covered in the first year, optics and astronomy in the second. As his textbook, Garland used an English translation of a French textbook, Adolph Ganot's *Eléments de Physique*,[19] and 'often asked students to recite, and even to come forward to the front of the room and illustrate their replies with the apparatus at hand.' In these demonstrations the students were assisted by Charles Schott, Sr., the same man who had sold Braid the brass spyglass tube that had been used to fashion Barnard's first telescope. Garland and Schott wore small skull caps and long laboratory coats and resembled 'old alchemists.'[20] Rumored to be too bibulous, Schott served as curator of instruments at Vanderbilt between 1876 and 1894.[21]

As a special student whose purpose was to develop his potential for astronomical work rather than to qualify for a degree, Barnard was allowed to register for classes weeks after a semester opened, and he often took reduced course loads. In addition to the coursework in physics and astronomy on which he concentrated during his first two years, he studied mathematics under William J. Vaughn, the chair in mathematics, who had been one of Garland's students at the University of Alabama, while W. M. Baskerville, McTyeire's son-in-law and a noted scholar in Anglo-Saxon and Middle-English, helped Barnard fill in the gaps of his rudimentary education by teaching him English literature and a smidgeon of French and German.

5

In many ways, Barnard was an unusual student. At twenty-five, he was older than most of the undergraduates. He was, moreover, a married man and had an invalid parent to care for. Not for him were the usual temptations of student life. He was altogether too moralistic for any of that, even if his demanding work ethic and domestic obligations had granted him any leisure time. He was not unsociable, but his social life always came a distant second to his work, and he clearly let it be known, if called upon to visit friends, 'Well, if it rains or is cloudy, I will come.'[22] Otherwise they knew they were not to expect him. To his fellow students he appeared awkward and shy, and he was notorious for rushing into classes 'at the last minute and out of breath.'[23]

Once he was settled on campus, Barnard's foremost priority was to resume his

comet-seeking. For this he preferred his own 5-inch Byrne refractor to the 6-inch
Cooke refractor, for the simple reason that it was portable rather than being mounted
permanently in an observatory and thus could be moved about readily as needed to
reach relatively inaccessible parts of the sky. He used the 6-inch refractor for transit
measurements with the meridian circle and observations of planets and double stars, for
which its greater power and stability were advantages. Though during his first year at
Vanderbilt he failed to discover any comets, he succeeded in making several other
important discoveries as by-products of these searches.

6

The first of these discoveries was made on July 17, 1883. While he was sweeping for
comets southwest of the Trifid Nebula (M20) in Sagittarius, he encountered at celestial
coordinates RA 17h 56m, declination $-27°$ 51', a 'small triangular hole in the Milky
Way, as black as midnight. It is some 2' in diameter, and resembles a *jet black* nebula.'[24]
He later gave a somewhat fuller description of it: 'a most remarkable small inky black
hole in a crowded part of the Milky Way . . . with a bright orange star on its n[orth]
p[receding] border and a beautiful little cluster [NGC 6520] following.'[25]

   The hole was so striking that Barnard was surprised that it had never been recorded
before. Though he knew of 'other dark holes and gaps in the Milky Way' – including
the apparent void which William Herschel had encountered long before in the body of
the Scorpion; on seeing it, Herschel had exclaimed to his sister Caroline, 'Da ist
wahrhäftig ein Loch im Himmel' (there truly is a hole in the heavens) – none, Barnard
wrote, was as 'small and decided' as this. Barnard's description of the object as
resembling a 'jet-black nebula' was certainly prescient, and anticipated his later view
that there were dark obscuring masses between the stars. But for a long time – awed by
the authority of William and John Herschel, both of whom had believed that the
apparent voids in the heavens were true vacancies – he could not bring himself to take
seriously the implications suggested by his own analogy.

7

Barnard was sweeping for comets again on October 4, 1883, when he paused for a
moment to take his eye from the telescope for a brief rest. He was at once 'struck with
the appearance of a large hazy mass of light south of the zenith and near α Pegasi. At
first I thought it might be a slight cloud,' he wrote, 'but it remained pretty constant.'[26]
The skies became overcast and no further observations could be secured that night, but
the following night Barnard found the mass of hazy light again. He continued to follow
it for the next few days. His observations showed that the cloud was moving eastward
along the ecliptic at the rate of about one degree per day, and he had no idea what it

could be unless it was 'an immense comet.' He sent Swift the coordinates of its position along with his surmise as to its cometary nature. Swift, by return post, replied: 'Letter just received . . . The light you see is no comet but a zodiacal light phenomena always opposite the sun called Gechenchine. I don't think I have spelled it just right. I never could see it.'[27]

What Swift was recalling was the phenomenon known as the gegenschein, from a German word meaning 'counterglow.' Perhaps the greatest tribute to the power of Barnard's sight (and the inky blackness of the Nashville skies at the time) was the fact that it had never been seen by Swift, whose eye was exceptionally sensitive to faint nebulous objects.

The gegenschein had, it turned out, been independently discovered several times before Barnard did so. It had been described by the English astronomer T. W. Backhouse, who had published a paper in the *Monthly Notices of the Royal Astronomical Society* in 1876 with the title 'On the Aspect of the Zodiacal Light Opposite the Sun.' It had also been seen in 1853 by the German astronomer Theodor J. C. A. Brorsen, and in March 1803 by the German explorer Alexander von Humboldt during a voyage from Lima to Mexico. From the tropics the zodiacal light, a cone of light best seen from the tropics after sunset or before sunrise, showed 'in a splendor which I have never beheld before,' Humboldt wrote, and on March 16, when the zodiacal light was at its brightest, 'the gegenschein with its gentle light became visible in the east.'[28] Still earlier, the gegenschein had been seen by the Jesuit priest Esprit Pezenas, who published a note about it in the *Memoirs* of the Parisian Academy in 1731.

Nevertheless, for all these independent discoveries, before 1883 the gegenschein had attracted scant attention. Few had seen it, and it was known to few. Over the next forty years Barnard took an almost paternal interest in it, making it an object of special study. From his first observations he described it as '20' in diameter, roundish, resembling a patch of dense haze or fog, pretty even in light, ill defined borders; very dense by averted vision, dimming the clouds [of stars] over which it passed . . . I should say the object was as noticeable as the Milky Way near Orion . . . I have seen dim hazy clouds at night which resembled it in appearance.'[29] Later observations established that it always appeared on the zodiac, exactly opposite the Sun; at certain times of the year it appeared small and roundish, at other times it was large and elongated. As seen from the northern hemisphere, it was brightest in September and October, and it was invisible in December and June when it crossed the Milky Way.[30]

Barnard's rediscovery of the gegenschein stimulated renewed interest in the phenomenon, and various attempts were made to explain it. The English astronomer John Evershed suggested that the Earth might have a tail of hydrogen and helium atoms that would, rather like the tail of a comet, point away from the Sun because of light pressure. In his view, a perspective view of this tail was what was seen as the gegenschein. Barnard, however, considered the absence of any observed parallax fatal to Evershed's theory.[31] Instead, Barnard convinced himself that the gegenschein was a purely atmospheric phenomenon, 'due in some way to concentration of the Sun's light by refraction in the atmosphere (opposite the Sun) as if the atmosphere acted as a spherical lens,' as he concluded in his last paper on the subject, published in 1919.[32]

Here, however, he was mistaken. The true explanation, as we now know, is that the gegenschein consists of dust lying beyond the Earth's orbit and faintly glowing by backscattered sunlight. The dust is part of the same cloud surrounding the Sun which in its denser parts appears as the zodiacal light.

8

Barnard made one last notable discovery in 1883. Whenever a star is occulted by the airless Moon, its light ought to snap out abruptly. So it usually does, as had been observed on countless occasions. On November 5, 1883, Barnard was watching such an occultation, of the star β^2 Capricorni, with the 6-inch Cooke refractor. This time, instead of snapping out abruptly, he was startled to find that 'first about nine-tenths of the light instantly disappeared, and for the space of one second there remained in its place a minute point of light, estimated of the 10th magnitude. This also instantly disappeared.'[33] Something similar was seen at the same occultation by Edwin F. Sawyer of Cambridge, Massachusetts, and by Professor David Peck Todd of Amherst College, at Amherst, Massachusetts; both were skilled observers, but neither could fully explain the observation. Barnard alone grasped its true signficance. 'Before the remaining point of light had disappeared the explanation of the phenomenon flashed upon me,'[34] he recalled. As he realized, his observation had shown that the star was in fact a very close double, one component of which was much brighter than the other.

For a time confirmation proved elusive. Under usual observing conditions, Barnard could find no trace of the fainter companion with the 6-inch Cooke refractor, even with the highest magnifying powers. Eventually, however, S. W. Burnham, the world's leading double star observer, succeeded in glimpsing it with the $18\frac{1}{2}$-inch refractor at the Dearborn Observatory in Chicago. Burnham showed that β^2 Capricorni was indeed a close and unequal double star (separation $0''.8$, magnitudes 6.1, 10.0), but he added that it 'taxed the powers of that splendid instrument to show it even to the trained eye.'[35] As with the gegenschein, it was yet another triumph for Barnard's eyesight, which was fast becoming legendary.

The novelty of the discovery – only once before had a double star been discovered in this way – made it especially noteworthy. Though Barnard would eventually be credited with the discovery of thirty-six double stars, J. T. McGill later noted that Barnard thought 'more of the discovery of this one than of the other thirty-five together.'[36]

9

The year 1883 ended with a series of extraordinary sunsets, and Barnard, who had since earliest childhood watched with delight every aspect of the great story that plays continuously overhead, kept careful notes about them, which he later sent to W. W. Payne, the editor of the *Sidereal Messenger*. First commenting on an unusually 'glowing' sunset on October 29, 1883, Barnard continued to record gorgeous displays of roseate twilights well into the following year.[37] Around the world, other observers gave similar reports.

The cause of these startling effects was clearly the scattering of sunlight by some fine dustlike material in the upper strata of the atmosphere. According to a theory first advanced by the English astronomer Sir Norman Lockyer, this dust had come from the Krakatoa explosion – the most violent terrestrial explosion of modern times – in which the island volcano in the Sunda Strait, between Sumatra and Java in Indonesia, had virtually destroyed itself on August 26–28, 1883. The sound of the blast was heard around the world – in Turkey, Australia, the Philippines and Japan – and in the ensuing tidal wave tens of thousands of people in western Java were drowned. The dust, ashes and smoke from the tremendous explosion were carried to a height of seventeen miles and soon encircled the globe.

In mid-December 1883, Barnard saw a large number of 'small bright bodies close to the Sun' in his telescope. In appearance they resembled 'little stars, many as bright as the first magnitude,' and he had the impression that they were 'small particles drifting with the air currents at considerable altitude.'[38] When Barnard reported what he had seen to Swift, the Rochester astronomer suggested that these particles were probably 'nothing more than thistledown.'[39] Barnard, however, concluded in a brief note in the *Sidereal Messenger*: 'If the phenomenon of our red skies is due to the volcanic eruption in Java, and the proof seems to point to that as a cause, my frequent observations of bright rapidly moving particles near the Sun can readily be explained, on the supposition that they were minute particles of ashes drifting through the atmosphere, which had their origin in the awful catastrophe of Krakatoa.'[40]

It is doubtful whether in later years he would have regarded the identification with ashes from Krakatoa as certain, for after he went to the Lick Observatory, he often saw such objects in the sky near the Sun, and realized that they could be caused by a great variety of common things – in the summer by 'small seeds, pollen, and particles of dust flying with the wind,' and in the winter by 'snow or frost crystals.' He added the following observation:

From the same cause – a small angular distance from the Sun – I have frequently seen with the naked-eye spider-threads floating across the sky, drifting with the wind, and brilliantly illuminated by the Sun. I remember one day in November of 1911, while at the Solar Observatory on Mount Wilson, California, Mr. [W.S.] Adams and Mr. [F.] Ellerman were silvering the great mirror in the dome of the five-foot reflector. The observing slit was open in the direction of the Sun. By hiding the Sun with the dome, we could see great numbers of spider-webs floating across the sky to the south-east and shining very brilliantly in the Sun as they twisted in the wind. It was a very interesting sight. I do not recall that any spiders were seen on these floating threads, although I have heard that they sometimes so transport themselves.[41]

Parenthetically, Barnard himself would later see what remained of Krakatoa as he steamed past it en route to Sumatra for the 1901 total solar eclipse, and must then have thought back to the red skies of 1883–4, caused by an explosion on the other side of the world.

1   Fanny D. Nelson, 'Edward E. Barnard – Fellow of Vanderbilt University,' *Our Day*, 2:2 (1883), 40

2   ibid. Similarly, Mary R. Calvert later explained to Robert G. Aitken, 'He, evidently, had no thought even then of remaining in Nashville if he should ever have the chance to go to some observatory, for when I asked Aunt Rhoda why he refused to accept the gift she said that he felt that that would have tied him to Nashville.' MRC to RGA, November 11, 1927; SLO

3   E. E. Barnard, *SM*, 1 (1883), 290. Barnard also published an account of the transit in *AN*, **105** (1883), 231

4   O. H. Landreth, 'Barnard at Vanderbilt University Observatory,' *JTAS*, 3:1 (1928), 16

5   ibid., pp. 16–17

6   ibid., p. 17

7   H. N. McTyeire to EEB, March 1, 1883; VUA

8   E. E. Barnard, 'A Few Unscientific Experiences of an Astronomer,' *Vanderbilt University Quarterly*, 8 (1908), 273–88:278. Barnard knew from his own observations of the flimsiness of comets. On October 21, 1881, he watched Encke's comet, which he found 'very bright and round, with a moderate condensation,' pass in front of a 9th magnitude star. 'The star,' Barnard reported, 'from the time of first contact with the comet until free from it, suffered no diminution of light save that due to contrast from a dark to a light background. The star remained so remarkably distinct during the entire process of occultation that it formally impressed me with the idea of a transit of the star across the comet – a pearly point floating between me and the bright mass of Vapours.' E. E. Barnard, 'Remarkable Transparency of Encke's Comet,' *AN*, **102** (1881), 207

9   P. B. Calvert, 'Reminsicences of Barnard,' *JTAS*, 3:1 (1928), 14. For information about the Calvert family, I am indebted to Barnard's great nephew James R. Calvert for taking the time to answer my inquiries. Also useful was an article based on the recollections of Ross Calvert, son of Peter Ross Calvert, Barnard's friend at the Photograph Gallery: Sara Sprott Morrow, 'A Portrait of the Calverts,' *Nashville!*, August 1974.

10   Quoted in Edwin Mims, *History of Vanderbilt University* (Nashville, Vanderbilt University Press, 1946), p. 20. My account of the early years of Vanderbilt University and conditions there during Barnard's years there is drawn mainly after Mims, as supplemented by Robert T. Lagemann, *The Garland Collection of Classical Physics Apparatus at Vanderbilt University* (Nashville: privately printed, 1983).

11   As Bishop McTyeire described it for the Commodore, quoted in Mims, *History*, p. 38.

12   It was renamed Barnard Observatory in 1942, but ten years later was dismantled to make way for the present Rand Hall. Many of the instruments were placed in storage, though bricks from the building were used to construct a house for the director of the new Dyer Observatory located near Radnor Lake, about six miles south of Vanderbilt.

See Lagemann, *Garland Collection*, pp. 70–1. The house on campus in which Barnard lived was also torn down – not a single one of Barnard's many residences in Nashville has survived.

13 Detailed descriptions of these instruments are given in Lagemann, *Garland Collection*.

14 Quoted in ibid., p. 46

15 Quoted in ibid., p. 87

16 J. T. McGill, 'Edward Emerson Barnard,' *JTAS*, 3:1 (1928), 36

17 Mims, *History*, p. 134

18 EEB to Joseph S. Carels, September 21, 1892; VUA

19 E. Atkinson, *Elementary Treatise on Physics, Experimental and Applied, for the Use of Colleges and Schools*. Translated and Edited from Ganot's *Eléments de Physique* (New York, William Wood and Co., 7th ed., 1875)

20 Lagemann, *Garland Collection*, p. 61

21 In a letter from Garland to EEB, July 21, 1887, Garland worries that 'he will fall back into his old habit of drunkenness.' VUA

22 P. R. Calvert, 'Reminiscences,' p. 13

23 R. H. Hardie, 'The Story of the Early Life of E.E. Barnard,' *Leaflets of the Astronomical Society of the Pacific*, Nos. 415 and 416 (1964), p. 7

24 E. E. Barnard, Editorial Note, *SM*, 2 (1883), 259

25 E. E. Barnard, 'On some Celestial Photographs made with a large Portrait Lens at the Lick Observatory,' *MNRAS*, 50 (1890), 311. This small black hole was later designated B86 in Barnard's catalog of these objects.

26 E. E. Barnard, 'Gegenschein,' Editorial Note, *SM*, 2 (1883), 254

27 LS to EEB, October 11, 1883; VUA

28 Alexander von Humboldt, *Kosmos*, 1st ed., 1845, vol. I, 145ff; quoted in Willy Ley, *Watchers of the Skies* (New York, Viking, 1966), pp. 275–6

29 Barnard, 'Gegenschein,' p. 254

30 E. E. Barnard, 'The Gegenschein or Zodiacal Counterglow,' *PA*, 7 (1899), 169–79

31 ibid., 173–4. See also J. Evershed, 'The Parallax of the Gegenschein,' *PA*, 7 (1899), 289–90 and Barnard's rebuttal, p. 290

32 E. E. Barnard, 'The Gegenschein and its Possible Origin,' *PA*, 27 (1919), 109–12. Barnard noted that one drawback to the atmospheric theory was the fact that the gegenschein sometimes appeared much elongated. However, from careful observations in fall 1920, which Barnard reported at the Twenty Fifth Meeting of the American Astronomical Society in Chicago in 1921, he found that 'the apparent change in form is due to its projection on to a zodiacal band in this part of the sky. There seems to be a permanent part of a zodiacal band that extends along the ecliptic from between the Pleiades and Hyades to just south of the eastern part of the square of Pegasus . . . Apparently the change to an elongated form occurs every year, when the Gegenschein is projected on this band, and is not real, but is due to the mingling of its light with that of the band.' E. E. Barnard, 'Probable Explanation of the Apparent Elongation of the Gegenschein,' *PA*, 29 (1921), 149

33 E. E. Barnard, Editorial Notes, *SM*, 2 (1883), 254

34 E. E. Barnard, Editorial Notes, *SM*, 3 (1884), 120

35 S. W. Burnham, 'The Early Life of E. E. Barnard,' *PA*, 1 (1894), 343; reprinted from *Harper's Magazine*, August 1893

36 J. T. McGill, 'Edward Emerson Barnard,' *The Vanderbilt Alumnus*, 7:3 (1922), 73.
    The once before was the discovery of the companion of Antares, by the Viennese
    astronomer Burg, during the occultation of April 13, 1819. The bright star reappeared
    at the dark limb of the Moon, and Burg noted 'the emersion of a star 6.7 mag, which,
    about 5 seconds after suddenly appeared to me like a star of the first magnitude, and it
    is from this transition that I have set the time of emersion. Perhaps Antares is a double
    star, and the first observed small one is so near the principal star that both, viewed even
    with a good telescope, do not appear separated.' Quoted in Robert Burnham, Jr.,
    Burnham's Celestial Handbook, (New York, Dover, 1978), vol. 3, p. 1664. Since
    Barnard's discovery, there have been a number of other double stars discovered during
    occultations, including μ Arietis and Electra, one of the Pleiades.
37 E. E. Barnard, 'The Red Sunsets,' *SM*, 2 (1883), 293–6
38 ibid.
39 LS to EEB, December 30, 1883; VUA
40 E. E. Barnard, 'The Red Skies,' Editorial Note in *SM*, 3 (1884), 222–3
41 E. E. Barnard, 'A Curious Observation,' *Observatory*, 37 (1914), 416–17

6

In the realm of the nebulae

1

Barnard discovered his first comet at the Vanderbilt University Observatory on July 16, 1884. While sweeping with the 5-inch Byrne refractor in the constellation Lupus, he encountered 'a suspicious object, which from its absence from my memory and the catalogues at once suggested the probability of its being a comet.'[1] So slow was the object's motion that it was two days before Barnard was able to verify that it was indeed a comet and not a new nebula. The comet, 1884 II, proved to be of interest in that it was later shown to have a short period of only 5.4 years. However, it was not recovered at its expected returns in 1890 (unfavorable for observation) and 1895 (favorable), and has never been seen again – it is quite possible that it no longer exists.

Incidentally, by this time there had been significant improvements in the way astronomical information was disseminated – a development which Barnard, who had lost two comets during his first two years of seeking, undoubtedly welcomed. When he had started comet-seeking in 1881, the Smithsonian Institution in Washington, DC, had been the official American bureau in charge of receiving and sending astronomical information, but it was often slow and inefficient, and in any case only communicated announcements to American observatories about comets discovered in Europe. The Smithsonian made no provision for announcements of American comets, and by default this responsibility devolved to Swift in connection with the Warner prizes and to John Ritchie, Jr., of the *Science Observer*, who 'made arrangements to telegraph certain of our correspondents, whenever a comet is discovered in this country, the expense of the telegram being paid by them.'[2] Ritchie relied on the cooperation of the Harvard College Observatory, under the direction of E. C. Pickering, for observations and computations, most of which were provided by S. C. Chandler and O. C. Wendell.

As there was no uniform code for transmitting such reports, the information often arrived in garbled form, and this meant that a good deal of time might be wasted in trying to puzzle out the correct information. A claim of priority or a needed confirmatory observation might thereby be lost. After the establishment of the Warner prize, this could mean the loss of money as well as prestige. Noting the sorry state of affairs before 1881, Chandler and Ritchie groused: 'The requirements of mercantile pursuits have long since lessened the chances of a mistake in a message of business importance, and the buying of fifty hogs or the sale of one hundred barrels of flour, can be done with ten-fold more accuracy than the announcement of the position of a comet.'[3]

That year, the two men proposed as a solution an ingenious cipher code based on the use of a standard dictionary (Worcester's Comprehensive Dictionary, edition of 1876), by which the elements of orbits and ephemerides of positions of newly discovered comets could be telegraphed quickly and reliably.[4] Following the massive confusion surrounding the discovery of the Great Comet of 1882, the reformers gained ground, and decisive action was finally taken in January 1883, when Spencer Baird, secretary of the Smithsonian, offered to turn over to Chandler, Ritchie, and the Harvard College Observatory the department of international exchange of astronomical information.[5] With the adoption of the *Science Observer* code, an astronomer could reasonably hope that critical information about a newly discovered comet, on being cabled to Harvard, would be quickly – and intelligibly – communicated to other observatories in the United States and thence, by transatlantic cable, to those in Great Britain and Europe. From this time forth, Barnard never again lost credit for a comet.

2

As Barnard's reputation grew, so did his correspondence. The men with whom he was now corresponding – Swift, Brooks, Chandler, Brashear, E. C. Pickering – were illustrious names to him. In some cases he had portraits to go along with the names, but so far he had met none of these famous men. His first opportunity to do so came in September 1884, when he made his first extended trip away from Nashville to attend the American Association for the Advancement of Science meeting in Philadelphia. Always practicing the strictest economy, he avoided unnecessary hotel bills by traveling at night, sleeping on the hard wooden benches of the cheaper railroad coaches.[6]

Given his shyness at the earlier meeting of the Association in Nashville in 1877, when he had trembled before the great Newcomb, Barnard no doubt started out for Philadelphia with similar feelings of awe. But the more worldly-wise Anson Nelson, the friend who had secured for him his introduction to Newcomb on that earlier occasion, warned him against possible disappointment:

> The meeting you are attending is an important one to you, for you will have the opportunity of coming in personal contact with most of the best scientific minds in this country. My opinion is, that you will find all these gentlemen at the meeting *below* your expectations; for this reason: they are on a sort of holiday excursion, with their bodies and minds relaxed from the toil and drudgery of everyday pursuits, and will not, therefore, seem to be so studious and thoughtful as they usually are at home. And yet you will see them, and form a personal acquaintance with many of them that will be of immense service to you.[7]

Heartened by Nelson's advice and wanting to make the most of his opportunity, Barnard decided on a grand tour – an indirect route to Philadelphia which would allow him to visit the observatories at Cincinnati, Washington, Cambridge, Allegheny, Albany, and Princeton. He also made forays into Boston and New York. His fondest memories were of his stopover in Pittsburgh, where he met Brashear. In a letter to Brashear written many years later, he recalled the circumstances of their first meeting:

> I had just arrived on a very early morning train on which I had neither dined nor slept, for the Diner and Sleeper had been a special terror to me in all this journey. After a vaguely directed search I finally found myself on an unpaved road cut in the clay soil of a steep hillside, hunting for Number 3 Holt Street (this street and number have remained in my memory all these years) – a street that seemed to have no existence in the memory of any one I had met.
>
> A man passed by with his dinner bucket – and this shows how early I was abroad that morning, and how hungry I was to notice the man's dinner bucket, which at that moment, could I have had my way without bloodshed, would have become a 'breakfast' bucket. This man said he didn't know of any street of that name. When I asked him, however, if he knew where John A. Brashear lived, he said, 'Why he lives right there in that house,' pointing to a small cottage not fifty feet away, on the down side of that road. 'Then,' said I, 'this must be Holt Street, and that house must be Number 3.' I felt then that it was a mighty fine thing to be a man who was better known than the street he lived on! But you mustn't feel too proud, Mr. Brashear, for Holt Street at that time, as I remember it, was a mighty mean street.[8]

On reaching the small cottage, Barnard was at first too timid to knock at the door, and paced back and forth until he finally gathered the courage to make his presence known.[9] Once inside, he had a hearty reception. Brashear's wife Phoebe asked him if he had had any breakfast, and 'the admission on my part that I hadn't anything to eat since the previous morning was taken instantly as a challenge, and soon a delicious breakfast of ham and eggs and coffee – and such a biscuit! – greeted my appreciative soul.'[10] Thus began a friendship that would become so intimate that the two men would be 'Johnnie and Eddie to each other.'[11] Brashear later that same day showed Barnard the nearby Allegheny Observatory, though neither the director, Samuel Pierpont Langley, nor his assistant James E. Keeler were in. When Barnard left Pittsburgh late that afternoon for New York, it was raining, and he recalled 'how dark and gloomy everything looked from the smoke and grime of the city.[12]

At the meeting in Philadelphia, Barnard, though by now a Fellow of the AAAS, still could not bring himself to approach Newcomb. However, other astronomers, equally eminent, were less off-putting. Pickering, for instance, whom Barnard had expected to be 'a formal and aloof dignitary,' proved to be 'comparatively a young man, and strongly resembled a simple countryman. Had anyone shown him to me on the street and told me that that was the famous director of the Harvard College Observatory, I should not have taken his word on oath. I should have been positive there was a mistake. [H]e is the most unassuming man you can imagine, and I admired him very much indeed.'[13] Among other astronomers at the meeting was England's John Couch Adams, who as a young man had won an immortal name in astronomy for his calculations of the position of then unknown Neptune from its perturbations on Uranus and was now a gray-bearded Cambridge don; Ireland's Sir Robert Ball, who would later call Barnard 'the greatest observational astronomer in the world';[14] and the distinguished retiring

Barnard, left, at the AAAS meeting in Philadelphia, 1884. Yerkes Observatory

president of the AAAS, Charles Augustus Young. Barnard also for the first time met his friend Lewis Swift, who talked on 'the Nebulae.'[15] In fulfillment of Anson Nelson's prediction, the meeting was a great success for Barnard, who returned to Nashville feeling that he had been warmly accepted into the fraternity of astronomers.

3

On December 1, 1884, Barnard's mother died. She was almost seventy and had lived long enough to see her son in the first blush of his fame – though with 'clouded mind,' she can hardly have had much real comprehension of his achievements. 'His mother's influence,' Barnard's niece Mary Calvert later wrote, 'must have strongly affected his whole life, but the impression I have . . . is that he must have been almost "on his own" from, say, early in his teens.'[16] For a while he had lived alone in a room in the Hotel St. Charles, but with the move to Patterson Avenue, he was finally able to give his mother shelter under the same roof, and she had followed Barnard and his wife to 'comet house' and finally to the little house on the Vanderbilt campus. In later years, Rhoda was the one who assumed most of the responsibility for taking care of her, and from what little is known of the situation, her cares must have been very heavy indeed.

Elizabeth Barnard was buried at Mt. Olivet Cemetery in Nashville, with all the expenses being generously paid by Vanderbilt University. Somewhat surprisingly, when thirty years later information was being sought for the family monument at Mt. Olivet, Barnard, whose memory for astronomical observations was phenomenal, could no longer recall even the year in which she had died.[17] But then the actual date was of little importance, for the end had been so long expected that he had inured himself to it long before. He no doubt shed his share of tears for his poor old mother, but it was Rhoda on whom he relied now, and before long he was back to his classes and to scouring the skies for his beloved comets.

4

Stimulated both by the desire of fame and the hope of winning Warner prize money, competition for new comets was very intense during the 1880s. Swift, who was now in his sixties, was still very much in the hunt. For a time his observations were threatened by electric lights being erected in front of the Warner Observatory, but he managed to get up a petition from the citizens of Rochester which led to a compromise with the electric company, who agreed to place hoods above the lights, 'regardless of cost,' so as not to interfere with his observations – possibly the first successful campaign against light-pollution.[18] Barnard's chief cometary rival was, however, not Swift but the fast-rising William Robert Brooks of Phelps, New York, who at that moment was just starting to be heard from but who would prove to be, after Pons, the most energetic, single-minded and successful searcher for comets that the world had ever known.[19]

The son of a Baptist minister, Brooks was born at Maidstone, England, on June 11, 1844. As a small boy he went with his family on a voyage to Australia, and when he was thirteen settled in New York State, where he remained for the rest of his life. The next year Brooks's passion for comets began when he viewed the splendid Donati's Comet through a homemade spyglass. As a young man, he spent three years in Buffalo working for the Shepherd Iron Works, and in 1870 married and settled in Phelps, in the Finger Lakes district of New York. Like young Barnard, he made his living as a photographer. The Phelps address was to become well known to astronomers for the reports which were received regularly from the 'Red House Observatory' – actually no more than a small observing platform built in an apple orchard on the property, from which Brooks observed with several portable telescopes.

Brooks began his career as a comet-seeker at almost the same time as Barnard, and also like Barnard, his first discovery was not officially acknowledged. At 2:45 a.m. on October 4, 1881, he was sweeping the skies with a homebuilt 5-inch reflector when he came across a suspicious object in Leo. Unfortunately, he had it in view for only five minutes before clouds moved in, so he could do no more than roughly estimate its position. Wanting to be completely sure of himself before going public – 'I have a *great dread* of making mistaken announcements,'[20] he explained to Barnard, who after his own experience the previous May no doubt appreciated his caution – Brooks kept the discovery to himself until the following morning, when he hoped to obtain a confirmatory observation. The sky being again cloudy, however, he decided to telegraph Swift without further delay. He was actually in his office composing the telegram when a foreign one, announcing the comet's discovery by William F. Denning at Bristol, England, was handed to him. Though Brooks had seen the comet nearly a day before Denning, it became known as 'Denning's Comet' (its official name was 1881 V), and Brooks lost out on his first bid for the Warner prize.

As with Barnard's lost comet, Brooks's near-brush with success only whetted his appetite, and he built a $9\frac{1}{4}$-inch reflector for more effective searching. In February 1883, he found Comet 1883 I near β Pegasi, and on September 1, he recovered the long-awaited Comet Pons of 1812 on its first observed return. At about this time, the three Americans – Swift, Brooks, and Barnard – formed their cometary triumvirate, dividing the skies into search zones among themselves. But afterwards when Swift, who was harassed with running a public observatory and verifying other people's discoveries, fell behind the pace, the field was left to the two younger men, each of whom was too ambitious to share the skies with anyone.

5

One of Barnard's most productive years for comet-seeking was 1885, which began with an independent recovery of Encke's Comet, which has the shortest period of any comet known – only 3.3 years.[21] Barnard first glimpsed it with the 5-inch refractor on January 2. At the time he did not know that it had already been seen elsewhere – by Wilhelm Tempel at Arcetri on December 13 and a few nights later by C. A. Young at Princeton, who had captured it with a 23-inch refractor. Young wrote graciously to Barnard: 'I do not know but *you* could have seen the comet with your 5-inch on Dec. 17, but it was very faint in the 23 inch, and I could not see it at all in the $9\frac{1}{2}$ [inch].'[22] The fact that Barnard was able to see it at all with such a small glass when it was still a difficult object in much larger telescopes was yet another testimonial to the excellence of his vision.

Barnard found a comet (1885 II) on July 8, another (1886 II) on December 3, and yet another (1885 V) on December 27 – though, as he was disappointed to learn, this last comet had already been discovered by Brooks on December 26. Of these comets, 1886 II was the most remarkable. Barnard published the following account of its discovery in the *Vanderbilt Observer*:

> On the first of December a comet was discovered [by Fabry] at Paris, France, and the telegram from Boston announcing the discovery was received at the Vanderbilt on the night of December 2d, and the comet was observed for position after mid-night. The next night the comet could not at first be conveniently observed, and, whilst waiting for it to attain a more suitable altitude, the writer, to utilize the time, began a careful search for comets in the region near and west of the

constellation of Orion. After an unsuccessful search of something like two hours, a
faint nebulous object appeared in the field of view. A thorough knowledge of that
region of the heavens at once convinced the observer that the object was an utter
stranger and doubtless a new comet . . . [23]

When Barnard found it, the comet still lay 250 million miles distant from the Sun –
well beyond the orbit of Mars – and was five months from perihelion, when it was
expected that it would become a brilliant object. It appeared to be intimately related to
the comet that Fabry had discovered three days earlier. The orbits of the two comets
were almost identical, and they followed almost parallel paths across the sky, with
Barnard's comet trailing Fabry's by about a month. Fabry's comet reached naked-eye
visibility in early March and passed perihelion on April 6. At the end of April, just
before it swept into the southern skies, its nucleus was brighter than the first magnitude
and it had developed a 10° long tail. By then Barnard's comet was just brightening to
naked-eye visibility. It passed perihelion on May 3, and for several days was a 4th
magnitude object with a 3° long tail, though unfortunately, as its altitude in the sky was
then decreasing rapidly for northern hemisphere observers, it had become a rather
difficult object – Barnard himself, observing from the Vanderbilt campus, barely
managed to make it out with the naked-eye as a hazy ill-defined spot, almost lost in the
heavy smoke over the city. The orbits of both comets, Fabry and Barnard, are believed
to be hyperbolic, which means that they will never return.

6

By the end of 1885, Barnard had been awarded the Warner prize for the fifth time –
more than anyone else. When asked the secret of his success, he once answered that
what was needed in comet-seeking above all else was patience. 'Very few observers
possess this requisite patience,' he remarked, 'and they soon tire out, or become so
disappointed that they abandon the struggle without making a single discovery. The
army of comet-seekers is constantly recruited from the ranks of amateurs, but as
constantly falls off again from the causes mentioned.'[24] 'Patience, care and endurance'
had been the qualities that had set Barnard off from the other small boys who had tried
to guide Van Stavoren's solar camera. Because of these qualities, he had succeeded
where all the others had failed, and the same qualities now made him a successful seeker
of comets.

Comet-seeking required intense concentration which made it, Barnard acknowledged, 'tedious in the extreme, requiring the eye to be constantly on the alert and
strained to the utmost for the detection of any faint object.' Because of the tedium and
the time it took away from other observations, Barnard would more than once resolve to
give it up altogether 'as a waste of time.'[25]

But if in certain moods Barnard was contemptuous of comet-seeking, in other moods
he saw it as the best possible preparation for other kinds of work, for the comet-seeker
alone made it his business to familiarize himself with everything in the whole heavens.

Thus, Barnard could take pride in the fact that there was 'not an object in the heavens which may be viewed by means of a five-inch telescope that I have not seen perhaps a hundred times during my search for comets.'[26]

Among the objects that most excited the comet-seeker's wonder were the nebulae, those strange objects which passed in such great numbers through the field of the eyepiece and were at the time wrapped in almost impenetrable mystery. At first the comet-seeker's interest in them extended no farther than to distinguish them quickly and reliably from comets, for as mentioned earlier, the typical nebula looked identical in appearance to the average telescopic comet – so much so that even professional astronomers were not immune to the embarrassing blunder of confusing the one for the other.

As a young astronomer, Barnard had undergone a thorough apprenticeship, spending many hours in preliminary sweeping of the skies in order to learn how to recognize nebulae which 'not a man living could tell from [comets by] the physical appearance alone.' How he came to acquire this uncanny ability even he could not fully explain:

> In the search for comets these nebulae are constantly encountered, and with one of them appearing in the field every now and then it might be inferred that the astronomer cannot always be certain of it when he has really discovered a comet. There are, however, several ways by which he can decide . . . The lighter the object the easier it is for him to tell. The vast majority of nebulae are so faint as to be invisible in the small telescopes usually employed for comet-seeking, so these will not bother him much, and as for the others, the comet-seeker, through a constant and vigorous prosecution of his work, soon becomes familiar with most of them. The location, the appearance with reference to little stars in the field – possibly some kind of indefinite sixth sense – impresses the astronomer with the instant conviction that this is a known nebula and not a stranger comet! In nine cases out of ten it is not possible to explain how he recognizes this or that nebula as one he has seen before, yet he can do it; nor does he himself know how it is done, unless by instinct. A dim, hazy object will enter the field of view as the telescope is swept across the sky. It looks very much like a comet. A quick glance at it . . . and he passes on. Perhaps in the next minute an object exactly similar in every respect enters the field. A glance at it; it does not look familiar.

The sixth sense referred to by Barnard is, parenthetically, rather reminiscent of the chess grandmaster's ability to glance at a crowded board with twenty to twenty-five pieces on it and, as if by intuition, to get to the heart of the game and see the solution immediately.

Barnard and Swift often shared information about nebulae in their voluminous correspondence. In one of his first letters to Barnard, Swift had invited the budding astronomer to search for the nebula surrounding Merope, one of the Pleiades. It had first been reported in 1859 by the German astronomer Wilhelm Tempel, from Venice – this 'breath stain' on the vault of the heavens would, Tempel later declared, perpetuate his name. Ever since, opinion had been sharply divided as to the nebula's existence or non-existence, some seeing it readily, others being unable to get a glimpse of it. Among the doubters were G. W. Hough and S. W. Burnham, who despite being armed with the powerful $18\frac{1}{2}$-inch refractor at the Dearborn Observatory could find no trace of it. Tempel, on the other hand, had claimed it with a 4-inch refractor, and Swift with only a 2-inch!

On commencing his own search from the yard of his house on Patterson Avenue, Barnard found the nebula 'plainly visible' in his 5-inch Byrne refractor. He later published an account of his observations, along with a drawing, in the *Sidereal Messenger*: 'My chart of Merope and stars near it is identical with Temple's drawing, although the latter was not seen until after my sketch was made.'[28]

In his comet sweeps, Barnard was before long coming across new nebulae that were not listed in any of the catalogs – by the time he left Nashville, he had discovered about a score. Most were excessively faint. As Swift described one of them after verifying it in his 16-inch refractor, it was 'about as near spiritual as anything conceivable here.'[29] This was, of course, only to be expected; by the 1880s the skies had already been thoroughly scoured with powerful telescopes for over a century, and it was not likely that any prominent objects could have been missed. Nevertheless, a few surprises remained.

Thus, on January 23, 1883, Barnard encountered a large and fairly bright nebulous object near the star 12 Monocerotis, and entered in his observing book: 'Found a large nebulous object, [near] a scattering cluster of bright stars; it is elongated southwest and northeast. Larger than the field of view.'[30] Finding no such object in any of the catalogs, Barnard assumed, naturally enough, that he had found a comet, and he kept it in his sights much of the night looking for the telltale signs of its motion. But it remained stationary. Several more nights of observation put its nebular identity beyond doubt, and on writing to Swift, Barnard learned that his mentor had himself seen the nebula many years before but had never published an account of it. As Swift explained: 'I did not know for a certainty that it was never recorded but took it for granted from its immense size that it was although I never could find a description that fitted it.'[31] He finally published a note about it in the *Sidereal Messenger* in 1884. He there described it as 'a nebula of the dumb-bell type, and . . . very large, one of the largest visible from this latitude,' and added that 'it is unaccountable to me that so conspicuous a nebula should have so long been overlooked. About eighteen months ago Mr. Barnard, now connected with the Vanderbilt Observatory . . . in his sweeps for comets, picked it up and called my attention to it, thinking it might possibly be a comet.'[32] Barnard's discovery was, then, entirely independent. In later observations with the 12-inch refractor of the Lick Observatory, Barnard would discover that what he and Swift had seen 'was simply a brightish knot in a vast nebulous ring that entirely surrounded the cluster.' The object would become even more impressive in long-exposure photographs, showing an intricate wreath-like structure that would earn for it the name the 'Rosette.'[33] (The nebula is no. 2237 in the NGC catalog; the associated bright cluster, which was first recorded by John Herschel, is NGC 2244.)

Among Barnard's other nebular discoveries at Vanderbilt was a large and faint nebula in Sagittarius which he swept up with the 5-inch Byrne refractor in 1884 – it is the interesting object now known as 'Barnard's Galaxy' (NGC 6822) though at the time, of course, its nature was entirely obscure.[34] Yet another was the large diffuse nebula near ξ Persei which he found with the 6-inch Cooke refractor on November 3, 1885[35] – the 'California nebula' (NGC 1499). As with the Rosette, its full splendor was first revealed in long-exposure photographs.

7

From the late eighteenth century when William Herschel had catalogued 2,500 nebulae and clusters with his large reflectors, their number had steadily risen. The *General Catalogue* published by John Herschel in 1864, which included nebulae he had discovered in his four years at the Cape of Good Hope, brought the list to 5079, of which all but 450 were discovered by his father or himself. Other skilled observers, such as William Lassell, Heinrich d'Arrest and Wilhelm Tempel, added hundreds more.

Swift took up the quest for new nebulae in earnest in 1883, soon after taking charge of the 16-inch refractor of the Warner Observatory. On July 9, he selected for his field of observation 'the region between the head and third coil of Draco.'[36] On that short summer's night he found fourteen new nebulae, and the next night's searching yielded seven more. After not quite a month, Swift confided to Barnard that 'there was more undiscovered neb. than I supposed. I see no reason why I can't find a thousand if I devote time enough to the work.' He confessed, however, 'It is very exhausting and tells on my eye. It requires looking with the intensity of despair.'[37]

Despite interference from the red skies which hid some of the faintest nebulae from sight, Swift by November 1884 had discovered 197 new nebulae – of which seven were discovered by his thirteen year old son, Edward. 'It may be of interest to state in this connection,' he wrote in summarizing his observations up to that time, 'that nearly three-fourths of the nebulae discovered at this observatory are in Draco, and that the constellation is not yet half explored.'[38] The numbers continued to mount; by June 1887, he had reached 600. As far as anyone could tell there was no end to the nebulae in sight – especially in regions like Draco near the celestial pole and well away from the 'Zone of Avoidance' of the band of the Milky Way which, as William Herschel had noted long before, the fainter nebulae tended to avoid.

But what were these nebulae which were turning up in such vast abundance? The nature of most of them remained shrouded in mystery. At the AAAS meeting in Philadelphia in 1884, Swift reminded his audience of the 'many questions which suggest themselves regarding the origins and design of these immense masses of matter strewn with such lavish abundance in some regions, and so sparsely scattered in others.' And noting that the efforts of astronomers to solve their mysteries had thus far proved largely futile, he added: 'For obvious reasons this is, and for ages to come must be, true.'[39]

At the time Swift uttered these words, the nebulae had been separated into two broad classes by the spectroscopic observations of Sir William Huggins, who with his wife Margaret had studied about seventy of them, beginning in the 1860s, from his observatory at 90 Upper Tulse Hill, London. Some of them, including the so-called planetaries, which showed pale planet-like disks in the telescope, had a bright emission-line spectrum, which was telltale of the presence of a hot, low-density gas. The Orion nebula was also of this type. These came to be known as 'green nebulae' – they were true nebulosities. The rest, which were called 'white nebulae,' showed faint, continuous – indeed star-like – spectra characteristic of hot gas under high pressure. The Great Nebula of Andromeda was the outstanding example of the type.

To some, the Andromeda nebula was the vision of an immense star system grown misty through excessive distance, one lying perhaps well beyond the Milky Way, and grandly evoking the 'island universes' about which the philosopher Immanuel Kant had speculated as early as 1755. In this view, the other white nebulae which observers like Swift were turning up in great numbers were also presumably island universes, vast galactic systems in their own right. Alternatively, the Andromeda nebula might be a relatively nearby cloud of gas and dust, swirling around its already condensed center – a prototype of the nebula from which the Solar System itself was thought to have been spawned according to the nebular hypothesis of Kant and Laplace. In that case, it no doubt lay well within the Milky Way. Unfortunately, at the moment there seemed to be no way of choosing decisively between these alternatives. Rev. T. W. Webb summed up the situation as it appeared in 1882 in calling the Andromeda nebula 'a mystery in all probability never to be penetrated by man.'[40]

8

Within three years of Webb's pessimistic pronouncement came a clue, if only it could be deciphered, into the nature of the Andromeda nebula. A nova or 'temporary' star appeared only sixteen seconds of arc from its star-like nucleus, thereby producing 'one of the strangest sights in the heavens,' in the words of one of those who witnessed it.[41] E. W. Maunder of the Greenwich Observatory remarked that the 'strange and beautiful object has broken silence at last, though its utterance may be difficult to interpret.'[42]

Only once before had a nova appeared in a nebula – in May 1860, when Arthur Auwers at Königsberg discovered a nova in M80 in Scorpio. By then M80 was no longer strictly speaking in the nebular class, however, having in 1785 been resolved by William Herschel into a globular cluster, 'the richest and most condensed mass of stars which the firmament can offer to the contemplation of astronomers.' In any case, the event had been more or less forgotten until the somewhat parallel eruption occurred in the Andromeda nebula a quarter of a century later.

The new star, 'S Andromedae' – the designation 'S' signified that, for want of a better category, it was considered a 'variable' star, the letters from R onwards being assigned to the variables in each constellation – had definitely not been present on the evening of August 16, 1885, when the nebula was examined by Max Wolf at

Heidelberg, Wilhelm Tempel at Florence, and R. Engelmann at Leipzig. It was first seen by Ludovic Gully at Rouen on August 17, who however dismissed the novel appearance as some kind of telescopic defect and thought no more about it. The first person to see it and to grasp its significance was Ernst Hartwig of the Dorpat Observatory in Estonia, who perceived it on August 20 as a star of at least 7th and possibly as bright as 6th magnitude. However, the observatory's conservative director would not allow Hartwig to announce his discovery until he had obtained a confirmatory observation in a moonless sky. Because of a run of clouds, this was not until August 31. Only then Hartwig announced the discovery to A. Krüger, editor of *Astronomische Nachrichten* and director of the Central Astronomical Telegram Office in Kiel. By then scores of other observers had independently discovered the new star – one of the first was a Baroness de Podmaniczky of Hungary.[43] The earliest Americans to see it, apparently, were J. C. McClure of Red Wing, Minnesota, on August 29, and H. S. Moore of McKinney, Texas, on August 30.[44]

A telegram from Kiel reached E. C. Pickering at Harvard on September 1, and Pickering immediately telegraphed the news to other American observatories, including Vanderbilt. Barnard decoded the message on the evening of September 1, and that night, through breaks in cloud, made his first observations of the new star with the 6-inch Cooke refractor:

> The first glimpse through the telescope showed at once the bright stranger beaming forth in the midst of the nebula. It was a bright eighth magnitude star – perfectly stellar – of a slight yellowish cast. The light from this object was quite intense and very steady. Upon the application of high powers it retained its star-like form as distinctly as did the neighboring stars. The object illuminated the nebulosity surrounding it like a star shining through fog . . .[45]

This curious illusion – that the star was illuminating the nebulosity surrounding it – led Barnard to conclude that the star was either '*in* the nebula or on the other side of it, certainly not on this side.'[46]

In the following weeks Barnard carefully documented the star's drop from prominence by comparing its light with that of other stars in the field. By the end of October the star had already faded to the 11th magnitude, and Barnard made the curious observation that

> since the star has diminished to a small point its light seems to come and go every few minutes. I have at almost every observation lately been struck with this apparent ebb and flow of light. For a few seconds the star will almost totally disappear; strain your eyes as best you may, you cannot see it clearly; but presently it will become quite bright and distinct; remaining thus a short time, it will again fade away, even while other stars in the field are fairly steady.[47]

Barnard considered but rejected the notion that these rapid fluctuations in brightness indicated changes in the star itself. Instead he realized that 'a very slight disturbance in the atmosphere – not enough to disturb the light of the neighboring stars – might be sufficient to blend together the light of the star and nebula, and thus cause an apparent fading of the stellar point.'[48] He was no novice to the effect; he had noted similar changes in the brightness of stars seen through the heads of comets and in the star-like centers of some of the nebulae he had encountered while comet-seeking. Of one such nebula, he recalled:

> I have been struck, when looking at this nebula, with the wonderful changes that apparently go on in an ordinary night. There seems to be an almost constant battle going forward between the nebula and the bright stellar nucleus; for a moment the nucleus will be overcome, and its light blending with that of the nebulosity, it will seem to have faded out; but presently you will see it flash forth brightly and steadily and perfectly stellar in form; a few moments thus and it is gone again . . .
>
> The fluctuations of the light of the *Nova* are precisely similar to those in the nebula referred to. While the *Nova* was bright there was great contrast between it and the nebulous background, therefore slight disturbances in the air would not so sensibly affect the star; but bring the light of the star and the nebulosity to near an equality, and the equilibrium is easily destroyed by a feeble disturbance that might not sensibly affect another star in the same field . . .[49]

Parenthetically, when Barnard some years later began to suspect the variability of the central star of the planetary nebula NGC 7662 in Andromeda, he long hesitated before finally declaring that the star was indeed variable. 'Familiarity with the changing aspect of a star involved in nebulosity, due to moonlight, bad seeing, etc., has made me extremely cautious in the matter,' he wrote in explaining his circumspectness.[50] Ironically, even for all his caution, it seems that in this case he was probably mistaken – the current opinion is that the star is not variable.[51]

The light from S Andromedae continued to fade. When Asaph Hall saw it for the last time on February 7, 1886, it had fallen to the 16th magnitude, the limit of visibility with the 26-inch refractor at Washington. Then it vanished forever.[51]

The wondrous apparition had, however, reignited the debate about the nebula's nature. Germany's Hugo von Seeliger regarded the nova as evidence of the stellar composition of the nebula, thus as supporting the 'island universe' theory. So did France's Camille Flammarion, who enthused: 'Probably the Andromeda Nebula is a cluster of stars, and the star which has flared up close to the center is one of its suns of which the photosphere has undergone a sudden conflagration . . . [It is] a huge sun, without doubt thousands of times more gigantic than our own! The heavens are immense; man is insignificant!'[53] Others, including England's Richard A. Proctor, claimed that the star's very brightness – in late August and early September it had just reached naked-eye visibility[54] – proved conclusively that the nebula could not be excessively remote. Writing eight years after the eruption, the judicious Agnes M. Clerke gave her assessment of the controversy, noting that two outbursts of such uncommon character as those observed in M80 and the Andromeda nebula could hardly have occurred accidentally just on the line of sight between the Earth and the central portions of those two objects. Thus, she argued, they had to be associated with the objects in the midst of which they appeared; but if so she deemed it eminently rash to conclude that [these nebulae] are really aggregations of sun-like bodies . . . For it is practically certain that, however distant the nebulae, the stars were equally remote; hence, if the constituent particles of the former be suns, the incomparably vaster orbs by which their feeble light was well-nigh obliterated must, as was argued by Mr. Proctor, have been on such a scale as the imagination recoils from contemplating.[55]

In 1902, Clerke expanded on her argument, suggesting that if, as she believed, the Andromeda nebula was indeed stellar in constitution, it must be 'made up of stars smaller and closer together than we have supposed' – a position which, as she pointed out, was not without its difficulties.[56] There were dim intimations that in the course of the 'intricate circlings' of these small stars some of them might be 'jostled together,' converting some of the energy of their motion into heat to produce the observed conflagration. Such explanations were far from satisfactory, but other processes that might be operating were even more dimly apprehended. At the time, the shining of the stars was still understood to be the direct result of the slow gravitational contraction proposed by Helmholtz in 1853 (with the energy derived from this source being at best augmented by the collisions with swarming meteorites, as had been proposed by Sir Norman Lockyer). Astronomers of the late Victorian era simply had no way of imagining an energy source capable of producing an outburst on the scale required if the new star had appeared in a stellar system beyond our Milky Way, so that, as Kenneth Glyn Jones has noted,

> This one item of astronomical evidence remained an obstacle in the way of progress to a firmer grasp of the scale of the universe for nearly forty years, and the recognition of the 'island universe' status of the great spiral nebula in which the nova had appeared, was perennially hampered by the incredible implication that a single star could, if only temporarily, equal in brightness a sizeable fraction of a huge system of millions of stars.[57]

The result was that, in Clerke's words again, 'the conception of the nebulae as remote galaxies . . . began to withdraw into the region of discarded and half-forgotten speculations.'[58]

The red sunsets due to Krakatoa, which had erupted only two Augusts earlier, had hardly died out when the new star in Andromeda had appeared, marking an explosion that was on an inconceivably greater scale – such, indeed, as the imagination might 'recoil from contemplating.' The event has now been identified as a Type I supernova. A massive star destroyed itself in a cataclysmic explosion which caused it, for one brief moment of glory, to flash with the intensity of four billion suns.[59] Barnard and others who watched it could not begin to grasp the nature of this striking apparition; the physics of the stars had not yet advanced far enough. What they saw they could not interpret, any more than Tycho Brahe or Johannes Kepler three centuries earlier could interpret the stellar cataclysms which they witnessed within our own star system.

1   E. E. Barnard, Editorial Notes, *SM*, 3 (1884), 188

2   John Ritchie, Jr., 'Editorial,' *Science Observer*, 3 (1881), 84–85:85

3   S. C. Chandler, Jr. and John Ritchie, Jr., 'On the Telegraphic Transmission of Astronomical Data,' *Science Observer*, 3 (1881), 65–77:65

4   The plan of the Science Observer Code, as it was called, was described by Chandler and Ritchie as follows: 'The dictionary in question has more than 390 pages, and has, generally, 100 words or more on a page. It is manifest, then, that any integral number up to 39,000 can be represented by a single word, by making the two digits on the right (the tens and units) of the proposed number correspond to the numerical order of the word on the page, and the three places on the left, to the number of the page. Thus, for example, 16,718 can be represented by the 18th word on page 167 [electrize]. In the same way, 349° 12′ can be represented by the 12th word on page 349 [proportionableness]. April 14d. 10h. 48m. = April 14d.45, being the 134th day of the year (135th of leap year), can be represented by the number 134 45, or the 45th word on page 134 [crush], and so on. In a similar manner each position of an ephemeris can be represented by two words, one for the right ascension and one for the north polar distance, which is to be preferred to declination, as the distinction of plus and minus is thereby avoided.' ibid., p. 66.

5   See John Ritchie, Jr., 'Announcement with Reference to the Astronomical Code,' *AJ*, 8 (1888), 189–90 and Bessie Zaban Jones and Lyle Gifford Boyd, *The Harvard College Observatory: The First Four Directorships, 1839–1919* (Cambridge, Mass., The Belknap Press of Harvard University Press, 1971), 194–8 and 331–4. The Harvard College Observatory continued in this role until 1966, when the bureau was transferred to the Smithsonian Astrophysical Observatory.

6   E. B. Frost, 'Edward Emerson Barnard,' *BMNAS*, 21 (1926), 3

7   Anson Nelson to EEB, September 4, 1884; VUA

8   EEB to JAB, November 22, 1915; VUA. The letter was written on the occasion of Brashear's seventy-fifth birthday and is quoted in *John A. Brashear: An Autobiography* (Boston and New York, Houghton, Mifflin Co., 1925), 62–3

9   J. A. Parkhurst, 'Edward Emerson Barnard,' *The Journal of the Royal Astronomical Society of Canada*, 17 (1923), 98

10   EEB to JAB, November 22, 1915; VUA

11   Parkhurst, op. cit., p. 99

12   EEB, undated memorandum; VUA

13   Frost, op. cit, p. 3

14   Parkhurst, op. cit., p. 98

15   Lewis Swift, 'The Nebulae,' in *Proceedings of the AAAS*, 33 (1885), 73. Swift said, in part: 'The assigned limits of this paper forbid entering largely into the many questions which suggest themselves regarding the origins and design of these immense masses of matter strewn with such lavish abundance in some regions, and so sparsely scattered in others.

   'The efforts of astronomers to solve this mystery, as well as those of distance from our system, their proper motion and their variation in size, shape and brightness, have thus far proved futile and have ended only in vague and contradictory conjectures. For obvious reasons this is, and for ages to come must be, true . . .'

16   MRC to RGA, November 11, 1927; SLO

17   EEB to J. T. McGill, October 10, 1921: 'I have been trying to find the date of my mother's death. I recorded it in a notebook but that notebook cannot now be found. There must have been some notice of it in the Vanderbilt Quarterly or some other publication at that time. I am unable to assign the year. It was probably 1884–5 or 86. It could not have been later. I came to the Vanderbilt in March of 1883 so I do not think it was in 1883 . . . I may eventually find it here but it is wanted now for the monument.' VUA. Concerning Barnard's memory, G. E. Hale later wrote: 'Although his observation of every class of celestial objects, stars, nebulae, comets, planets, satellites, meteors and many others, were numbered by tens of thousands, he could always recall the day and often the hour of one of them, as well as the exact details recorded.' Quoted in Helen Wright, *Explorer of the Universe: A Biography of George Ellery Hale* (New York, E.P. Dutton & Co., 1966), p. 126. Strange that Barnard, who could remember in such detail his astronomical observations, should have failed to recall the date of his mother's death.

18   LS to EEB, October 11, 1883; VUA

19   For information on Brooks, see Joseph Ashbrook, 'Harvester of the Skies,' *The Astronomical Scrapbook: Skywatchers, Pioneers, and Seekers in Astronomy* (Cambridge, Mass., Sky Publishing Corp., 1984), pp. 92–7

20   W. R. Brooks to EEB, October 24, 1881; VUA

21   The comet was, in fact, discovered by Pierre Méchain in 1786, by Caroline Herschel in 1795, and twice by Pons, in 1805 and 1818, before its orbit was worked out by J. F. Encke of the Berlin Observatory.

22   CAY to EEB, January 9, 1885; VUA

23   E. E. Barnard, 'Second Comet Barnard, 1885,' *The Vanderbilt Observer*, vol. 3, no. 5 (February 1886)

24   E. E. Barnard, 'How to Find Comets,' *San Francisco Examiner*, February 5, 1893

25   EEB to ESH, May 7, 1886; SLO

26   Barnard, 'How to Find Comets,' op. cit.

27   LS to EEB, August 28, 1881; VUA

28   E. E. Barnard, note, *SM*, 1:2–3 (May 1882), 33

29   Swift was describing a new nebula Barnard had found near η Cassiopeia. LS to EEB, August 22, 1883; VUA

30   E. E. Barnard, 'The Great Nebula in Monoceros,' Editorial Note in *SM*, 4 (1885), 313–14.

31   LS to EEB, February 19, 1883; VUA

32   Lewis Swift, 'A New and Remarkable Nebula,' Editorial Note in *SM*, 3 (1884), 57–8

33   For Barnard's visual observations with the 12-inch refractor of the Lick Observatory, see E. E. Barnard, 'The cluster GC 1420 and the nebula NGC 2237,' *AN*, **122** (1889), 253; for his photograph and a discussion thereof, see 'Photographic Nebulosities and Star Clusters Connected with the Milky Way,' *AA* 13 (1894), 177–83 and 'Photograph of Swift's Nebula in Minoceros [sic.], N.G.C. 2237,' *AA*, **13** (1894), 642–4

34   It is an irregular 'dwarf' galaxy, one of the closest objects of its kind. It is generally regarded as a member of the Local Group of Galaxies. See Robert Burnham, Jr., *Burnham's Celestial Handbook: An Observer's Guide to the Universe Beyond the Solar System* (New York, Dover, 1978), vol. 3, p. 1616.

35   E. E. Barnard, 'An Excessively Faint Nebula,' *SM*, **5** (1886), 27. Barnard writes of the

discovery: 'On the night of November 3, I found an excessively faint, but rather large nebulosity that is not in G.C. or supplement. This nebula lies a short distance following the south of a 9th magnitude star. It seems to be extended somewhat. It is probably about $\frac{1}{2}°$ long . . . This requires the lowest power and cannot be seen by direct vision. It is only by directing the vision slightly to one side of its place that it is possible to see it, it then flashes out feebly.'

36   Lewis Swift, 'The Nebulae,' *SM*, **4** (1885), 1–4:4
37   LS to EEB, August 6, 1883; VUA
38   Lewis Swift, 'The Nebulae,' p. 4
39   Lewis Swift, 'The Nebulae,' *Proceedings of AAAS*, **33** (1885), 73
40   Quoted in Gérard de Vaucouleurs, 'Discovering M31's Spiral Shape,' *Sky and Telescope*, **74** (1987), 596
41   Thomas Fraser, Diary, September 20, 1885; SLO
42   E. W. Maunder, 'The New Star in the Great Nebula in Andromeda,' *Observatory*, **8** (1885), 321–5:321
43   N. de Konkoly, Letter, *Observatory*, **8** (1885), 331
44   McClure wrote to Barnard about it the following day, but the letter did not reach Vanderbilt until September 3.
45   E. E. Barnard, 'The New Star in the Nebula of Andromeda,' *SM*, **4** (1885), 240
46   ibid., 243
47   E. E. Barnard, 'Nova Andromedae,' *SM*, **4** (1885), 308
48   ibid.
49   ibid., p. 309
50   Barnard first suspected the variability of the star in observations with the Yerkes 40-inch refractor in 1897, as noted in *Ap J*, **14** (1901), 153. He did not publish his observations in detail until 1908, however, when he announced: 'I have . . . established conclusively the fact that the nucleus of 7662 is actually variable to an extent of upwards of three magnitudes. At times it has appeared as a bright yellowish star of about the 12th magnitude and at other times, equally favorable, it has been either entirely invisible or excessively faint.' See E. E. Barnard, 'The Variability of the Nucleus of the Planetary Nebula N.G.C. 7662,' *MNRAS*, **68** (1908), 465–81. Ironically, his conclusion about the variability of this star was questioned by C.R. O'Dell in 1963 on the grounds that the apparent brightness of a star surrounded by nebulosity is critically dependent on the seeing. Echoing Barnard's own argument, O'Dell wrote: 'As the seeing varies the ability to discern the star will change because of the superposition of the nebula, while nearby comparison stars will not be affected.' Quoted in Robert Burnham, Jr., op. cit., vol. 1, p. 157.
51   D. Osterbrock to W. Sheehan, personal communication, April 24, 1992
52   Asaph Hall, 'Nova Andromedae,' *American Journal of Science*, **31** (1886), 301–4; 'Parallax of Nova Andromedae,' *Observatory*, **9** (1886), 204. Parenthetically, between September 29, 1885 and February 7, 1886, Hall made a series of careful measures of the star's position relative to another preceding the nova by about 2' of arc in an attempt to ascertain whether the star would show any appreciable parallax and thus reveal its distance. His conclusion was that these measures 'gave no certain indication of any parallax. There was indeed a slight apparent diminution of the distance of the nova from the comparison star but the increasing difficulty of accurate measurement, as the

Nova faded away, might fully account for the slight discrepancy.' Hall was almost certainly the last to see the star; the claim of Rev. T. E. Espin, who reported that he glimpsed the star with a $17\frac{1}{4}$-inch reflector on March 6, 1886, is probably spurious.

53   C. Flammarion, 'Apparition d'une Étoile dans la Nébuleuse d'Andromède,' *L'Astronomie*, **4** (1885), 361–7:366–7

54   See, for instance, W. R. Brooks, Editorial Note, *SM*, **4** (1885), 246. Brooks noted that on September 2 the new star 'appeared to me . . . almost of the 6th magnitude and was distinctly seen by the naked eye.' Incidentally, on the night of August 31, 1885, Brooks had discovered a new comet, in Canes Venatici.

55   A. M. Clerke, *A Popular History of Astronomy During the Nineteenth Century* (London, Adam & Charles Black, 3rd ed., 1893), p. 484

56   A. M. Clerke, *Problems in Astrophysics* (London, Adam & Charles Black, 1903), p. 451

57   Kenneth Glyn Jones, 'S Andromedae, 1885: An Analysis of Contemporary Reports and a Reconstruction,' *Journal for the History of Astronomy*, **7** (1976), 27–40:27

58   Clerke, *A Popular History*, p. 505

59   See G. de Vaucouleurs, 'The Supernova of 1885 in Messier 31,' *Sky and Telescope*, **70** (1985), 115–18

7

Go West, young man!

1

By the time he came to Vanderbilt, Barnard was already in the habit of nervously waiting for every fragment of every night in which there was a patch of clear sky. His dedication bordered on fanaticism. Olin Landreth, seeing this in him, nicknamed him 'Enthusiastically Energetic Barnard.'[1] He was impelled by some blazing internal force which harried him on. But he was not alone in his fanaticism, and every night he could be sure that his arch rival Brooks was out scanning the sky with as keen intensity as his own.

Unlike Brooks, however, who could still speak of feeling '*almost* electrified'[2] with the discovery of each new comet, Barnard was growing increasingly resentful of the claims on his time made by comet-seeking. He went so far as to tell E. S. Holden in May 1886 that as of the previous August he had actually given up regular comet-seeking, as it was 'a great waste of time that might be valuable if put to other use.'[3] Such comments cannot be accepted at face value – the comets he discovered in December 1885 show that at least at times his resolve against comet-seeking must have wavered. Meanwhile, Brooks remained as relentless as ever, and in less than a month between April 27 and May 22, 1886, swept up three new comets, thereby setting a record for rapidity of discovery.

Barnard's relative slackening of commitment to comet-seeking spared more time for other researches. He began experimenting with photography with the 6-inch Cooke refractor, and he also renewed what had become a somewhat dormant study of Jupiter, entering the following observation of the planet in his notebook on April 1, 1886:

> At $12^h 45^m$, three . . . dark projections ranged along the inner edge of the belt and just south of the equator. I noticed that from the summit of each there extended for a short distance in a following direction a dusky streak, looking like smoke. I was strongly impressed with the resemblance to what might be called a silhouette view of three volcanic peaks, ranged in a line and vomiting smoke, which a strong wind was carrying eastward.'[4]

2

Even without the demands of regular comet-seeking, Barnard gave himself little quarter, and his tendency to push himself so hard amazed some and alarmed others. In

William Robert Brooks. Mary Lea Shane Archives of the Lick Observatory

August 1886, Bishop McTyeire – himself no stranger to compulsive overwork – had to write Barnard a letter in which he strongly advised him to take a vacation, 'for two or three weeks, even longer if you find yourself away, and doing well':

> You need rest; if not now – you need rest to get strength for the future. And, certainly, you deserve it; you are entitled to it.
> The Stars and comets will keep on their way, and be found in the right places

when you return. Forget them for awhile. Don't look up, except to say your prayers, for the next month. Rest, *rest*.[5]

Barnard could hardly refuse McTyeire, with the result that at the end of August he did indeed take a brief rest from the telescope in order to attend (with Landreth) the American Association for the Advancement of Science meeting in Buffalo. At the meeting he saw (but still did not dare approach) Newcomb, and he also met Brooks for the first time, the latter having made the short trip over from Phelps. And Barnard himself gave a brief talk on 'Telescopic Observations of Meteor Trains.'

The rest – or perhaps the meeting with Brooks – seems to have had the effect on Barnard of stimulating a renewed quest for comets. On October 5, 1886 he discovered his sixth, one day ahead of Hartwig in Germany and Pechüle in Copenhagen. This comet was captured 'almost by chance' while Barnard was sweeping the constellation Sextans with his 5-inch Byrne refractor 'in an open space between the observatory dome and a large mass of trees.'[6] Barnard's eastern horizon at Vanderbilt was cut off to a considerable altitude by trees, and it was while sweeping in a narrow gap between the trees and the dome that the comet was found. 'Had it not been seen in this gap,' he explained, 'it would not have been found by me, for when its altitude is sufficiently great to bring it above the trees, it cannot be seen, being blotted out by dawn.'[7] Barnard scarcely had time to run to the observatory and turn the 6-inch refractor upon the suspected comet in order to secure a position before it faded from view. 'Though positive it was a comet,' he wrote, 'yet not having seen it long enough to detect motion, I feared to risk the announcement as a comet but at once gave the usual notification to Dr. Swift of a "suspected comet." '[8]

Later that same day Barnard had to stand for a five-hour examination in English. His teacher, W. M. Baskerville, afterwards recounted how 'Barnard remarked timidly as he handed in his paper that he had worked under great difficulties; he thought that the night before he had discovered in the early morning hours a new comet, and that he must now go back to finish his calculations and announce the result to astronomers!'[9] Next morning Barnard indeed successfully recovered the object, which had now moved perceptibly toward the northeast.

This comet (1886 IX) later became a fine object – in fact, it proved to be the best performer of all of Barnard's many comets. It reached naked-eye visibility by early November and in mid-December was a 3rd magnitude object with a 15° long tail. As it retreated from the Earth, it swept into the southern skies, where it continued to be followed by astronomers until June 1887. Since its orbit is hyperbolic, it will never return again.

3

Instead of savoring his triumph as most men would do, Barnard, in a letter to E. C. Pickering written shortly after this discovery, complained bitterly of the comet-seeker's lot:

The discovery of a comet is not so slight a matter as one might think and the discoverer very seldom gets his just dues; the discovery of a single comet represents months and sometimes years of patient searching, at the risk of health from exposure to the biting cold of winter and the damp dewey nights of summer. I have many a cold bleak winter night searched ceaselessly from sunset until dawn, almost frozen and completely worn out and nothing to show for my labors and not being able to make up any lost sleep in the day time, before I came here I had to work every day, and since being here I have had to attend the classes in the day and get my lessons as any other student would.[10]

Such was Barnard's emptiness, the depth of his earlier privation, that no matter how hard he tried to fill it with achievements, he remained insecure and unsatisfied. The need for recognition and approval always motivated him most; this was what his soul, deeply scarred by his earlier life of crushing poverty and hardship, hungered for so desperately. The desire for fame was a spur that prodded him to so many sleepless nights of toil.

In Nashville, he had always received complete adulation. Gradually, however, he had come to the depressing realization that despite the unsophisticated public's enthusiasm for his achievements, most professional astronomers did not regard the discovery of a new comet as a matter of great importance, and it can hardly have helped matters any that Warner had discontinued his prizes as of the end of 1885, so that comet-seeking was no longer even lucrative. Some astronomers – the more snobbish among them – were even resentful of the recognition that the discoverer received. A well-known professional astronomer wrote on the death of another prolific amateur discoverer of comets some years later: 'A discoverer is much better known to the public and the newspapers than in the profession . . . I may say that the spirit of such discoveries is not always welcomed in an observatory.'[11] This attitude was shared by many professional astronomers of Barnard's day, for whom, as Donald E. Osterbrock has written, 'computing an orbit mathematically counted for far more than finding a new comet.'[12] As we shall see, Barnard did eventually compute an orbit or two, but remained a slow and error-prone computer. Though his observations were always first-class, his reductions of them were often wrong, so that, as E. S. Holden later noted, 'the computers learned to check [them] before using them.'[13]

Since it was the recognition and approval of professional astronomers that he always craved the most – his own dream having always been to get on at one of the great observatories – Barnard was facing rather urgently, at the time he wrote to Pickering, the whole question of his future.

Barnard was approaching thirty, and he would soon be finished with his studies at Vanderbilt. One of his options was to stay put; he knew that he would be very welcome to remain at Vanderbilt. But he also knew that he had long since absorbed what Garland could teach him, and that the modest equipment of the Vanderbilt University Observatory would only allow for a rather pedestrian program of research. The most glorious possibility would be a position on the staff of the Lick Observatory, then fast abuilding on Mt. Hamilton, a 4200 foot peak near San Jose, California. When completed the observatory would boast a 36-inch refractor, the largest in the world. The possibility of such a position must have long haunted the regions of his dreams, but by 1885, he was not only dreaming about it, he was actively pursuing it by courting the observatory's director-to-be, Edward Singleton Holden.

4

Holden, for better or worse, was destined to become one of the most important figures in Barnard's life.[14] He was born in St. Louis in 1846. His mother died of cholera when he was three, and a few years later his father sent him to Cambridge, Massachusetts, to live with an aunt and to attend a private school taught by one of Holden's cousins. Another cousin was the wife of George Phillips Bond, son of William Cranch Bond, the first director of the Harvard College Observatory. After William died in 1859, George succeeded him as director, and the teen-aged Holden, a frequent visitor to the observatory, was often given the chance to peer through the 15-inch refractor. In time he developed 'a boundless admiration for the science to which the Bonds, father and son, devoted their life.'[15]

At sixteen Holden returned to St. Louis to attend Washington University Academy, a prep school, and two years later was admitted to Washington University itself, where he found a mentor in William Chauvenet, professor of mathematics and astronomy and well known as the author of a then widely used textbook on mathematical astronomy. Holden later married Chauvenet's daughter, Mary. Following his graduation in 1866, Holden continued his education at the US Military Academy at West Point, and after a brief stint in the artillery, he resigned from the Army in order to take a position at the US Naval Observatory in Washington, DC. He arrived in 1873, an exciting time at the observatory as it was just putting into operation its 26-inch Clark refractor, then the largest in the world. Simon Newcomb was placed in charge of it, and Holden became his assistant.

Even then, however, plans were afoot for an even larger telescope, to satisfy the wish of James Lick, the eccentric San Francisco piano builder and real estate millionaire, to build as a monument to himself a telescope 'superior to and more powerful than any telescope yet made.' In 1874, Newcomb was approached for advice about such a telescope by Darius O. Mills, president of the Bank of California and a member of the first Lick Board of Trust, and Holden, as Newcomb's assistant, was included in these discussions. When the question of a director for the future Lick Observatory came up, Newcomb did not hesitate to recommend Holden despite the fact that the latter was then only twenty-eight years old. When Newcomb soon afterward realized that the observatory would not be finished for many years, he withdrew the recommendation as premature,[16] but Holden remained intimately involved in the planning for the observatory and indeed 'the trustees and Newcomb considered Holden the probable director from that time on.' So did Holden, and henceforth all of his career decisions were made with this in mind.[17]

Edward Singleton Holden. Mary Lea Shane Archives of the Lick Observatory

For two years Newcomb and Holden had the 26-inch refractor more or less to themselves, and they tried to make discoveries with it. Newcomb searched in vain for the 'dark companion' of Procyon, whose existence had been deduced by its disturbing effect on the bright star's motion, and both astronomers were interested in seeing whether the large refractor would reveal any new satellites. By 1875, Newcomb was getting tired of the night work. Wanting to concentrate more on his mathematical studies, he turned the telescope over to Asaph Hall, and Holden became Hall's assistant. One of the first things Hall did on assuming control was, as he later told Seth Carlo Chandler, to 'find out what my predecessors had been doing.' He learned that both Newcomb and Holden had used the 26-inch to search for faint satellites of Uranus and Neptune, and in addition he noted: 'I found in a drawer in the Eq[uatorial] room a lot of photographs of the planet Mars in 1875. From the handwriting of dates and notes probably Holden directed the photographer, but whoever did the pointing of the telescope had . . . satellites under his eye.'[18]

Hall was interested in searching for Martian satellites himself, convinced that earlier astronomers had failed to scrutinize space close enough in toward the planet. In August 1877, with Mars in a favorable position for such a search, he decided to launch his own investigation. Naturally, he wanted his assistant Holden out of the way, so that he would receive sole credit in the event of a discovery, and later confided to E. C. Pickering, 'By the greatest good luck Dr. Henry Draper invited him to Dobb's Ferry at the very nick of time. He could not have gone much farther than Baltimore when I had the first satellite nearly in hand.'[19] Hall announced the discovery of two satellites on August 18. Meanwhile, at Dobb's Ferry, Holden and Draper heard the news and immediately began searching for the new satellites with Draper's 28-inch reflector. For a while at the end of August they thought that they had discovered yet another satellite.[20] Hall tried to confirm its existence but was unsuccessful – it was later shown that one of Holden and Draper's observations was actually of Hall's own outer satellite, Deimos, while two others were of faint field stars.

After making these observations, Holden travelled from Dobb's Ferry to Nashville for the AAAS meeting,[21] and on his return to Washington, he found yet another 'satellite' on September 24. This one, Hall noted with disgust, did not even obey Kepler's laws: 'Its existence was therefore a mathematical impossibility.'[22] By now Hall had spent so much time trying to verify Holden's reports that he was falling behind in his own work, and with growing irritation wrote to Pickering: 'If I were to go through this experience again other people would verify their own moons.'[23] Rumors about Holden's spurious 'moon' continued to circulate in the astronomical community for years, and Holden became known as the man 'who had set all Washington astronomers laughing by detecting a third satellite of Mars with an impossible period and distance, and remaining deceived by it for months!'[24]

More creditable to Holden's reputation than the Martian satellite debacle was his *Monograph on the Central Parts of the Orion Nebula*, a large tome of 221 pages which was the most massive assault to date on the long-standing question of whether changes of form took place in the Orion nebula – a claim first made by William Herschel. Holden, who was a born cataloguer, began his study by collecting together every important drawing and description of the Orion nebula made since the seventeenth century. In addition he contributed his own study of the nebula, made in 1878–80 with the 26-inch refractor and a device designed for him by physicist Charles S. Hastings. Holden's method was to compare the brightness of different regions of the nebula with the image of a piece of paper simultaneously projected through the telescope – basically, a primitive attempt at photometry. The paper was illuminated from behind by an oil lamp whose brightness could be adjusted until it just matched the brightness of part of the nebula, and from these observations Holden believed that he was able to establish slight changes in the nebula's brightness. Unfortunately, the study was flawed, probably because Holden was unable accurately to control his lamp, and his results have never been confirmed.[25] For that matter, his *Monograph*, based as it was entirely on visual observations, became obsolete even before it appeared. In March 1882, Henry Draper obtained a portrait of the nebula, a 137-minute exposure on a gelatino-bromide dry plate which showed it, as Holden himself admitted in a hastily written appendix to his *Monograph*, 'for nearly every purpose . . . incomparably better' than any drawing made by a visual observer. It was abundantly clear that photography and photography alone would provide data reliable enough to settle the question of changes in the form of the nebula – and now a century later, definite changes have yet to be detected.

Holden lacked the great observer's patience and willingness to tinker at the telescope until he got it right. Temperamentally he was always more drawn to library work, and in 1880 he very gladly gave up his duties on the great telescope in order to concentrate on the, for him, far more congenial task of reorganizing the Naval Observatory's library – a task which he seems to have carried out with aplomb.

When in that same year, 1880, the directorship of the University of Wisconsin's Washburn Observatory became vacant through the sudden death of James Craig Watson, Holden was offered the job, and he accepted – though with prompt assurances to Captain Richard S. Floyd, the president of what was by now the third incarnation of the Lick Board of Trust, that he regarded this step as only a needed preparation for the eventual Lick directorship. Indeed, he continued to receive information about the observatory and to dispense advice as needed through a voluminous correspondence with Floyd. By 1881, a winding road had been built up Mt. Hamilton, the top of the peak had been blasted away in order to create a level surface for the observatory, and a 12-inch Clark refractor, formerly Draper's, had been mounted in the dome that had by then mushroomed on the mountain. It was used that November by Holden and S. W. Burnham to observe a transit of Mercury, and Holden also travelled to Mt. Hamilton for the transit of Venus the following year. In 1883 he sailed to Caroline Island, in the Pacific between Hawaii and Tahiti, for a total solar eclipse, using the precious minutes of totality to search – unsuccessfully – for intra-Mercurial planets.

5

Though Barnard may have heard Holden present a paper on the 'Trifid Nebula' at the AAAS meeting in Nashville in 1877, there is no evidence that they then met. The first documented contact between the two men took place shortly after Holden's return from Caroline Island in 1883, when Barnard sought the opinion of the author of the massive *Monograph* about a suspected band of nebulosity west of the Orion nebula. Several letters were exchanged on the subject – the tone of Holden's is dignified and formal, in marked contrast to the chatty and often humorous style of Swift. Moreover, unlike Swift, who was always quick to settle such questions by direct recourse to the telescope, Holden's first impulse always led him to the library – thus he referred Barnard first to Heinrich d'Arrest's and then to Lord Rosse's drawing of the nebula. On September 12, 1883, he told Barnard, 'I shall be very glad to look at the nebula and tell you what I see, if it will be any advantage to you.' On September 21 he repeated his offer: 'I will look at the nebula myself later on and let you know what I find, if you wish.'[26] Presumably Barnard so wished, but there is no evidence that Holden ever did attempt to settle the matter by direct observation. Holden also begged off Barnard's request for help in sorting through the new nebulae that he was picking up in his comet-sweeps, excusing himself on the grounds that he had lost his copy of John Herschel's *General Catalogue* on Caroline Island.[27]

Barnard probably interpreted this as a brush off, for he allowed the correspondence to lapse for the time being. Indeed, except for one rather minor inquiry from Barnard about eclipses of Jupiter's satellites – Holden suggested that Barnard direct his inquiry instead to David Peck Todd of Amherst College[28] – it seems that there were no further communications between the two men until September 1885.

That month, Barnard recaptured Holden's attention by his detailed observations of the Andromeda nebula and its new star. Barnard had reported in the *Sidereal Messenger* seeing 'a well-defined but faint nebula' involved in the preceding end of the great nebula.[29] Holden wrote to Barnard that it might be new, but he advised him to go back and study the old drawings, such as the careful engraving made by G. P. Bond in 1847, to make sure.[30] A few days later he followed up with a postcard: 'In order to say if [your observations] indicate any new thing it is necessary to *know* what is old.'[31] As it turned out, what Barnard found was not new – his 'nebula,' actually a 'condensation' of the great nebula, had already been catalogued by John Herschel.[32]

At about this time, Holden had been offered the presidency of the University of California at Berkeley so that he might be close at hand during the final stages of the Lick Observatory's preparation. He accepted on condition that when the observatory passed from the Lick trustees to the Regents on its completion, he would resign the presidency and move up to the directorship.[33] The condition was accepted. Thus the directorship that Holden had coveted for so many years was finally placed securely within his grasp. (Ironically, not long before this, Newcomb himself had thrown his support to Todd, but it was too late.) Though Floyd made it clear that Holden would not actually be allowed to take charge, or even to take up residence, on the mountain until the Regents were in control – terms under which Holden soon chafed – there was still much that could be done from Holden's office in Berkeley. Above all there was the question of assembling a staff, and Holden began considering the talented young Tennessean for a position.

In the course of their correspondence about the Andromeda nebula, Holden casually mentioned to Barnard his imminent plans to leave Wisconsin for California in order to take charge of the University and, eventually, the Lick Observatory. Barnard replied immediately and warmly:

I received your card this morning, and I am very greatly delighted to hear that you were going to California to take charge of that grand observatory. I am so glad; and I am sure no man in the country is so well suited for that great charge as you are. I'm sure I feel that the University of California deserves the highest congratulations for its success in securing one so eminently fitted for the position. I am not surprised for I have all along said you would be the man for the Lick Observatory.

You have my heartiest wishes for your welfare and I hope your health may always be good so that you can enjoy the grand instrument which under your charge I know will reveal wonders that man hath never dreamed of. Please accept my humble but most sincere congratulations, and may you live long to honor our land.[34]

Holden resonated to such flattery, for he responded in kind: 'Thank you heartily for your letter of congratulation. I appreciate it – and you are right. There is going to be a grand chance on that mountain for every man who is there – there will be four or five of us.'[35] By referring to 'us,' Holden at least hinted at the possibility that Barnard might be among the happy few. Barnard's subsequent behavior indicates that he indeed read thus between the lines. Together with taking every opportunity to flatter Holden – including asking him for his portrait, in exchange for a portrait Barnard took the liberty of sending him – he made a concerted effort to impress the future director with his own qualifications. Thus Holden began receiving detailed information about Barnard's comets and most recently discovered nebulae.[36] He also learned of the young astronomer's aspirations for celestial photography – 'no doubt the Astronomy of the future,' Barnard called it[37] – which Holden shared. Barnard notified Holden that during spring vacation, 1886, he had succeeded in determining the photographic focus of the 6-inch Cooke refractor – 'it was 0.17 inch outside the visual focus,' he discovered[38] – and at such focus he had taken some tolerably sharp pictures of the Moon.

While Barnard was making every effort to put his best foot forward, Holden was seeking out confidential opinions of him from other astronomers. He received the following ringing endorsement from Ormond Stone, an expert on nebulae at Virginia's Leander McCormick Observatory:

Barnard is 29 I think; he is married; I think he has only a wife; at least he had no children . . . a year ago; . . . he gets about $600 salary; I doubt whether he has much future prospects there [at Vanderbilt], but I may be mistaken; I think they have but little appreciation of his real ability, tho they are glad to have the advertising his comet hunting gives them. I consider him one of the most promising men I know, and in every way a thoroughly good fellow. He is an active member of the Baptist Church.[39]

Meanwhile, by the spring of 1886, James E. Keeler, the first staff astronomer hired for the Lick Observatory, was already on Mt. Hamilton setting up the observatory's time service, and Holden informed Barnard of his intention to select the rest of his associates as soon as the lens of the 36-inch was finished and delivered. And he added:

My idea has been that we can probably have two or three astronomers who have already attained a high reputation who will be in charge of such work as Meridian Circle Observations, Equatorial Observations, &c., and that some of the younger men in the country would be willing to come (as Keeler is) to take up certain routine work, and to use the leisure which I shall be able to give them, for their own special researches.[40]

Barnard could not fail to catch the encouragement in the phrase 'some of the younger men in the country,' and did not waste time in clarifying his own situation for Holden in a letter dated May 7, 1886 – the one in which he first announced that he had given up comet-seeking. He noted that he would 'graduate in mathematics here at the University in June if my studies have a successful termination.' In other words, he would be available for another position should one be offered. Then, in what was as Osterbrock notes 'the closest approach to an explicit job application that the customs of those days permitted,'[41] he offered some background information about himself:

Up to my coming here I had been a photographer from the time I was about 7 or 8 years old, most of that time as a photo printer and later as an operator. So you see I am somewhat prepared for the difficulties that will occur in my photo experiments here . . . I wished to let you know that I am trying to do the best I can and yet give sufficient time to my studies in the University which go sometimes rather hard with me as I had never previously gone to school except about two months before I went to work in a photograph gallery.

Barnard promised to make some of his prints of the Moon for Holden, and also told him of his resolve to begin experiments with stellar photography, despite the fact that his equipment at Vanderbilt was in some respects wanting for such a purpose. 'I cannot see how our telescope can be driven sufficiently steady,' he explained, 'but possibly by the aid of a high power on finder I may help the clock keep the image steady for a long enough time.' At last, just in case Holden missed his point, he added: 'It seems to me that the Lick Observatory, with the proper instruments, would be the best place in the world to do wonders in Celestial Photography.'[42]

6

At least in part, Holden's interest in Barnard as a future employee of the Lick Observatory was because of 'the advertising,' in Ormond Stone's phrase, that any new comets discovered from Mt. Hamilton would give the new observatory. Being savvy enough to recognize the importance of keeping the wider public interested in the work of the observatory, Holden did not share the snobbish attitude of many professional astronomers about such 'discovery work.' As a means of promoting his interests, Barnard's discovery of October 1886, mentioned earlier, could not have come at a better time; moreover, it was only the first in a new crop, which could not fail to impress Holden and – so Barnard must have hoped – clinch a position for him.

With the new year, others followed in quick succession. On January 22, 1887, Barnard learned by telegram of the discovery of Brooks's latest comet. The next night it was, he recorded, 'raining heavily at dusk.' Conditions remained unpromising until midnight, when it suddenly turned very cold and clear. (How many observers, seeing it raining heavily at dusk, would have kept up a vigil until midnight waiting for the sky to clear? This Barnard did routinely.) Barnard went to the observatory and took several positions of Brooks's new comet with the 6-inch refractor, before bringing out the 5-inch refractor for comet sweeping. 'At about 5 o'clock a.m.,' he noted, 'I ran upon a dim hazy object several degrees south-west of β Cygni which I recognized as a stranger.'[43] The stranger indeed proved to be a comet, and as it had already passed perihelion in late November 1886, it received the designation 1886 VIII – the eighth comet to come to perihelion that year.

Barnard found another comet among the stars of Puppis at 10:30 p.m. on February 16, 1887. 'It was,' he noted, 'very faint and moving with astonishing rapidity to the northwest.'[44] This comet, 1887 III, was noteworthy in that Barnard himself, perhaps with help from Landreth and Vaughn, calculated a preliminary orbit from his observations of February 16, 22, and 28. The orbit was published in both the *Astronomical Journal*, the leading American journal of record, and the German *Astronomische Nachrichten*, and he later published a revised orbit which was based on further measures of the comet's position.[45] Significantly, after this *tour de force*, he never again published another orbit of comet – or of anything else for that matter. Though he no doubt hoped that this feat would impress professionals, telling Holden, 'I have lately taken up the computation of comet orbits but my experiance in this has not been great from a lack of time, and I am anxious to pursue it further,'[46] computing was not, and never would be, in his line.

On the evening of May 12, 1887, Barnard discovered his last comet at Vanderbilt, a 10th magnitude object near the borders of Hydra, Scorpio, and Centaurus. This faint comet (1887 IV) raised Barnard's official total to nine – one more than Brooks and two more than Swift had at the time. Indeed, he was now fifth on the all-time list, behind only Pons, Messier, and the still active Wilhelm Tempel and F. A. T. Winnecke.

On the Vanderbilt campus, such achievements had earned Barnard the kind of acclaim usually reserved for football players. When in that same year 1887 Vanderbilt's first annual appeared, it was named the *Comet* in his honor. And in a debate in the pages of the *Nation*, which had begun when an editorial writer (in referring to the earlier dismissal of a Vanderbilt professor of geology who had championed the theory of evolution) had made the inflammatory charge that 'Vanderbilt University, which ere this should have been a Cornell or Ann Arbor, is now not much more than a large theological seminary where the free discussion of scientific truth will not be tolerated,' W. M. Baskerville retorted: 'In practical astronomy . . . Vanderbilt has in the last three years done more for the world than Cornell has accomplished during its whole history.'[47] What he did not say was that everything accomplished in practical astronomy at Vanderbilt in the last three years had been the work of one remarkable man – E. E. Barnard.

7

By March 1887, Holden had made up his mind to try to hire for the Lick staff, in addition to Keeler who had already been hired, the noted double star observer S. W. Burnham and George C. Comstock, Holden's former assistant at the Washburn Observatory. He thought that he would probably have the money to hire two more observers, John Martin Schaeberle of the University of Michigan and Barnard. In mid-July he officially offered Barnard a job: 'Astronomer at $1200 a year with quarters &c. . . . the duties to commence Oct. 1/87 *about*. I hope you will accept this and I ask you to telegraph your decision to me.'[49] Burnham and Schaeberle also received offers of positions, but Comstock withdrew himself from consideration, having earlier decided to remain in Wisconsin in Holden's old position as director of the Washburn Observatory.

For Barnard, it was not a question of whether he would accept, but rather of how quickly he could settle his affairs in Nashville and head out West. He was no doubt 'recklessly overeager,'[50] but also perhaps, given his massive insecurities, he may have been afraid that if he delayed Holden would change his mind. In any case, he hastened to Tullahoma, where Chancellor Garland was spending the summer, to discuss his resignation from his position as Instructor at Vanderbilt University (a title he had assumed on taking over from Garland responsibility for teaching all the coursework in practical astronomy). At the same time he and Rhoda began making plans to dispose of their household goods. After receiving Garland's acquiescence to his resignation, Barnard on July 26 telegraphed Holden his acceptance.[51] In a follow-up letter written two days later he expressed his hope that 'it would be early enough if I leave say about the middle of September – I cannot see how I can get away earlier.'[52]

Holden had undoubtedly counted on Barnard to wait for definite instructions before setting out as any reasonable man would. He now urged him to delay his departure until he received further word, but too late. Barnard and his wife had already disposed of all of their household goods except what they were planning to carry along – books, the 5-inch telescope, and a few articles of table furniture. Moreover, Barnard was now without a position – he had resigned from Vanderbilt effective as of September 1.[54] His resignation was accepted by McTyeire 'with deep regret at parting with one whose connection with the University has been so honorable to it.'[55] Barnard could claim to be a graduate of Vanderbilt's School of Mathematics, but his formal studies had always been rather irregular, and he left without taking a degree. Despite his haste, it is to his credit that he did manage to see to it that his financial support was transferred to one David Spence, another poor student with an interest in astronomy.

Barnard chose to ignore Holden's instructions to him to remain in Nashville until he was called for, writing on August 20 that 'we cannot well alter our time of coming as it would entail unnecessary expense.' Then he added with what must have seemed in his own mind a rather magnanimous gesture: 'If it is a matter of delay on account of the beginning of the salary, don't let that interfere, as I am anxious to begin work.'[56] As he would soon find out, even these liberal terms were not in Holden's power to grant.

Holden had earlier requested from Barnard a brief account of himself – 'your academic and other life – and mention how many comets you have discovered &c &c'[57] – which he thought might prove useful whenever the newspapers requested information about the Lick Observatory and its astronomers (as well as, presumably, in presenting a case for hiring Barnard to the Regents). Barnard had written what amounted to a résumé, carefully documenting what he considered to be his most important discoveries and the publications in which they had appeared through August 1887. Then, on a more personal note, he had added:

> I had made up my mind this summer to cease *comet seeking* and to get into a higher class of work. My deep interest in comets, however, would have made my work lie in a cometary direction. I hope you will aid me in this determination. My associations here have been the pleasantest, and I leave only because of the superior advantages that the Lick Observatory will afford me. Another strong inducement is (for I have always had the highest admiration for you personally) that I shall be immediately under your charge, and shall receive the benefits of your training. It will be my earnest endeavour to aid in the progress of the Observatory to the very best of my ability. I am perfectly temperate, neither smoke chew nor use intoxicating drinks.[59]

Despite his heartening claim to temperance, Barnard would give Holden more trouble than any other member of his staff, and the part of this letter about his wanting to be under Holden's charge would prove to be, if Holden ever recalled it, singularly ironic.

After hearing from Barnard that he was willing to start work without a salary, Holden did not hear from him again until September 15, 1887, when he was startled by the young enthusiast's wire from Kansas City: 'On the way and will arrive about the twenty sixth.'[60] By this time, Barnard (in company with Rhoda, whose thoughts on the momentous occasion do not seem to have been recorded) was already embarked on what would be, emotionally, seven tumultuous years, and intellectually the greatest adventure of Barnard's life.

1   O. H. Landreth, 'Barnard at Vanderbilt University Observatory,' *JTAS*, 3:1 (1928), 15

2   W. R. Brooks to EEB, May 10, 1886; VUA

3   EEB to ESH, May 7, 1886; SLO

4   E. E. Barnard, 'Observations of Jupiter with a Five-inch Refractor, during the years 1879–1886,' *PASP*, 1 (1889), 99

5   H. N. McTyeire to EEB, August 6, 1886; VUA

6   E. E. Barnard, *SM*, 5 (1886), 275

7   ibid.

8   ibid.

9   Edward Mims, *History of Vanderbilt University* (Nashville, Vanderbilt University Press, 1946), p. 152

10 EEB to ECP, October 21, 1886, HCO

11 Harlow Shapley, replying in confidence to the American Unitarian Society's inquiry

about a memorial tablet to honor Joel H. Metcalf, who died in 1925. Quoted in John W. Briggs, Letter to the Editor, *Sky and Telescope*, **78** (1989), 341.

12   D. E. Osterbrock, 'The Rise and Fall of Edward S. Holden, Part 1,' *JHA*, **15** (1984), 81–127:94

13   ESH to E. B. Knobel, June 16, 1893; YOA

14   For biographical details about Holden and his stormy directorship, see Osterbrock, 'The Rise and Fall of Edward S. Holden,' op. cit., and Osterbrock, John R. Gustafson, and W. J. Shiloh Unruh, *Eye on the Sky: Lick Observatory's First Century* (Berkeley, University of California Press, 1988).

15   E. S. Holden, *Memorials of William Cranch Bond, Director of the Harvard College Observatory 1840–1859 and of his son George Phillips Bond, Director of the Harvard College Observatory 1859-1865*, (San Francisco, A. Murdoch, 1897), p. 1

16   SN to Darius O. Mills, October 8, 1874; SLO

17   Osterbrock, 'The Rise and Fall of Edward S. Holden,' p. 85

18   Asaph Hall to SCC, March 7, 1904; quoted in Owen Gingerich, 'The Satellites of Mars: Prediction and Discovery,' *JHA*, **1** (1970), 109–15:113

19   Asaph Hall to ECP, February 14, 1888; HCO

20   ESH to Rear Admiral John Rogers, August 28, 1877; USNO

21   Holden presented his paper, 'On the Proper Motion of the Trifid Nebula,' in which from a comparison of the drawings of the nebula with later drawings, Holden concluded that an actual change in the nebula had taken place. This paper is summarized in *Proceedings of the AAAS*, **26** (1885), 120–3. Compare his later investigation, discussed below, in which he reached similar conclusions about the Orion nebula.

22   Asaph Hall to ECP, October 30, 1877; HCO

23   ibid.

24   Richard A. Proctor, 'Note from Mr. Proctor,' *SM*, **6** (1887), 259–62:260. This was the first published account of Holden's blunder. For background to Proctor's animosity toward Holden, see Osterbrock, 'The Rise and Fall of Edward S. Holden,' Part I, 87–9. See also: W. W. Payne, 'The Holden–Proctor unpleasantness,' *SM*, **6** (1887), 192; Holden, 'President Holden's reply to Professor Proctor,' ibid., 210–12.

25   Osterbrock, Gustafson, Unruh, *Eye on the Sky*, p. 65

26   ESH to EEB, September 12, 1883 and September 21, 1883; VUA

27   ESH to EEB, September 26, 1883; VUA

28   ESH to EEB, April 27 and May 2, 1885; VUA

29   E. E. Barnard, 'Small Nebula Involved in Preceding End of Great Nebula,' *SM*, **4** (1885), 247

30   ESH to EEB, September 28, 1885; VUA

31   ESH to EEB, October 13, 1885; VUA

32   It was, he later informed Holden, 'none other than GC 106. I am very sorry I did not know this.' EEB to ESH, October 31, 1885; SLO. This object is actually a rich star cluster in M31. Its modern designation is NGC 206.

33   ESH to John S. Hager, July 1, 1885; SLO

34   EEB to ESH, October 26, 1885; SLO

35   ESH to EEB, October 27, 1885; VUA

36   EEB to ESH, October 31, 1885, November 12, 1885; SLO. Also: ESH to EEB.,

October 27, 1885; VUA

37 EEB to ESH, December 10, 1886; SLO

38 EEB to ESH, May 7, 1886; SLO

39 Ormond Stone to ESH, December 14, 1885; SLO

40 ESH to EEB April 26, 1886; VUA

41 Donald E. Osterbrock, *James E. Keeler: Pioneer American astrophysicist* (Cambridge, Cambridge University Press, 1984), p. 68

42 EEB to ESH, May 7, 1886; SLO

43 E. E. Barnard, 'Comet 1886 (Barnard c 1887, Jan. 23),' Editorial Notes, *SM*, **6** (1887), 114–15

44 E. E. Barnard, 'Comet d, 1887 (Barnard, Feb. 16),' Editorial Notes, *SM*, **6** (1887), 161

45 E. E. Barnard, *SM*, **6** (1887), 161; *AN*, **117** (1887), 59; *AJ*, **7** (1887), 95

46 ESH to EEB, August 20, 1887; SLO

47 The *Nation*, August 6 and August 20, 1885. The episode is described in Mims, *History*, pp. 100–5. The geology professor was Alexander Winchell, who in 1878 had published a book, *The Pre-Adamites*, and at the Vanderbilt commencement of that year had talked on 'Man in the Light of Geology.' Even Professor Thomas O. Summers of the Biblical Department was forced to admit that the latter was one of the most beautiful lectures he had ever heard, noting, 'It made us almost sorry that we could not accept the nebular theory and evolution, its corollary,' quickly adding, however, that 'nothing is clearer to our minds than that "special creation" is taught in the Scriptures, and we must abide by that.' Unfortunately, matters did not end there. Winchell was criticized in the Nashville *Christian Advocate* and other Methodist publications, and Bishop McTyeire and the Vanderbilt Board of Trust, caving in to pressure, finally had Winchell's lectureship abolished. Following this state, the *Advocate* proclaimed, 'Vanderbilt is safe,' while the Tennessee Conference of the Methodist Church at their meeting the following year declared that in an age when scientific atheism walked abroad, 'our University has had the courage to lay its young, but vigorous, hand upon the mane of untamed speculation, and say we will have no more of this science.'

48 ESH to EEB, March 19, 1887; VUA. The letter was not actually mailed at the time.

49 ESH to EEB, July 14, 1887; VUA

50 Osterbrock, Gustafson, Unruh, *Eye on the Sky*, p. 71

51 ESH to EEB, July 26, 1887, telegram; SLO

52 EEB to ESH, July 28, 1887; SLO

53 ESH to EEB, August 13 and August 20, 1887; VUA

54 EEB to H. N. McTyeire, September 8, 1887; VUA. The letter reads in part: 'In handing in this my resignation, I would state my reasons for such action. Having been appointed one of the astronomers on the staff of Professor Holden, director of the Lick Observatory in California, I wish to sever my connection with the University to fill that position.

'You will readily perceive that the advantages there offered to a young man persuing the profession of Astronomy, are such as cannot be offered elsewhere in this country or even in Europe, it therefore behooves one to not neglect such an opportunity for promotion in the science which I have chosen as my life profession.

'I would state that the inducements there offered are not so much in the way of Salary as in the superb instruments and wonderful atmospheric conditions that promise to make singularly successful all kinds of astronomical researches.

'In leaving the Vanderbilt University I experience a feeling of the deepest regret mingled with a sensation of pleasure at the new life that is opening up before me. My entire connection with the University has been unmarred by any unplesantness whatever, indeed my intercourse both with professors and students has been singularly pleasant throughout.'

55 H. N. McTyeire to EEB, September 9, 1887; VUA. Nevertheless McTyeire gave him his best wishes: 'Following your motto, ad astra, you go to the Pacific Coast and the great observatory there. Your former official associates, and your many friends, cannot help but feel pleasure at your deserved promotion. They also feel sure that you will give the best account of your increased facilities for adding to the world's knowledge, and your own fame.'

56 EEB to ESH, August 20, 1887; SLO

57 ESH to EEB, August 13, 1887; VUA

58 'I have a clear record to the discovery of ten (10) comets, one of which was an independent discovery – being one day late, though no public announcement had been made until my announcement of it . . .

'Following is the list which you will verify by reference to the *Astronomiche Nichrichten* [sic]:

	Comet	Discovery
1)	1881 VI	1881 Sept. 17
2)	1882 III	1882 Sept. 13
3)	1884 II (periodic 5.3 yrs)	1884 July 16
4)	1885 II	1885 July 8
5)	1886 II	1885 Dec. 3
6)	1885 V (Independent)	1885 Dec. 27
7)	1886 IX	1886 Oct. 4
8)	1886 VIII	1887 Jan. 23
9)	1887 d	1887 Feb. 16
10)	1887 e	1887 May 12

'Besides these I have discovered with my 5 in. and the 6 in., some 23 new nebulae, which are recorded in Sidereal Messenger Science Observer, Observatory, etc . . .'

In addition, Barnard mentioned his discovery of the double star β^2 Capricorni, his observations of the break up of the nucleus of the Great Comet of 1882, and his observations of Jupiter. EEB to ESH, August 20, 1887; SLO

59 EEB to ESH, August 20, 1887; SLO

60 EEB to ESH, September 15, 1887, telegram; SLO

8

Hanging fire

1

The great observatory of which Holden was to be the first director was endowed by James Lick, who wanted 'a powerful telescope, superior and more powerful than any telescope yet made.' Lick, whose eccentric personality and eventful life have been described in detail elsewhere,[1] spent his early years as an apprentice wood joiner in his father's shop in Pennsylvania, and after a disappointment in love which resulted in the birth of a son whom he never recognized, he sought a new life for himself as a piano builder in South America. The business proved to be very successful, but after twenty-six years of self-imposed exile, he sold out and sailed for San Francisco, carrying with him a strong box full of Peruvian gold coins, his workbench and tools, and six hundred pounds of chocolates made by Domingo Ghiradelli (who later took his advice and followed him to San Francisco). Lick arrived in San Francisco in January 1848, only a month before the Treaty of Guadalupe Hidalgo was signed, in which Mexico ceded California to the United States, and began buying up real estate on a large scale after James Marshall's discovery of gold at Sutter's sawmill, when many of the residents proved only too glad to part with their property on the cheap in order to buy mining gear and raise a stake in the gold fields. When thousands of '49ers began streaming through the Golden Gate, San Francisco grew rapidly, and Lick was able to sell back land he had acquired so cheaply at handsome profit.

In 1873, Lick suffered a stroke, and began pondering the question of what to do with his immense fortune. Still bitter about his disappointed love, he refused to the end to legally recognize his son. Instead he wished to build a lasting monument to himself. At first he dreamed of commissioning colossal figures of himself and his parents bestriding San Francisco Bay, but he gave up the idea as soon as someone pointed out that such figures would be subject to bombardment in the event of a war. He then turned his thoughts to building a pyramid larger than that of Cheops at the corner of Fourth and Market Streets.

Eventually, however, Professor George Davidson, the head of the Pacific Branch of the US Coast and Geodetic Survey and president of the California Academy of Sciences, was able to persuade Lick that a large telescope would be a more fitting monument. Lick therefore drew up a will setting aside for that purpose the sum of $1 200 000, though he later reduced the sum to $700 000.

By July 1874, Lick had set up the first Lick Board of Trust for the purpose of carrying out this bequest. The president was Thomas O. Selby, the mayor of San Francisco, and another prominent member was Darius O. Mills, the president of the Bank of California. The first question to be decided was whether the telescope was to be a refractor or reflector. Lick, wanting the 'largest' as well as the 'most powerful' telescope in the world, at first refused to consider anything under 72 inches, the size of Lord Rosse's by then long dormant reflector (the 'Leviathan') at Parsonstown, Ireland. But the reflector was not then in fashion; Rosse's telescope, sneered at as 'the largest and most useless telescope in existence,' had in fact been something of a disappointment – the discovery of the spiral form of some of the nebulae had been the only important discovery ever made with it. Disappointing, too, had been the 4-foot reflector erected at Melbourne, Australia, in 1867. The problem with such instruments was that, just as in William Herschel's day, the mirrors were still being fashioned of speculum metal, a brittle alloy of copper and tin which was difficult to figure accurately and even when polished reflected only half the incident light, considerably less when tarnished. Despite these disadvantages, David Gill, a Scottish astronomer, strongly advocated a reflector to the Lick trustees, and so, for a time, did Newcomb.[2] Davidson, however, was firmly in the refractor camp, and Newcomb later changed his mind, so that in the end Lick opted for a refractor.

At the time the largest refractor in the world was still the 26-inch Clark at the US Naval Observatory in Washington, DC, and no one was willing to attempt anything larger than 36 inches. The Clark firm was the obvious choice to make the lens for the Lick instrument, too, but Lick was scared off by the asking price of $180 000 in gold, and resolved to have nothing more to do with the Clarks. Instead, in December 1874, Simon Newcomb set sail for Europe on behalf of the Lick trust in order to solicit offers from firms there.

Another important question being discussed was where to put the telescope when finished. At first Lick wanted it at Fourth and Market Streets in downtown San Francisco. Davidson, however, succeeded in persuading him that a high altitude was essential for sensitive astronomical observations, and himself favored a site in the Sierra Nevada, altitude 10 000 feet. For a time Lick was of the same mind, and thinking the matter settled, Davidson left on an expedition to observe the 1874 transit of Venus. During his absence Lick came to favor a different site – Mt. Hamilton, a 4200 foot summit located east of San Jose in the Diablo Mountain Range, whose appeal as the site for an observatory had first been noticed by Thomas Fraser, Lick's trusted foreman, who had visited it on horseback in August 1875. When Davidson returned from the transit and found that Lick had abandoned the Sierra Nevada site, he became so disgusted that he 'declined further conference with him.'[3] By then it no longer mattered. The officials of Santa Clara county had agreed to build at county expense a road from San Jose to the summit of Mt. Hamilton, and it was nearing completion when Lick died on October 1, 1876.

After Lick's death, his son sued to be recognized as heir to the fortune. The legal wrangles delayed progress on the observatory for several years. Finally, John Lick settled for a sum of half a million dollars, and only then, in the spring of 1879, did the

Board of Trust feel confident enough of their legal position to move forward again. The president of the trust was now Richard S. Floyd, a onetime cadet at the US Naval Academy who had served as an officer in the Confederate Navy during the Civil War. Floyd, Fraser, Newcomb, and Holden met in Washington, DC in the summer of 1879 to discuss their plans, which now went far beyond the 'telescope superior to any' that had originally been envisaged and called for nothing less than a fully appointed research institution.

There was still one nagging question – the seeing conditions on the mountaintop had never been formally tested. What if the atmosphere proved too unsteady for the most exacting kinds of astronomical work? Newcomb, in particular, had 'grave doubts on the subject,' as he later explained:

> A mountain side is liable to be heated by the rays of the sun during the day, and a current of warm air which would be fatal to the delicacy of astronomical vision is liable to rise up the sides and envelope the top of the mountain. I had even been informed that, on a summer evening, a piece of paper let loose on the mountain top would be carried up into the air by the current.[4]

If the seeing turned out to be as poor as Newcomb feared, the situation would be embarrassing, to say the least, with the road up the mountain already built at great expense to the citizens of Santa Clara county. The trustees hired S. W. Burnham, the celebrated double star observer from Chicago, to make the tests, and in the late summer of 1879 Burnham brought his 6-inch refractor to Mt. Hamilton for two months of observing. The results surpassed all expectations. Whereas at the US Naval Observatory in Washington, DC, a night watchman once counted only two 'first class' nights in a whole year,[5] Burnham experienced forty-two 'first class nights' between August 17 and October 16 and discovered forty-two new double stars.[6] He wrote enthusiastically to Floyd: 'If the proof of the pudding is in the eating, these discoveries ought to be eloquent on the subject of Mt. H for the site of an Obsy. They will certainly be much more satisfactory to outsiders than the mere opinion of anyone.'[7] Burnham's glowing report put all doubts about the site firmly to rest, and under Fraser's supervision, construction on the summit began in earnest in 1880. The first dome, for a 12-inch refractor acquired from Henry Draper, was in place by November 1881.

Meanwhile, with Lick no longer around to protest, the trust gave the Clarks the contract for the 36-inch lens. By way of background, a refracting telescope uses a lens to form an image by bending the rays of light to a focus. However, a simple lens brings light of different colors to different foci. In the case of a bright object, this results in an image that is surrounded by unfocused colors (chromatic aberration). By combining two closely spaced lenses made of different kinds of glass, the chromatic aberration produced by the one lens can be made to nearly cancel that of the other for a certain band of wavelengths, usually the yellow and green region of the spectrum to which the eye is most sensitive. Thus such a lens is said to be achromatic – light comes to nearly the same focus across this visual spectral region, though as one moves beyond it, say into the blue and violet regions to which photographic emulsions are most sensitive, the focus begins to vary with wavelength again, with the result that even achromatic refractors are not perfectly free of chromatic aberration.

One lens is made of ordinary crown glass and the other of flint glass (also known as leaded crystal). Obviously the glass needed for such a lens has to be very clear and pure, free of bubbles, fractures and other defects, and there was a long delay while the Paris firm of Charles Feil and Son struggled through numerous mishaps and to the verge of bankruptcy in order to produce a perfect enough crown disk. At last, in September 1885, such a disk was delivered to the Clarks' Cambridgeport, Massachusetts workshop, and it took the Clarks only fourteen months to finish the lens. Fraser himself personally escorted it aboard a special railroad car from Cambridgeport as far as San Jose, and up the winding mountain road, to the summit, by wagon. Upon its arrival it was carefully locked away in a safe to await completion of the rest of the telescope. On January 9, 1887, James Lick's body, which had temporarily reposed since his death in the Masonic Cemetery in San Francisco, was trundled up the mountain and reinterred at the base of the pier in fulfilment of his last wishes. A metal plate placed there bears the simple inscription: 'Here lies the body of James Lick.'

The massive iron dome for the great telescope, seventy-six feet broad, was manufactured at the Union Iron Works in San Francisco. The same firm was responsible for the dome's movable floor, whose iron framework was covered over with hard wood. The floor was designed to travel up and down a distance of seventeen feet in order to accomodate the observer in all viewing positions, thereby eliminating the need for the observer to dangle precariously in the darkness from a tall ladder. The power to raise it and to turn the dome was to be supplied by hydraulic pressure from reservoir tanks located at various elevated points on the mountain. Meanwhile the tube and mounting of the great telescope were under construction at the firm of Warner and Swasey in Cleveland, Ohio. Their job was 'to manufacture a cylinder of metal 56 feet long, 4 feet in diameter, and 6 tons in weight, and to make it move with the precision of a Swiss watch.'[8] When S. W. Burnham paid a visit to the Cleveland factory in late October 1887, he reported to Holden on the progress:

> I asked Capt. F[loyd] when he expected to be able to turn over the obsy and after estimating the several things of furnishing the mounting, transporting it, getting it on the hill and set up, and things straightened out generally, he said the minimum time would be Jan. 1. My private opinion is that practically it will be some considerable time beyond that, making only moderate allowance for the unexpected.[9]

2

Barnard and his wife, as we have seen, had set out from Nashville on the morning of September 14, 1887, in blissful ignorance of how far the observatory remained from completion. The route west led first to the small town of Mackenzie, Tennessee, where they switched railcars for Memphis. That afternoon they passed through a country 'parched and baked with the intense heat of the sun.' Clouds of dust from the dry soil rose as the train passed, 'stifling the nostrils and filling the eyes with an impalpable powder,' and Barnard wondered how people could bear to live in such a country: 'The stations were burried in dust, and a more discouraging outlook could scarse be imagined.' There were cotton fields, white with the ripe balls of cotton, in which pickers were busily gathering in the crop, 'with that intolerable sun pelting down upon them its firey beams.'

> One scene that struck us with a peculiar sadness was witnessed in the afternoon when the dust and heat were greatest; in a large cotton field, with no house in sight and no other living being in view, was a poorly clad woman gathering cotton, in the sweltering sun and near her, sheltered by an old cloth supported on two sticks, was her infant, lyi[n]g in a rude wooden 'crib'; they were in view but for a moment as the train dashed past, but the scene was indelably impressed on my memory, and it was several hours before the landscape seemed to brighten again.[10]

After an hour layover in the depot in Memphis, which the Barnards found deplorably filthy, they boarded the train for Kansas City. As they crossed the Mississippi, into the cool night air, with the stars sparkling above them, Barnard recorded: 'What a change! Not a speck of dust – not a thought of heat; we opened the car window and let the cool fresh air in. Later the sleeping car porter came by and asked, "Whar' you been, mister, t'git so much dus," and persuaded us to go to the back of the car and shake ourselves off.' To Barnard, the dust flying back in the wake of the train resembled 'a comet's tail.'[11]

When they awoke, they were in the Ozarks. As the Sun rose higher into the sky, the train chugged on through fields of grain, now ripe for the harvest, through Fort Scott to Kansas City, where Barnard telegraphed Holden of his impending arrival. By the next morning, as they were nearing Denver, they passed through a 'desert-like' country that was mainly remarkable for its colonies of prairie dogs. Barnard found these creatures 'singularly amusing in their attitudes and inquisitive stare.' Later that same morning they caught their first breathtaking view of the Rocky Mountains.

Unfortunately, Barnard carried his account only as far as Denver. He left no surviving impressions of the rest of the journey on the great transcontinental railway through Salt Lake City to San Francisco. Nor does there seem to be any documentation of what happened when he first presented himself to Holden at the offices of the Lick trust, at 120 Sutter Street in San Francisco. Holden, presumably, was less than thrilled, while one can only imagine Barnard's bitter disappointment when he learned that the observatory was still months away from completion. He had been more than willing to start work without a salary, but the possibility that he would not be allowed to take up residence on the mountain had never crossed his mind. Now the unpleasant reality stared him in the face. He and Rhoda would have to rent a room in San Francisco – they found one at 1500 Taylor Street – and they prepared themselves for several months of hardscrabble existence.

Barnard still had some money left from the Warner prizes, and faced with the sudden need for additional cash, he sold to the University of Southern California his beloved 'pet' – the 5-inch Byrne refractor that had served him so well and with which he had made so many wonderful discoveries. Though he does not seem to have recorded his feelings on this occasion, one can well imagine that he must have been close to heartbreak.

Even so, Barnard's resources would have been soon depleted, had not Holden managed to find Barnard a job copying legal documents in the law office of Jarboe and Harrison, located at 230 Montgomery Street. The firm's senior partner, John R. Jarboe, was Holden's personal friend. Though Jarboe seems to have treated Barnard well, the job did not pay very well, and Barnard continued to dip steadily into his savings. The next few months must have been dreary indeed for someone who wrote in an almost illegible hand and was a painfully slow typist.

Though Holden had many things on his mind, he did his best to keep Barnard informed of possible diversions, and often went out of his way to provide complimentary tickets to theatrical and operatic performances. Rhoda, however, usually begged off because of illness. For a time Barnard was contemplating a cruise to Hawaii (then known as the Sandwich Islands),[14] but Rhoda again became ill and it had to be postponed, while still later Barnard himself got cold feet, realizing that if he sailed out of San Francisco in early December as planned he would be unable to return to San Francisco until late January at the earliest, and he was afraid that he might miss the opening of the observatory.[15] (On that account, at least, he needn't have feared.) Perhaps significantly, almost immediately after Barnard had made up his mind definitely not to go, Rhoda began to recover.[16]

What must have made Barnard's life during these months almost unendurable was the fact that he no longer had a telescope. Whether or not one accepts Gerrit L. Verschuur's diagnosis of Barnard as an 'addict of the heavens,'[17] Barnard probably often experienced acute symptoms not unlike those of withdrawal whenever the stars were shining. Under the circumstances, it is no surprise that astronomically speaking his months in San Francisco were barren indeed. He borrowed from Holden the textbook written by Holden's father-in-law, Chauvenet,[18] and used George Davidson's 6.4-inch refractor for a few nights to search for some of Lewis Swift's latest nebulae – the only astronomical observations he made in San Francisco.[19]

3

Slogging on in the law office, Barnard struggled to keep from falling prey to depression. From the distance, and receiving information only at second hand from Holden, he had difficulty keeping abreast of the latest developments on Mt. Hamilton. Captain Floyd, as president of the Lick trust, was still firmly in charge there, and intended to remain so until the observatory had been turned over to the Regents – or, as he put it in an apt nautical metaphor, until he had safely 'launch[ed] this Observatory into the Ocean of Science.'[20] Except for Keeler, no other astronomer, not even Holden, was allowed to take up residence on the mountain – according to Holden, for no better reason than that the president of the Lick trustees 'didn't want to be pestered.'[21] Barnard did, however, manage a brief visit in early December, though strictly in the role of tourist.[22] By then, the mounting and tube of the great telescope, having been carried up the mountain in parts by teams of horses, were being assembled in the giant dome.

At last, on New Year's Eve 1887, the 36-inch lens was removed from the safe in which it had been kept for a full year and installed in the telescope. Alvan Clark was now dead, but his son Alvan Graham Clark, a 'terrible old blow and grumbler,' as Keeler found him,[23] was present for the occasion, as were Ambrose Swasey, Floyd, and Keeler. Before the last of the Old Year expired, they hoped to have a first look through the new telescope. But suddenly foul weather rolled in – a 'rattling South Easter,' Floyd called it, swept over Mt. Hamilton with strong winds, drifting clouds, and freezing cold. The skies again cleared on the evening of January 2, 1888. Then, because of the cold, the dome was found to be frozen solid in its track. With great effort, workmen managed to pry open the shutter, and the great telescope was swung toward the bright star Aldebaran. The moment of truth had arrived.

With a sinking feeling, the four men found that they were unable to bring the image of the bright star to a focus. For one despairing moment it seemed that the telescope would prove to be absolutely useless, in spite of all that Lick, Floyd, the Clarks, and Swasey had done. The source of the problem, fortunately, proved to be readily correctible – it turned out that the Clarks had misstated the focal length of the lens by six inches, with the result that the tube had been made too long. Swasey simply sawed off a six-inch segment from the end and remounted the lens. By the next clear night, January 7, the telescope was ready for another test.

Again Floyd, Clark, Swasey, and Keeler gathered in the dome, joined this time by Floyd's niece, Cora Matthews. Since the dome was still frozen in place, the observers had to wait for various objects to transit the open shutter. First across was the white star Rigel – to everyone's relief, it came to a perfect focus. Next came the Orion nebula. In the Trapezium, the famed group of stars embedded in the nebulosity, Clark detected a faint star never seen before, about which S. W. Burnham later wrote to Floyd:

> If the discovery had been made anywhere else, I should have had very little faith in it, because the Trapezium is one of the most ancient of mare's nests for new stars; and I suppose some . . . venerable humbugs will claim this star, and any others which may be found. But I think we can get ahead of them in somethings after awhile.[24]

Finally, at just past midnight, Saturn was captured within the great blue circle of the eyepiece. Keeler described it as 'beyond doubt the greatest telescopic spectacle ever beheld by man. The giant planet with its wonderful rings, its belts, its satellites, shone with a splendor and distinctness of detail never before equalled.'[25] The seeing was close to perfect, and examining the planet with a magnification of × 1000, Keeler discerned for the first time an incredibly fine division, 'like a spider's thread,'[26] in the outer ring. There could no longer be the slightest doubt as to the telescope's excellence.

Three days later, the Regents formally approved Holden's recommendations for the members of the Lick staff, and Holden at once telegraphed Burnham, Schaeberle, Keeler and Barnard, and also Charles B. Hill, who had worked with Davidson at the US Coast Survey and was hired as assistant astronomer, secretary, and librarian, news of their official appointments: 'Regents appointed you to day to date from day of transfer of observatory.'[27] In the event, that would not be until June 1, 1888.

At the same meeting, the Regents adopted Holden's *Articles on the Government of the Lick Observatory*, of which Article Six would later become the most contentious: 'No communications to journals purporting to emanate from Lick Observatory, or relating to the work of the Lick Observatory are to be made by officers or employees without the formal approval of the Director.'[28]

Before turning the observatory over to the Regents, Floyd wanted to have a complete inventory of the property done. He offered the job to Barnard 'for . . . the same compensation that [the trustees] allow to Professor Keeler, that is $75 per month and your board and lodging here.'[29] Barnard was eager to get to Mt. Hamilton in any capacity, and expressed an interest in the job, though he did not feel comfortable with Floyd's further request that he 'arrange the photographic and other apparatus.'[30] Floyd, however, assured him that nothing more had been intended by this than 'the general way of putting things in places most convenient for their after permanent arrangement,' and that the requirements of the job were entirely 'extra-professional.'[31] With this clarification, Barnard offered his services for $25.00 plus expenses, preferring 'a definite sum to being employed by the month.'[32] One would not guess from the modest sum asked for that by this time Barnard was nearly broke; his savings were exhausted except for about $100.[33]

On March 28, 1888, Barnard took a room for the night at the St. James Hotel in San Jose, and at 7:30 the next morning caught the Mt. Hamilton stage.[34] He arrived at the summit in gale-like conditions. Keeler, who had met Barnard earlier in San Francisco and again on his brief visit in December, thought that Barnard 'look[ed] rather glum,' and guessed that it was because of the nasty weather. 'I don't imagine he minds it himself, but I suppose he is thinking of his wife,' he added.[35] Captain Floyd was also still on the mountain. Only forty-five years old, he was already in failing health. Soon after Barnard's arrival he suffered a heart attack and was forced to leave Mt. Hamilton for Kono Tayee, the family home on Clear Lake. He never returned.[36]

Though Barnard probably continued to look glum most of the time he was working on the inventory, he must have cheered up on at least one occasion. On the evening of April 5, Keeler, who had been observing Mars with the 36-inch refractor since mid-March, glimpsed the Martian satellites, and woke up Barnard to show them off. 'He was delighted,' Keeler wrote. 'It was his first view through the telescope.'[37]

But these were rare interludes. Working himself to the point of exhaustion, Barnard completed the inventory in only two weeks. Starting with 'The Main Observatory Building,' he listed items ranging from '1 36″ equatorial complete and in position,' 'Large Dome of 36 Inch Equatorial,' '1 Brashear star spectroscope complete, on carrier,' down to 'One single chicken house near Cottage 2,' '1 3 foot step ladder,' and '1

ladle, 1 skimmer, 1 egg turner.'³⁸ By mid-April he was back in San Francisco preparing
to make his final report, but the night after he arrived he was taken ill. Holden noted in
his diary that Barnard was 'ill from work – and from disgust at the frightfully dirty
condition of his Quarters.' And with resentment borne of his long exclusion from the
observatory he was to direct, Holden put the blame squarely on the trustees:

> [Barnard] has the choice of 2 sets [of quarters] – one with a kitchen and no furniture
> – the other with bedrooms and no kitchen – both filled with bedbugs! – This is
> simply a sample of the utter disregard of the needs of the Obsy. – All the work
> having been put on the Main building.³⁹

When Barnard, still sick in bed two days later, fretted to Holden that he was not yet able
to complete the inventory,⁴⁰ Holden wrote back: 'All right about the Inventory – you
get well and cheer up – It will go fast now – I have got hold.'⁴¹

Of course Barnard did eventually recover, and he was able to complete the inventory
– it ran to ninety-eight pages. Meanwhile, Holden was only slightly premature in
claiming to have 'got hold.' He was able to win permission from the trustees for himself
and Barnard to move onto the mountain as of May 1, 1888, 'to take possession of the
quarters intended for [us] and to fit the same for [our] use, but so as not to interfere with
the work now going on.'⁴² Thus on May 1, Holden moved into one side of the so-called
Astronomers' House, a three-story brick building located near the Main Building. On
the following day Barnard and Rhoda – and also Elizabeth Calvert, who had come out
to California from Tennessee in order to provide her older sister companionship and
support – also arrived on the mountain.⁴³ The Barnards set up their household in one of
the small wooden cottages previously used by workmen and located in the saddle just
below the Astronomers' House. (Because of its shape, this cottage was later – after the
Barnards had left – nicknamed 'the Ark.') One of the first priorities, obviously, was to
have it fumigated so as to make it fit for human occupancy.⁴⁴ In order to make it feel
more like home, the Barnards planted a small garden outside with geraniums, violets,
and mignonette.

Finally settled upon the mountain, Barnard wasted no time resuming regular
observing. On May 9, 1888, he made his first entry in his observing book: 'Field of 12
in. lowest power. Found 4 new nebulae within 1° of Castor.'⁴⁵ Eight dreary months of
near total telescope deprivation were finally at an end!

Holden, meanwhile, was writing on Barnard's behalf to Andrew S. Hallidie, who
with Timothy Guy Phelps and Judge John S. Hager served on the Regents' Lick
Observatory Committee: 'I . . . beg to ask you that you will endeavour to have Mr.
Barnard's pay commence May 1 at the same time that mine did. He is doing the same
work here that I am – and observing at night – May 9 – he discovered 4 new nebulae etc.
etc. He has been without pay for 8 mos. and it would be a deserved recognition of his
talents and of his patience if this were granted.'⁴⁶ In the same letter, Holden also asked
for money for needed repairs, and took another shot at what he regarded as the
negligence of the Lick trustees: 'As this is a personal and not an official letter I may say
that the reservation is very like an old time Southern plantation – The front door knob

Lick Observatory on Mt. Hamilton, about 1888. The Main Building is in the background, with the 36-inch refractor dome on the left and the 12-inch refractor dome on the right. In front of it is the Astronomers' House, and at the lower left of the photograph is the cottage occupied by Barnard and his wife from 1888 to 1894. Mary Lea Shane Archives of the Lick Observatory

shines with polishing and the parlor is clean – but everything else is dirty and shiftless.'[47] His gibe was clearly directed at Floyd, the scion of an old Georgia family. Hallidie, while not commenting on Holden's sarcasm, granted his requests, and was particularly glad to agree with the proposal about Barnard's salary.[48] Thus, with a sufficient source of income and observing once again, Barnard could finally write with satisfaction: 'It is quite delightful up here now, and all are well.'[49]

When, on May 19, 1888, Captain Floyd signed at Kono Tayee a deed turning over all the property Barnard had inventoried to the University of California, almost fourteen years had passed since James Lick had executed the first deed of trust mentioning the great Lick telescope. The document was signed by the rest of the Lick trustees on May 31. Holden had once fumed about those same trustees: 'How this *great* thing is temporarily belittled by the little men.'[50] Now the 'little men' – Floyd, Fraser, and the rest, who despite their detractors and with little fanfare had done their work so splendidly – got out of the way, and the observatory was formally turned over to the Regents. A new era had begun. On June 1, 1888, Holden announced: 'The observatory begins its active existence tonight.'[51]

1   See, for example, Donald E. Osterbrock, James R. Gustafson, W.J. Shiloh Unruh, *Eye on the Sky: Lick Observatory's First Century* (Berkeley, University of California Press, 1988), and Helen Wright, *James Lick's Monument: The Saga of Captain Richard Floyd and the Building of the Lick Observatory* (Cambridge, Cambridge University Press, 1987). Other useful accounts are: F. J. Neubauer, 'A Short History of the Lick Observatory, Part 1,' *PA*, **58** (1950), 201–22; Part 2, 318–34, and Part 3, 369–88; Michael Criss, 'The Stars Move West: The Founding of the Lick Observatory,' *Mercury*, 2:4 (1973), 10–15 and 2:5 (1974), 3–8

2   D. E. Osterbrock, 'The Quest for More Photons: How Reflectors Supplanted Refractors as the Monster Telescopes of the Future at the End of the Last Century,' *The Astronomy Quarterly*, **5** (1985), 89

3   Quoted in Neubauer, 'Short History,' p. 206

4   Simon Newcomb, *Reminiscences of an Astronomer* (Boston and New York, Houghton, Mifflin & Co., 1903), p. 188

5   Jan K. Herman, *A Hilltop on Foggy Bottom*, (Washington, DC: Naval Medical Command, Department of the Navy, 1984), p. 52

6   S. W. Burnham, 'Report of Mr. Burnham,' *LOP*, **1** (1887), 13. 'By first class seeing,' he wrote (p. 15), 'I mean such a night as will allow of the use of the highest powers to advantage, giving sharp, well-defined images, and when the closest and most difficult double stars within the grasp of the instrument can be satisfactorily made.'

7   SWB to RSF, November 9, 1879; SLO

8   Osterbrock, et. al., *Eye on the Sky*, p. 56.

9   SWB to ESH, October 14, 1887; SLO

10   EEB, unpublished typed MS in VUA. There is legal correspondence on the other side of each sheet. Barnard no doubt wrote this account in his spare time when he was working at the law office of Jarboe and Harrison in San Francisco.

11   ibid.

12   EEB to ESH, October 26, 1887; SLO

13   ESH to EEB, October 7, 1887; VUA

14   EEB to ESH, November 18, 1887; EEB to ESH, November 28, 1887; SLO

15   EEB to ESH, December 2, 1887; SLO

16   EEB to ESH, December 5, 1887; SLO

17   Gerrit L. Verschuur, *Interstellar Matters: Essays on Curiosity and Astronomical Discovery* (New York, Springer-Verlag, 1989), p. 30

18   EEB to ESH, October 26, 1887; SLO

19   E. E. Barnard, 'Ueber Nr. 14 und 15 des Swift'schen Nebelcatalogs Nr. 6 in A.N. 2798,' *AN*, **118** (1887), 173–4

20   RSF to George C. Comstock, December 24, 1887; UW

21   ESH, diary, March 31, 1888; SLO

22   JEK to EEB, December 24, 1887; VUA

23   JEK to ESH, January 6, 1888; SLO

24 SWB to RSF, February 28, 1888; SLO

25 J. E. Keeler, San Francisco *Examiner*, January 10, 1888.

26 J. E. Keeler, 'First Observations of Saturn with the 36-inch Equatorial of the Lick Observatory,' *Sidereal Messenger*, 7 (1888), 79–83. See also D. E. Osterbrock, 'Keeler's gap,' *Science*, **209** (1980), 444 and D. E. Osterbrock and D. P. Cruikshank, 'J. E. Keeler's Discovery of a Gap in the Outer Part of the A Ring,' *Icarus*, **53** (1983), 165–73

27 ESH, note; SLO

28 Neubauer, 'Short History,' p. 323

29 RSF to EEB, March 13, 1888; SLO

30 EEB to RSF, March 15, 1888; SLO

31 RSF to EEB, March 19, 1888; SLO

32 EEB to RSF, March 24, 1888; SLO

33 ESH, diary, March 31, 1888; SLO

34 EEB to ESH, March 28, 1888; SLO

35 JEK to ESH, April 1, 1888; SLO

36 RSF to H. E. Mathews, May 3, 1888; SLO

37 JEK to ESH, April 9, 1888; SLO

38 Osterbrock, Gustafson, Unruh, *Eye on the Sky*, p. 61

39 ESH, diary, April 16, 1888; SLO

40 EEB to ESH, April 18, 1888; SLO

41 ESH to EEB, April 20, 1888; VUA

42 H. E. Mathews to ESH, April 25, 1880; ESH to A. S. Hallidie, May 11, 1888; SLO

43 ESH, diary, May 2, 1888; SLO

44 ESH, diary, May 3, 1888: 'During this week all the cottages . . . were fumigated for bed bugs.' SLO

45 EEB, observing notebook; LO

46 ESH to A. S. Hallidie, May 11, 1888; SLO

47 ibid.

48 A. S. Hallidie to ESH, May 15, 1888; SLO

49 EEB to H. E. Matthews, May 20, 1888; SLO

50 ESH, diary, March 28, 1888; SLO

51 E. S. Holden, Letter, *AJ*, 8 (1888), 43–4:44

9

On Mt. Hamilton

1

When Holden arrived in Berkeley to become the first president of the University of California, its newspaper, *The Berkeleyan*, wrote of him: '[T]here was something . . . confident, yet not haughty . . . something in the low, even voice, something in the wonderfully frank individuality that won the listener's respect. That something was indicative of self–control, of decision, of the ability to mark out a line of action and of the will-power to stick to it.'[1] Other opinions were expressed by the undergraduates who saw him from afar as an Eastern sophisticate in manner and dress – 'the Dude,' or 'Champagne Eddy.'[2] Holden, in fact, would soon show a remarkable affinity for collecting epithets, most of them highly uncomplimentary, during his years as director of the Lick Observatory.[3]

Insofar as a man's writings give any sense of his personality, Holden comes across as a rather attractive figure. His works were always clearly written, and they were remarkably wide-ranging, covering such diverse topics as the bastion system of fortifications, early Hindu mathematics, the vocabularies of children under two years of age, cryptography, rattlesnake suicides, and earthquakes, to say nothing of his voluminous astronomical writings.[4] Holden's retentive memory and unusual abilities as an assimilator and organizer of information made him especially well suited to the work of cataloguer and bibliographer. As a conversationalist, he was regarded as 'entertaining to the point of brilliancy.'[5]

Despite his undoubted vast abilities, there was, however, something about Holden's personal manner that 'offended and antagonized everyone that he had to deal with personally.'[6] This is so well substantiated, and by so many witnesses, as to be well-nigh incontestable. He appears to have been personally cool, formal, and autocratic. To some extent these were the characteristics to be expected of a former West Pointer. But though his military manner, or what one of the local newspapers once referred to as 'an excess of dignity,'[7] was not likely to appeal to everyone, it alone cannot account for the contempt – and even loathing seems not too strong a word – that those who were his associates eventually came to feel for him, almost to a man, but no one more so than Barnard.

Even before he assumed the Lick directorship, Holden found himself embroiled in controversies. During his second year as president of the University of California, he quarreled with the popular astronomy writer Richard A. Proctor in the pages of the San

Francisco *Examiner*. Proctor had predicted that, like previous large telescopes, the Lick refractor was likely to prove a disappointment. Holden had attempted to refute this claim, which was only to be expected, but then went on to attack Proctor personally. Proctor replied at length in the same personal terms, Holden replied to his reply, and so on.[8] By the time the attacks and counterattacks died out, Holden had seen himself referred to in print as 'the telescopist who failed utterly at Washington,' 'a convicted utterer of untruths,' and as 'a man who had set all Washington astronomers laughing by detecting a third satellite of Mars with an impossible period and distance, and remaining deceived by it for months' – the first published reference to the embarrassing Martian satellite affair. Though Proctor died the following year – he succumbed to pneumonia after being thrown out of a New York City hotel into a raw, windy, rainy night by doctors who feared he had contracted yellow fever while living in Florida – the damage to Holden's reputation had been done.

A later controversy was even more damaging. On May 1, 1888, a month before he ascended to the Lick directorship, Holden presented the Lick trustees with a bill for $6000 for the 'expert services' he had provided to them since May 1876.[9] The trustees were outraged, and so was Newcomb, at that moment lying ill in a Chelsea, Massachusetts hospital bed.[10] Nowhere in his bill did Holden even mention Newcomb, despite the fact that Holden's initial involvement in the project had been exclusively as Newcomb's assistant and protégé, and the bill also tended to give the impression that he – Holden – had been the 'sole inspiration behind the observatory.'[11] The trustees resolved to fight Holden tooth and nail – in court, if necessary, for which purpose they set aside $10 000 of Lick's bequest as a hedge against future legal expenses. Meanwhile, the newspapers had got hold of the story, and printed it under banner headlines: 'High Priced Stargazer Meets With Well Deserved Rebuff.'[12] Holden, portrayed as 'avaricious,' 'reaching,' and 'greedy,' afterward argued rather unctuously that his position had been completely misrepresented in the newspapers and that his real object in presenting his claim had been solely to place on record the facts, which he believed showed clearly how extensive his services to the Lick trust had been:

> It is very repugnant to me that money affairs have anything to do with [it]. I do not desire to make any fight for money; nor to wantonly encroach on the funds provided by the generous founder of the Observatory . . . My claim was made in good faith after frequent conversations about it with Officers and employees of the Trust. No one of them ever hinted that it was not just nor gave me to understand that it was likely to be disapproved.[13]

The last point – that he had frequently discussed the bill with the trustees – seems unlikely at best, but regardless of Holden's deserts in the matter, from a public relations standpoint the whole affair was an unmitigated disaster. If, as he alleged, he had only wanted to 'place into the record certain facts which were in great danger of being forgotten' – in other words, his own self-serving version of the history of the Lick Observatory – his approach certainly backfired. Henceforth Holden bashing became a popular sport for the local newspapers, and though the legal wrangles continued for several years, in the end the bad publicity – and the fact that he later needed for observatory projects the $10 000 being retained by the trustees as a legal defense fund – made Holden withdraw his claim without collecting any of the money he had asked for.

2

For all his faults, Holden was a conscientious director, and on moving to Mt. Hamilton threw himself into his work with the energy of a demon. To give but one example, during his first month in office alone he wrote more than five hundred letters, and indeed throughout his tenure 'the heavy correspondence of the Lick Observatory was conducted by Professor Holden's personal pen . . . He did not utilize the services of a stenographer and only an occasional letter was copied with a typewriter.'[14]

His West Point background led him to insist on a strict military way of doing things. Thus, on the very first day when the observatory began its active existence, Holden issued detailed instructions of the regular work to be performed by each staff member – what he described as a 'complete working system of the whole observatory.' He underscored to the staff that 'when this [system] is once understood, everything will go on smoothly.'[15] In addition to the regular duties of each staff member, there was often extra work to be done, and Holden issued instructions for such work on cards, sometimes signed peremptorily: 'By Order of the Director.' The card was handed to the person who was to do the work, and when it was finished the card was to be returned to him. Holden conservatively estimated that close to two thousand such cards were issued during the first nineteen months of his administration.[16] In sympathy to Holden, there was a great deal of such work to be done. Conditions on Mt. Hamilton were rugged, and every necessity except water had to be brought by stage from San Jose, a distance of twenty-six miles. 'When [the stage] cannot arrive,' observed Mrs. S. D. Proctor, the widow of Holden's late nemesis who came on a visit to the observatory in 1891, 'the little colony on the mountain are as isolated as Arctic explorers.'[17]

The water came from a spring on the mountain, and was pumped by a windmill, located on 'Huygens Peak,' into the system of reservoirs. In the dry season, June to October, the reservoir capacity was often insufficient for the observatory's needs and had to be augmented by rain water collected during the previous winter. The water was needed not only for drinking but to provide hydraulic power to revolve the great dome and raise its floor, and since the water passed many times through the water-engines and hydraulic rams, it became covered with a heavy residue of oil, making it, Holden admitted, 'really unfit for [drinking], and produc[ing] more or less illness when . . . used. But it *must* be used. There is no other.'[18] The only fuel supply was wood which had to be collected from 1200 to 1400 feet below the summit, but the delivery was far from reliable – 'The procrastination of our neighbors,' Holden complained, 'has ceased to be annoying. It is majestic – colossal – like a great feature of nature. It must be reckoned with like the inexorable forces of heat, magnetism, gravitation.'[19] Even when there was enough fuel on hand, it was often a serious problem to keep the fires burning.

Despite numerous experiments with ventilating chimney-pots, Holden complained that

> out of six offices there are only two in which a fire can be lighted in all winds. In one of the brick dwellings fires will not burn in a southeast wind, and in the other a north wind is equally fatal. The wind sweeps up the deep cañons on either side, and blows almost vertifically down the flues, so that the flames are driven out into the room several *feet!* or else volumes of smoke make it simply impossible to remain in the apartment.[20]

Such, then, were the mundane problems which captured the attentions of the director. To someone like the unworldly Barnard, who lived only to observe, Holden's often lonely struggles must have seemed remote. Barnard had little sympathy with any of them. But Holden was right to insist, as he once did, that 'before the least scientific work can be done life must somehow be organized':

> If the shutters of the great dome are frozen tight together, the great telescope cannot be used. If there is no wood to burn in the office-stoves, no computations can be made, no matter how enthusiastic may be the computer. If the chimneys of the Observatory will not draw, it is beyond any man's power to work at his desk, be he never so devoted. These matters must be attended to somehow. The energy that is left over is available for the astronomical work.[21]

Much of the responsibility for seeing that these matters were attended to fell to Holden, and his heavy administrative responsibilities left him little energy for research.

3

Deprived of the warmth of a mother's love at an early age, Holden lacked the human touch which would have served him well in his relations with other men. But though he did not prove particularly adept at handling them once he got them, Holden did recognize good men when he saw them, and he succeeded brilliantly in the staff he brought together to live and work on Mt. Hamilton. 'One fact is indisputable,' Newcomb wrote later, 'and that is the wonderful success of the director in selecting young men who were to make the institution famous by their abilities and industry.'[22]

Probably Holden's greatest coup was to lure Sherburne Wesley Burnham, the world's greatest observer of double stars, back to Mt. Hamilton where he had spent two months in 1879 testing the seeing conditions there. At fifty, he was the oldest of the astronomers hired for the Lick staff.

A native of Thetford, Vermont and a graduate of its excellent Academy, Burnham as a young man had taught himself shorthand and at the age of nineteen had moved to New York City to earn his livelihood by it.[23] During the Civil War, he served as shorthand reporter to General Benjamin Franklin Butler at the Union army headquarters in New Orleans. After the war, he moved to Chicago, where he became a member of the firm of Ely, Burnham, & Bartlett, official reporters of the courts of

Chicago. By then, he was already keenly interested in astronomy, and he had a small, though not very satisfactory, telescope.

His Chicago residence, at 36th Street and Vincennes Avenue, was only two blocks from the Dearborn Observatory, in Douglas Park near the Lake Michigan shore. Its imposing ninety-foot tall stone tower, capped by a cylindrical dome, housed what was then the world's largest refractor – the $18\frac{1}{2}$-inch Clark. Burnham's proximity to the world's largest telescope naturally made him wish for a better telescope of his own, and in 1869 he arranged to meet Alvan G. Clark, who was then passing through Chicago on his way home from Iowa where he had observed the total solar eclipse of that year, and Burnham ordered from him, at a cost of $800, a 6-inch refractor. The only stipulation was that 'its definition should be perfect.' This telescope was later housed in Burnham's backyard in a small dome which the neighbors referred to as the 'Cheese Box,' and 'rumors had it that the "Star Gazer" was busy with fortune telling and necromancy. Some of the bolder couples even approached him about their future, but Burnham rejected fame and fortune and turned all inquiries aside with his own sense of humor.'[24]

From first to last, Burnham's chief astronomical interest was in double stars. At a time when many astronomers imagined that the field had been nearly exhausted, Burnham, working with his modest telescope almost in the shadow of what was then the largest refracting telescope in the world, discovered many new pairs. He published a first list of eighty-one in 1873,[25] and soon added many more – his final list of discoveries with the 6-inch came to 451. Needless to say, these results created a sensation, and Burnham was soon brought into correspondence with other leading experts on double stars such as Rev. T. W. Webb of England and especially Baron Ercole Dembowski of Italy who measured many of them for him. As his reputation grew, he was invited to use ever more powerful instruments. Professor G. W. Hough, the director of the Dearborn Observatory, invited him to use the $18\frac{1}{2}$-inch refractor on a regular basis, and in 1874 he was allowed one night on the newly erected 26-inch refractor in Washington. It was then that he first met Holden, who came away impressed with his energy and his keen eye. In 1881, when Holden was appointed director of the Washburn Observatory in Wisconsin, he hired Burnham to observe double stars with the $15\frac{1}{2}$-inch Clark refractor in Madison. Burnham was then eager to make a career of astronomy, and was keenly disappointed when this job ended after only five months and he had to return to court reporting.

Later, Burnham's attitude toward a career in astronomy changed. When in 1885 Holden first approached him about a position on the Lick staff, he declined. He was by then reluctant to leave his lucrative court reporter position, partly out of consideration for the needs of his wife and six children, for whom Mt. Hamilton offered the prospects of substandard housing and no schools, but partly also out of fear that if he took up astronomy as a fulltime profession it would ruin it for him as a hobby. 'I have thought of your proposition,' he wrote to Holden, 'but have come to the conclusion that the time for any change of life with me has gone by . . . It is a great accomplishment, to which I might lay some claim, to learn from experience and to know when one is well off. Looking at things in an unambitious and humble sort of way, I think I am pretty well off on both sides of the line between home and work.'[26] Holden, however, was not so easily put off. After much cajoling, he finally got Burnham to agree to come.[27]

Among the other astronomers Holden recruited for the Lick staff, John Martin Schaeberle was in some ways the most enigmatic. He was born in Germany in 1853 and came to the United States as an infant. His hardworking immigrant parents settled in Ann Arbor, Michigan, and as soon as he was old enough, Schaeberle was apprenticed in a machine shop in Chicago. There he became interested in astronomy and tried his hand at making small reflecting telescopes. Later he returned to Ann Arbor to complete his education. After getting his high school diploma, he studied astronomy under James Craig Watson at the University of Michigan, and remained there for a time as an assistant. Hired by Holden to take charge of the Lick Observatory's meridian-circle, an instrument used to obtain accurate positions of the stars, Schaeberle seems to have been entirely satisfied with such routine work. In personal characteristics he was a confirmed bachelor who found the isolated conditions on the mountain congenial, engaged in regular gymnastic exercises, and was a self-described 'fresh-air crank,' who avoided 'dust, smoke, smokers, fumes, etc as I would a person having a filthy disease,' and made a point of sleeping by an open window whenever possible.[28]

Keeler, at thirty-one, was the same age as Barnard. Whereas the other members of the staff had all been trained in the methods of classical astronomy, he was thoroughly forward-looking, a pioneer in the still infant science of astrophysics, which aspired to use the methods of physics in the study of the Sun, planets, stars, and nebulae. A graduate of Johns Hopkins University in Baltimore, where he was influenced by the pioneer American spectroscopist Henry Rowland, Keeler worked for several years as Samuel Pierpont Langley's assistant at the Allegheny Observatory in Pittsburgh and later studied physics in Germany under the great Hermann Helmholtz. He had been eagerly recruited by Holden for the Lick staff, and by the time the observatory opened, had already spent two years on the mountain assisting Floyd, setting up the time service, and designing a spectroscope for use with the 36-inch refractor. This spectroscope was built to Keeler's specifications by John Brashear, whom Keeler had known in Pittsburgh; it arrived on Mt. Hamilton in February 1888. In addition to his astrophysical interests, Keeler was a gifted visual observer. He was also the most talented draughtsman on the staff.

Finally Barnard had been chosen to add to his impressive record of cometary discoveries and to lend his experienced hand in the astronomical photography undertaken at the observatory. Compared to the others on the staff, who were somewhat specialized in their areas of research, Barnard was also by far the most omnivorous observer.

4

As director, Holden, of course, received the most generous salary from the Regents ($5000 a year, a $1000 a year cut from what he had been paid as president of the

University of California). Burnham was the next highest paid at $3000 a year, followed by Schaeberle at $2000, Keeler at $1400, and Barnard at $1200. The accommodations similarly reflected the pecking order. Holden's dignity as director gave him title to most of one half of the three-story brick Astronomers' House. Burnham, though maintaining a second household for his large family in San Jose, needed large quarters for those occasions when they joined him on the mountain, and so received most of the other half. Schaeberle and Keeler, both bachelors, occupied smaller apartments (two rooms each, with a shared 'mess') in the same building, while Barnard and Hill occupied makeshift cottages that had formerly been used by workmen.

'In order that the work of the Observatory may go on without friction,' wrote Holden, 'it is necessary to recognize the departments of work for which some one person is immediately responsible, and to place the corresponding instruments in charge of the various astronomers.'[29] Obviously time on the 36-inch refractor was at a premium. Holden reserved it two nights a week for himself, Burnham had it two nights for observing and measuring double stars, and Keeler two nights for his spectroscopic work, which involved mostly the measurement of the velocities of stars in the line of sight (radial velocities). Saturday nights between seven and ten o'clock the observatory was open to the general public, and visitors were given views through all of the telescopes, including the 36-inch. After the visitors left, one of the observers, usually Burnham, took over the great telescope for the rest of the night. Thus, under usual circumstances, Barnard was not allowed to use the 36-inch at all, though he could formally petition Holden for time on it for specific observations.

Most of Barnard's work was done with the 12-inch Clark, which he used every night except on the rare occasions when it was needed by Burnham. He was also placed in charge of the observatory's $6\frac{1}{2}$-inch equatorial refractor and a 4-inch comet-seeker with a so-called 'broken tube.' (The tube, that is, was bent at right angles in the middle so that the eye-piece end could remain horizontal, thus saving the observer strain when looking at objects near the zenith; a large prism at the right angle directed the light from the object-glass to the observer's eye.)

In addition to their observational routines, each member of the staff had other assigned duties. Keeler, for instance, was placed in charge of the time service and of the seismometers used to monitor earthquakes, while Barnard was made responsible for the astronomical telegrams. Hill was a sort of man of all trades of the observatory, and had the most various duties. When Burnham was using the 36-inch, he helped him with the physically exhausting work of swinging the telescope around to the different parts of the sky as required by his observing program, lent Schaeberle a hand on the meridian circle, and was Keeler's back up on the time service. He also was required to try the fire hose every Monday, to keep and issue all keys, to maintain a list of the books in the observatory library, and to look after the Visitors Book, the Request Book, and the Water in Reservoir Book.[30]

Lick Observatory's 12-inch Refractor. Mary Lea Shane Archives of the Lick Observatory

5

Though Barnard would eventually regard his exclusion from the great telescope as unfair, even unmerciful, at first he seems to have been reasonably satisfied with the role he had been given as 'junior astronomer.' Indeed, many years later he recalled the days immediately after he arrived at Lick Observatory as 'happy ones for myself and my wife.'[31] Keeler, somewhat weary after having spent two years on the mountain already, had written shortly before the observatory opened: 'I must say I rather envy the frame of mind in which the others will come here – all hungry for work.'[32] Of all the members of the staff, Barnard's appetite for work was arguably the most ravenous, and he wasted no time in resuming his research.

His first months on Mt. Hamilton were quietly productive. In early June he sketched Sawerthal's Comet, which had been a naked-eye object the previous April. With the 12-inch refractor, he commented on the 'peculiar wing-like appendages of the head [which] gave the comet a most ghostly and bird-like appearance.'[33] On June 9, he observed an occultation of the 6th magnitude star 47 Librae by Jupiter, which Holden and Keeler observed with the 36-inch.[34] A week later he searched, and successfully found, Olbers' Comet. Recovered the previous August by Brooks on its first return since 1815, Olbers' Comet was by now almost out of range of telescopic observation – before Barnard saw it, it had not been seen at any other observatory for three months.

Meanwhile, Barnard had also been conducting experiments to determine the chemical (photographic) focus of the 12-inch. On July 21, a Saturday evening, after two hundred visitors had queued up for a glance through the telescope, he stayed up until 4 a.m. making trial exposures on the Moon in preparation for the following night's lunar eclipse.[35] During the eclipse itself, he successfully photographed the partial phases, and contrasted the Moon's rather colorless appearance during totality with the bright reddish-orange cast it showed at the eclipse of June 11, 1881, which he had observed from Nashville.

6

The first summer in which the observatory was in operation, the public eagerly awaited any news of its work, and Holden was only too willing to oblige. During July and August the great refractor was frequently turned toward Mars – always a sure-fire subject to capture the interest of the public – and though the planet was by then well past opposition, a few of the delicate Schiaparellian 'canals' were glimpsed by Holden, Keeler, and Schaeberle, as the director announced to the newspapers. Holden also addressed the sensational reports of Henri Perrotin of the Nice Observatory, who claimed to have witnessed the submersion of a whole Martian 'continent' by a neighboring 'sea.' Holden wrote that as of July, when observations began with the 36-inch,

> The submerged 'continent' had reappeared . . . and was seen by us here essentially as it has always appeared since 1877. It was most unfortunate that the Lick telescope could not be used for this purpose until so late a date; but it has shown its great power in such work . . . and has conclusively proved that whatever may have been the condition of the 'continent' previous to July it was certainly in its normal condition from that time onward.[36]

Another object that came in for its share of attention was the Moon. James Lick himself had hoped that the great telescope would 'prove or disprove the existence of animals,' and early in his planning for the observatory Holden had listed among his highest priorities getting 'photographs of the Moon – large – for distribution.'[37] In August 1888, Burnham, at Holden's request, made a series of sixty-seven trial exposures with the 36-inch refractor. For this work a 33-inch corrector lens, purchased by the trustees at a cost of $14 000, was mounted in front of the main lens to change its best focus from yellow to blue light and thereby convert it from a visual to a photographic telescope. Holden was so delighted with Burnham's results that he later preempted the whole project for himself – as we shall see, the photography of the Moon would become his consuming research interest.

Realizing their potential public relations value, Holden distributed many complimentary sets of Burnham's finest prints of the Moon, and among the recipients was the Vanderbilt University Observatory. Bishop McTyeire, assuming that the set was a gift from Barnard who had not been heard from since heading out west, acknowledged them to the junior astronomer:

> We highly appreciate your thoughtful & generous kindness. – Don't forget your old friends, grovelling here below, while you are perched on high, communing with stars & skies.[38]

But Barnard, uncomfortable with the misattribution and aware of Holden's sensitivity in such matters, at once wrote dutifully to the director:

> Did the President of Vanderbilt University acknowledge the Moon pictures to you? I fear from his letter to me . . . that he has misunderstood . . . & approved *I* had sent him the pictures . . . . I am so sorry if the Bishop has not acknowledged them *to you* and I will write to him letting him know that you alone was responsible for . . . the pictures.[39]

Though Burnham's photographs were of quite respectable quality, many of the later exposures taken by Holden would suffer from blurring due to vibrations and imperfect focusing, and it would be no sacrifice to have others believe that Holden alone was responsible for them.

7

The inky skies of Mt. Hamilton were highly favorable for comet-seeking. As soon as the observatory opened, Barnard, after a year-long lapse, renewed his quest for them in earnest. In this work, he used both the 4-inch 'broken tube' comet-seeker and the $6\frac{1}{2}$-inch equatorial.

In the early morning hours of September 3, 1888, he recorded in the pages of his observing book the first fruits of these labors. After using the 12-inch for routine observations of Faye's Comet, then making its sixth return since its discovery by the French astronomer Hervé Faye in 1843, Barnard

> went in house and began [comet] seeking with the 4 in. broken tube [comet] seeker. Found a small hazy obj., which on examining with 12 in. looks like a [comet].[40]

This new comet was then a tailless object of the 8th magnitude moving among the stars near the Gemini–Monoceros border. Later that morning Holden recorded with evident satisfaction in his diary: 'This morning Barnard picks up a comet (the first discovered this side of the Rocky Mts.).'

Through the Lick Observatory's telephone office, Holden at once sent the telegram Barnard had composed to Harvard College Observatory, only to receive back from Chandler a few days later a mild admonishment as to the preferred protocol:

Your various telegrams (addressed to the Harvard Observatory, and not, as is the general practice . . . to Ritchie or myself, a course which saves much delay) were duly received and finally got around to my hands, and distributed to where they would do most good, namely to Washington and Albany, and the orbit cabled to Germany last night (Sept. 10, 9 P.M.) where it is now probably in print . . . If it had been addressed straight to Ritchie's or my address, it could have been cabled and distributed that same night.[41]

In the same letter, Chandler informed Holden that Brooks had claimed to have found the comet independently. However, he assured Holden that in his view Barnard's priority was firmly established: 'At best, [Brooks] was a day late in the discovery, and certainly late as regards the announcement, as my cable to Krueger [the director of the Central Astronomical Telegram office at Kiel] had gone before he had sent to Swift; I believe too, that his announcement to Swift was of a "suspected" comet. At any rate Barnard is safe.'[42] Brooks, it turns out, had no intention of disputing with Barnard the question of priority – even before Chandler's letter to Holden arrived, he had graciously acknowledged to Barnard: 'You are one day ahead.'[43]

The orbit of this comet (1889 I) was computed by Schaeberle. Though it was found that the comet would pass perihelion at the end of January 1889, the respective movements around the Sun of the comet and the Earth were such that the comet would remain favorably placed for observation for at least a year. Indeed, when in July 1889 it again became favorably placed for observation after an absence of several months, Swift announced the discovery of a new comet, only later realizing his mistake.[44] Barnard himself followed it longer than any other comet had ever been followed – his last observation of it, with the 36-inch refractor on September 8, 1890, was two years, five days after he had discovered it. The comet had by then reached a distance of 6.25 astronomical units from the Sun – further, he pointed out, than a number of short-period comets at aphelion.[45] It then disappeared forever; since its orbit is hyperbolic, it will never return again.

Barnard's next comet (1888 V) was discovered in the early morning hours of October 31, 1888, while he was sweeping the stars of Hydra. When found, the comet was already past perihelion and bound outward from the Sun in a highly elliptical path, which will not bring it around again for 2400 years. Before sending his telegram announcing the discovery, Barnard went to Holden's quarters in the Astronomers' House to get his approval. With his heavy daytime responsibilities, Holden seldom kept late hours on the telescopes, and Barnard knew that he would have long since turned in and would have to be awakened. Holden, however, was delighted. 'By all means, send [the telegram] off at once,' he told Barnard. '[C]heck the cipher message – but send it off – I congratulate you my dear fellow – & you were right to wake me of course.'[46]

The director was always glad for the publicity that Barnard's comets brought to the observatory – by helping to keep the observatory and its work in the public eye, it made his fund-raising efforts easier. Soon after Barnard made this second comet discovery from Mt. Hamilton, Holden asked Hill, who as an amateur astronomer in San Francisco had been an occasional correspondent to the San Francisco *Examiner*, to contribute to that newspaper a brief 'Account of Mr. Barnard's Latest Discoveries at the Lick Observatory.' Hill penned the following:

> The fact that the active life of the Lick Observatory had been but fairly commenced when two comets were discovered at that institution by E. E. Barnard of the astronomical staff, seems to demand more than the simple telegraphic announcment always published of such discoveries. Comet-seeking demands more patience and energy than almost any other branch of scientific work. Routine astronomical work is generally a constant accumulation of *results*, of more or less value, but comet-seeking carefully and tirelessly continued on every clear night possible may go unrewarded for years; and while the discovery of one of these 'celestial wanderers' is not an event of very great scientific importance (unless, indeed, the orbit should prove to be periodic), it is always, at least, of popular interest . . . And when the energy and ability of a single observer have attached his name to a list of twelve comets within less than eight years the feat thoroughly merits all the renown it has, in this case, received.[47]

Aside from general accounts of the observatory which had only mentioned him in passing, this was Barnard's debut in the local newspapers.

8

As a starry-eyed job-seeker, Barnard had earlier professed his admiration for Holden in the most flattering and obsequious terms. He had been no less effusive about how much he looked forward to receiving the benefits of his training once he had joined the staff. There is no reason to believe that any of this was insincere – at least any more so than such attitudes among job-seekers usually are. But if Barnard hoped to find in Holden a scientific mentor, he realized soon after his arrival on Mt. Hamilton that whatever else he might be (and Barnard would before long have some choice words about that, too), Holden was a mediocrity when it came to scientific research. Moreover, his personality repelled him.

He was, however, pleasantly surprised by Burnham, the member of the staff to whom he became closest during his years on Mt. Hamilton. To his near contemporaries Schaeberle and Keeler, Barnard would remain on cordial but somewhat formal terms, but when it came to the older man he would write: 'Somehow my life seems wrapped up in his. Besides my most sincere admiration for him as a man and astronomer there seems to be a love for him that surmounts all else.'[48] Always tending to gravitate to older men, Barnard found in Burnham, as earlier in Swift and Garland, someone who fulfilled his need for the father-figure he had never had.

At first, though, Barnard had dreaded the meeting with Burnham. 'Somehow I have always pictured to myself the astronomers of whom I have heard but not seen,' he later explained. 'These mental pictures almost always proved to be wrong when I finally met the man.' The picture of Burnham he had formed in his mind's eye was of 'a large, heavy-set man with a stern countenance – one you would not feel at home with. What

Sherburne Wesley Burnham. Yerkes Observatory

gave me this impression I do not know. I was young then and knew little of people in general, and I had associated a great name in any calling with a bearishness of manner that would be intolerable.' When Burnham arrived on Mt. Hamilton at the beginning of August 1888, Barnard found him 'covered with dust from the long stage ride,' yet noticed immediately 'his light blue eyes and cordial greeting. Instead of the large, austere man I had expected here was a thin, rather small man, one of the gentlest and kindest of men – more human than any man I had ever met. All fear that I had felt vanished when I came to know him and only love for him could take its place.'[49]

Barnard described Burnham as of 'slight, wiry build, [and] tough as iron.'[50] Unlike the temperate Tennessean, who neither smoked, chewed, nor used intoxicating beverages, Burnham smoked incessantly – cigars – and appreciated a good glass of Burgundy. And whereas Barnard was always impatient whenever the skies were cloudy, Burnham was more philosophical about cloudy weather and 'turned to a book or a game of whist or congenial talk over a glass of wine' instead of 'dashing in and out to see if there was any hope of clearing or if the wind had changed.'[51] Yet despite their differences, Barnard was drawn to the older man's quiet strength and homespun philosophy. Moreover, Burnham's own love of telescopic work made him sympathetic to Barnard's compulsive need to observe. He often allowed Barnard some of his own time on the 36-inch, and he was always ready and willing to verify his observations made with smaller instruments. With good reason, Barnard would always remember him as 'generous to a fault; his hand . . . open to all who were in need.'[52]

Both men worked throughout the whole night whenever the weather permitted, Burnham usually on the 36-inch, Barnard on the 12-inch or one of the smaller instruments. They usually got together 'near midnight, when one's vitality always gets low, . . . in his office or mine,' as Barnard later recounted, 'to have a little lunch consisting of coffee and crackers and cheese, the coffee being made by Burnham over a lamp chimney by a little device that would hold a small pot. With this stimulant the rest of the night would go by in work on the measurement of double stars, etc. His rapidity in measuring was remarkable. This was partly due to the fact that he always made three settings of the wires before recording the readings, keeping them in his memory in the interval.'[53] Burnham's ability to make out double stars was as legendary as Barnard's for picking up objects of excessive faintness. 'If a star disc deviated an almost infinitesimal quantity from the circular, his eye detected it at once,' Captain W. de W. Abney, president of the Royal Astronomical Society, noted in 1894 on awarding him the Society's gold medal.[54]

On one night in October 1888, Barnard joined Burnham in the dome of the 36-inch. The great telescope was then turned toward the Trapezium of Orion, where Alvan G. Clark had discovered a new faint star on the first night of observations with the 36-inch. Burnham had often measured it, though he noted, 'I have never seen with this or any other telescope so faint a star, or one so difficult.' He was measuring it again on this night when Barnard wandered into the dome. Burnham invited him to take a turn, and as soon as he peered into the eyepiece, Barnard

> detected another new star just preceding the trapezium, which had been missed by all who had examined it with this telescope. It is about the same magnitude as the Clark star . . . He also saw that this exceedingly faint star was itself double . . . As a double star it is quite unlike anything known in the heavens, and the severest possible test for the defining and illuminating power of the large telescope.[55]

Before the night was through, Barnard had discovered a still fainter star within the Trapezium. This one, however, Burnham had to take on faith – no one on the Lick staff except the keen-eyed junior astronomer was ever able to see it.

In addition to getting together for midnight lunches to break up their lonely watches at the telescope, Barnard often joined Burnham, who was an avid amateur photographer, on the excursions he made to capture on film the breathtaking vistas around Mt. Hamilton. Among Burnham's specialities were scenes in which the valleys below the observatory were filled with clouds and the tops of Mt. Hamilton and the adjacent mountains projected like islands above them. Oddly enough, however, despite his fascination and considerable skill in terrestrial photography, Burnham remained skeptical when it came to the astronomical uses of photography. 'Just now there is a sort of craze on the subject,' he told Holden, 'and a great many people will waste time which they might better employ.'[56] In another letter to Holden he opined: 'I think you will find photography is not going to revolutionize anything . . . Fifty years hence the situation may change.'[57]

Barnard later painted a romantic picture of the days he spent with Burnham in exploring the canyons around Mt. Hamilton:

> In the heat of summer everything becomes parched and dry on the mountain for the want of rain, and one feels frequently a sense of oppression in the eternal dryness and dust. Under these conditions we often went down into the canyons for relief, for a visit to these places, where spring seems never to cease, is, in the dry season, like a visit to fairy land. Here one finds the green foliage of trees, wild flowers and graceful ferns in abundance – among them the delicate maiden-hair fern. Through these canyons always runs a cool crystal stream, over whose rocky margins and steep banks are spread masses of velvety green moss. Here and there open spaces in the foliage overhead let the sun's rays sparkle on the water as it flows peacefully by or leaps over beautiful waterfalls into the deep silent pools below. Such scenes as these were eagerly sought by Mr. Burnham as subjects for his camera. The north slopes of the canyons high above the streams, are generally covered with dense chaparral, while the south sides are more or less free of it, the bare spaces being covered with wild oats. It was easy to descend these southern slopes, say into Sulphur Creek Canyon, and one could take long running jumps of twenty feet or more on the soft soil and reach the bottom in about twenty minutes. That was the pleasure of it! But in climbing up the same slopes it sometimes took hours to return, for the wild oats and the soft soil made one slip back almost as fast as he came up. When the exhausted one finally reached the top he usually registered a vow never to descend into the canyon again. But always in a few days he was ready to go once more, for the call of the canyon was more than fascinating – it was irresistible.[58]

In October 1888, the two friends with several others made an excursion to nearby Mount Santa Ysobel, from which Burnham hoped to get a photograph of the Lick Observatory from the south. The plan was for them to cross Sulphur Creek canyon in the afternoon and camp in an open and level space, free from chaparral, in a grove of trees near the summit of Ysobel, where they had been told there was a fresh water spring. Expecting their refreshment there, they took only enough water with them to make coffee. After a supper of salty ham, they began searching for the spring in the moonlight – but in vain. At first light the next morning Burnham clambered up the mountain to take his photographs, and as soon as he returned, the group started their return trip, which led through dense chapparal, back toward Smith Creek. They were now becoming desperate with thirst. Burnham suggested that chewing on acorns would alleviate it somewhat, but Barnard dryly remarked, 'I got no satisfaction out of them.' Finally, at five o'clock that day, they reached Smith Creek, and refreshed themselves by dunking their heads into the cool stream. By then Barnard's mouth was so dry that his tongue 'rattled about in it.'[59]

For the presidential election that fall, the two men decided to make the trip to the polls at Smith Creek Hotel, which had been built shortly after the completion of the Mt. Hamilton road in 1876. The hotel was located at a site two thousand feet below the summit. Burnham intended to vote a straight Republican ticket, Barnard a straight Democratic ticket; yet as Barnard later recounted the adventure,

> notwithstanding that we would thus mutually cancel each other's vote he thought we should do our duty anyway and go and vote. It was a very short and easy journey down the slopes of Mount Hamilton to the hotel . . . though it was seven miles by the stage road. But Mr. Burnham thought we should make a pleasure trip of it and suggested that we descend into Sulphur Creek canyon to the south of the observatory and wander through this canyon to the confluence of Sulphur Creek with Smith Creek and thence down Smith Creek to the voting place. It was quite a journey and, though delightful, was in some respects a difficult one for there were some dangerous points to be negotiated. Nevertheless, after many stops to photograph beautiful spots and to climb around precipitous corners, we got there all right just as the polls were closing. While we were congratulating ourselves on the fact that we were on time we were notified that we could not vote, as neither of us had registered. We then turned around and began the tedious ascent of the mountain, feeling satisfied that we had at least done our duty and tried to save the country once more.[60]

In the result, Barnard's candidate, Grover Cleveland, received the majority of the popular votes cast, yet the vote in the electoral college went to Benjamin Harrison, who became president. Parenthetically, Harrison would visit the Lick Observatory for two nights in April 1894 – one of many illustrious visitors who have made that pilgrimage over the years.

9

As seen from Mt. Hamilton at night, San Jose in the 1880s was still a relatively sparse settlement, its electric lights, gleaming thirteen miles off in the distance, being compared by Barnard to 'a wide and bright star cluster.' During the first winter of the observatory's existence, the heating of offices and living quarters was still by wood stoves, and electric lights had not yet replaced oil lamps.

One never knew quite what to expect in the darkness. One early dawn when he was crossing the saddle from Sulphur Creek Canyon to the North Canyon, Barnard encountered a mountain lion, and recalled later: 'We mutually regarded each other for a few moments from a distance of some 10 or 15 feet, then he leisurely passed on down the slopes of the canyon while I went home, each of us with a high respect for the other doubtless – though I think the respect was mainly on my side.'[61] Rattlesnakes, though more often gopher snakes, also lurked along the paths. But above all the mountain was celebrated for its lizards. 'Some of them are large and slender, like snakes on legs, and remarkably quick in their movements,' Barnard observed. '[O]thers are small and though shy, are apparently more friendly. All of them are beautifully graceful, with long slender tails. From the cottages on the saddle to the large brick residence of the Director, higher up near the summit, was a plank walk with cleats on it to aid in walking up. This was often referred to as the "chicken walk", and . . . was a favorite place for lizards which sought it for shade and protection.'[62] As for the flora of the mountain, there were, in addition to scanty grass, wild oats, and scrub oak or chapparal, a few specimens of Douglas oak, Digger pine, golden oak, stunted manzanitas, and Gowrias.[63]

Such was Mt. Hamilton in the early days of the Lick Observatory – a kind of monastery devoted to the pure religion of the stars. Far above the breathtaking canyons, where the howls of the coyotes pierced the lonely night, the pioneers of the world's first permanent mountaintop observatory lived and worked. Ralph Waldo Emerson, for whom Elizabeth Barnard had named her son, had once written:

> If a man would be alone, let him look at the stars . . . One might think the atmosphere was made transparent with this design, to give man, in the heavenly bodies, the perpetual presence of the sublime . . . If the stars should appear one night in a thousand years, how would men believe and adore; and preserve for many generations the remembrance of the city of God which had been shown! But every night come out these envoys of beauty, and light the universe with their admonishing smile.[64]

Nowhere had that smile been more admonishing (in Emerson's sense, of giving friendly encouragement) than on Mt. Hamilton – yet even in this peaceful scene there were troubles in the hearts of men.

1   Quoted in Donald E. Osterbrock, James R. Gustafson, W. J. Shiloh Unruh, *Eye on the Sky: Lick Observatory's First Century*, (Berkeley, University of California Press, 1988), p. 67

2   Donald E. Osterbrock, *James E. Keeler: Pioneer American astrophysicist* (Cambridge, Cambridge University Press, 1987), p. 62

3   D. E. Osterbrock, 'The Rise and Fall of Edward S. Holden, part 1,' *JHA*, **15** (1984), 81–127:81

4   For a complete bibliography of Holden's works, see: W. W. Campbell, 'Edward Singleton Holden,' *BMNAS*, **8** (1919), 347–72

5   ibid., p. 357

6   Richard H. Tucker to Mary A. Tucker, March 5, 1900; SLO

7 Santa Clara *County Official Paper*, January 26, 1889; clipping in SLO
8 San Francisco *Examiner* for February 27, 28, March 1, 27, and 28. The editor of the *Sidereal Messenger*, W. W. Payne, reprinted some of the choicest excerpts as 'The Holden–Proctor unpleasantness,' *SM*, 6 (1887), 192. See also: E. S. Holden, 'President Holden's reply to Prof. Proctor,' ibid., pp. 210–12; R. A. Proctor, 'Note from Mr. Proctor,' ibid., pp. 259–62
9 ESH to H. E. Mathews, May 1, 1888; ESH, 'The Lick Trustees to E. S. Holden Dr. For expert services rendered . . .,' May 1, 1888; SLO
10 SN to RSF, July 14, 1888; SN to H. E. Matthews, August 10, 1888; SLO
11 Osterbrock, Gustafson, Unruh, *Eye on the Sky*, p. 87
12 F. J. Neubauer, 'A Short History of the Lick Observatory, Part 2,' *PA*, 58 (1950), 323
13 ESH to the Lick Board of Trustees, July 5, 1888; SLO
14 Campbell, 'Edward Singleton Holden,' p. 357
15 ESH, Regulations for the Observatory, June 1, 1888; SLO
16 E. S. Holden, 'Address of the retiring President of the Society, at the Second Annual Meeting (March 29, 1890),' *PASP*, 2 (1890), 50–68:67
17 S. D. Proctor, 'Californian Observatories,' *Knowledge*, 14 (1891), 31–2:32
18 Holden, 'Address,' p. 62
19 ibid., p. 63
20 ibid.
21 ibid., p. 53
22 Simon Newcomb, *The Reminiscences of an Astronomer* (Boston and New York, Houghton, Mifflin & Co., 1903), pp. 192–3
23 For the life of S. W. Burnham, see E. E. Barnard, 'Sherburne Wesley Burnham,' *PA*, 29 (1921), 309–24 and E. B. Frost, 'Sherburne Wesley Burnham,' *ApJ*, 54 (1921), 1–8
24 Neubauer, 'Short History,' p. 326
25 S. W. Burnham, 'Catalogue of 81 Double Stars, discovered with a 6-inch Alvan Clark Refractor,' *MNRAS*, 33 (1873), 351–7
26 SWB to ESH, September 28, 1885; SLO
27 SWB to ESH, July 28, 1887; SLO
28 JMS to EBF, March 14, 1923; YOA
29 ESH, 'Draft of Regulations'; SLO
30 ESH, diary for the year 1888, SLO
31 EEB to J. T. McGill, January 30, 1922; VUA
32 JEK to ESH, April 1, 1888; SLO
33 EEB, observing notebook; LO
34 E. E. Barnard, 'Observations,' *AJ*, 8 (1888), 64; E. S. Holden, 'Note on the Occultation of 47 Librae by Jupiter, 1888 June 9,' ibid.
35 EEB, observing notebook, LO
36 E. S. Holden, 'The Great Telescope: What has been done by the Lick Equatorial – The Schiaparelli Canal Question,' *Common Advertiser*, New York, November 1888; from clippings book, SLO
37 ESH, diary for the year 1888; SLO
38 H. N. McTyeire to EEB, October 26, 1888; VUA
39 EEB to ESH, undated; SLO
40 EEB, observing notebook, LO

41   SCC to ESH, September 11, 1888; SLO

42   ibid.

43   W. R. Brooks to EEB, September 5, 1888; VUA

44   EEB, observing notebook; LO. Barnard entered the clipping from the San Jose
     Mercury for July 7, 1889: 'Rochester, July 6 – Professor Swift discovered a new comet
     this morning at 3:15 o'clock.' Immediately below the clipping he entered: 'Swift was
     deceived by the old comet e of Sept. 2, 1888. [comet] I 1889.'

45   E. E. Barnard, 'Comets 1889 I and II, and some suggestions as to the possibility of
     seeing the short-period comets at aphelion,' *AJ*, **10** (1890), 67–8 and 'Comet 1889 I,'
     AJ, **11** (1891), 51

46   ESH to EEB, October 31, 1888; VUA

47   C. B. Hill, 'Two Comets: An Account of Mr. Barnard's Latest Discoveries at the Lick
     Observatory,' San Francisco *Examiner*, November 15, 1888

48   EEB to HC, October 11, 1892; SLO

49   Barnard, 'Sherburne Wesley Burnham,' p. 315

50   ibid.

51   Helen Wright, *Explorer of the Universe: A Biography of George Ellery Hale* (New York,
     E. P. Dutton & Co., 1966), p. 125

52   Barnard, 'Sherburne Wesley Burnham,' p. 315

53   ibid.

54   W. de W. Abney, 'Address,' *MNRAS*, **54** (1894), 277–83:279

55   S. W. Burnham, 'The Trapezium of Orion,' *MNRAS*, **49** (1889), 352–8:354

56   Quoted in Osterbrock, Gustafson, Unruh, *Eye on the Sky*, p. 69

57   SWB to ESH, July 25 [?1886]; SLO

58   Barnard, 'Sherburne Wesley Burnham,' p. 318

59   ibid., pp. 319–20

60   ibid., pp. 318–19

61   EEB, MS draft for Atlas of the Milky Way; YOA

62   Barnard, 'Sherburne Wesley Burnham,' p. 321

63   Holden, 'Address,' p. 65

64   Ralph Waldo Emerson, 'Nature,' in S. Bradley, R. C. Beatty, and E. H. Long, eds.,
     The American Tradition in Literature, (New York, Grosset & Dunlap, Inc., 3rd ed.,
     1967), vol. 1, p. 1065

10

A year of wonders

1

On New Year's Day, 1889 occurred a total eclipse of the Sun. What Barnard once referred to as 'one of the most sublime and impressive sights that a human being can ever witness,'[1] a total eclipse was, apart from the sheer spectacle, of great scientific importance, for only under these conditions could nineteenth century astronomers study the ghostly corona, the outer atmosphere of the Sun, which ordinarily is overwhelmed by the much more brilliant airglow. Unfortunately the path of totality – the area where the Moon just blocks out the Sun's disk and allows the spectacle to be seen – is always a rather narrow swath of the Earth's surface, so that, as Barnard quipped, 'total eclipses of the Sun have a habit – from perversity alone, I believe – of occurring in the most out of the way places . . . often making the journey to see it a very great one.'[2] The path of totality of the eclipse of New Year's Day 1889 cut obligingly across northern California, offering a splendid opportunity for the astronomers of the newly founded Lick Observatory to see it without having to mount a major expedition.

Naturally the curiosity of the public was also greatly aroused, and Holden realized the tremendous opportunity it offered to encourage a wider interest in astronomy. He wrote a handbook, *Suggestions for Observing the Total Eclipse of the Sun on January 1, 1889*, and seized the moment to implement a plan he had had in mind since before his arrival at Lick, calling for a local organization dedicated to fostering the public interest in astronomy. Thus the Astronomical Society of the Pacific was born. The organization was, Holden stressed, to be 'popular in the best sense of the word,' and it would carry out this purpose in part by disseminating, through its bimonthly *Publications* – edited by Holden himself, who was also its most frequent contributor – knowledge of astronomy generally and especially of the Lick Observatory and its work.

In addition to all of this activity, Holden did everything he could to encourage amateurs to attempt to photograph the eclipse. Under the supervision of Charles Burckhalter of the Chabot Observatory, a public observatory in Oakland, amateur photographers belonging to the Pacific Coast Amateur Photographic Association were assigned positions along the path of totality in order to secure photographs of the corona.

Holden sent the official Lick Observatory expedition to Bartlett Springs, in the Coast Range northwest of Sacramento. The party was led by Keeler and joined by Barnard, Hill, and the observatory's first graduate student, Armin O. Leuschner, who had recently arrived on Mt. Hamilton from Ann Arbor. Holden and the two senior astronomers, Burnham and Schaeberle, would remain on Mt. Hamilton to monitor the (scientifically unimportant) partial phases. The public was informed that the observatory would be closed for visitors that day, a Tuesday – a poor public-relations decision that the San Jose newspapers fiercely criticized. 'The Emperor of Mt. Hamilton uses his power to deny the public a visit to a public institution supported by taxpayers' money,' fumed one editorial writer. 'Does he own the place?'[3]

Early in December 1888, the telescopes and other equipment needed for the expedition to Bartlett Springs were sent ahead, and on December 16, Keeler, Barnard and Leuschner – together with Holden's thirteen-year old son Ned – left Mt. Hamilton in fog and rain. Hill and his wife were to join the party at Bartlett Springs at a later time just before the eclipse. As the train started out from San Jose to San Francisco, Leuschner and Ned were left behind on the platform, but by taking a later train they were able to catch up with the others in San Francisco. Meanwhile Keeler and Barnard paid a visit to Holden's friend Jarboe, the lawyer, successfully appealing to him to take Ned with him for eclipse viewing at Cloverdale, a less isolated site than Bartlett Springs.

Thus relieved of their young charge, Keeler, Barnard and Leuschner continued their own journey by rail to Sites, the end of the line, where they arrived on the night of December 19. However, they discovered to their dismay that their telescopes and other equipment were nowhere to be found. Keeler went back to Colusa, the nearest town, to look for them, while Barnard and Leuschner waited two nights and a day for the next stage to take them the last forty-two miles to Bartlett Springs. After leaving Sites at 7 a.m., they rode 'the entire day through wind and rain in an open wagon – the "stage" – with no protection but our overcoats.' When they arrived at Bartlett Springs, it was after dark, and they were 'wet through with rain and chilled with cold.'[4]

At Bartlett Springs, Barnard and Leuschner learned that their freight was at Williams and would not go through Sites at all. They at once tried to convey this information to Keeler at Colusa, but found that the telegraph line was badly grounded. The next evening the freight finally arrived. Barnard was gloomy – he had caught a bad cold from riding all day in the rain, and the skies were hopelessly cloudy. 'We have tried to get a glimpse of a star or something so that we could get our meridian and get to work and put the instruments in place,' he grimly informed Holden, 'but it has rained all the time since our arrival – possibly we may get to see the sky tonight but it is very doubtful.'[5] Despite his bad cold, he maintained a constant vigil on the sky that night, and through breaks in the clouds he and Leuschner succeeded in observing Polaris and deriving their meridian.

Keeler arrived at 10 p.m. the next night, Christmas Eve, after enduring fifteen harrowing hours on the stage. 'The roads,' he complained, 'were muddy beyond anything I had ever seen before. We had to walk up the steep hills, and at short intervals it was necessary to stop and dig the mud off the wheels, in order that the four horses could drag the empty stage.'[6]

At that time Bartlett Springs consisted of little more than a resort hotel, whose fare was 'villainous' and whose only comfortable room was the bar, as Keeler reported to Holden.[7] For the observing site, Barnard selected a croquet ground about a hundred yards southeast of the hotel. He and Leuschner dug holes and erected posts in the middle of this ground on which they would later mount their telescopes and other instruments. 'We received but little outside aid in this work,' he grumbled. 'It was wet and raining most of the time.'[8]

On December 28, the rain finally stopped long enough to allow the men to get their instruments in position and ready.[9] Each observer was to be responsible for a specific project during the just less than two minutes of totality – Leuschner was to make photometric observations of the corona, Keeler, assisted by Hill, was to study the spectrum of the corona visually with a spectroscope, and Barnard was to attempt to photograph the corona. Barnard's photographic apparatus consisted of an 8×10 box camera, a 1-inch Voigtländer lens of nine inches focal length, and a $3\frac{1}{4}$-inch Clark lens, focal length forty-nine inches, which had been part of a telescope used on Mt. Hamilton for making readings of the water scales on the distant reservoirs. The Clark lens was found to give sharp photographic images if its aperture was stopped down to 1.75 inches, but its chemical (photographic) focus was unknown and had to be determined in the field. Barnard did this on the evening of December 29 'by wedging up the end of the box [containing the lens] so that it would point to [the star] Rigel [and making] a series of trails . . . on two sensitive plates, varying the distance of the objective before each exposure by a tenth of an inch. These trails were equally distributed on each side of the visual focus. Upon being developed they indicated a good chemical focus at one tenth of an inch outside the visual focus.' [10]

With this apparatus, Barnard hoped to photograph the delicate details of the inner corona, but all depended, of course, on the weather conditions during the eclipse. The incessant rain did not make for much optimism. On December 31, the night before the eclipse, when Barnard sat down to write his daily progress report to Holden, things looked very depressing indeed: 'It is densely clouded now,' he recorded at 11 p.m., 'and looks as if it would rain or snow before morning.'[11] He was convinced that the eclipse the next day would be lost in the clouds.

Though the next day dawned unexpectedly bright and clear, as the morning advanced the sky began to haze over again. At 11:30 a.m., an hour before first contact, the instant when the Moon would first touch the Sun's disk, Barnard noted with growing apprehension that the haze was becoming 'very dense in the south.' First contact was observed through haze, and once again things were beginning to look rather hopeless. However, the sky cleared around the Sun just before totality. 'So close was our escape,' Barnard afterwards mused, 'from total failure from clouds, that it seemed as if Providence had interfered in our behalf.'[12] Barnard had his three cameras arranged side by side, so that both ends of the boxes could be reached with as little loss of time as possible in uncapping the lenses and changing the plate-holders. He made exposures with two of the cameras simultaneously with his right and left hands; as soon as they were finished, he immediately began an exposure with the third camera, then changed the plate-holders and started another cycle. As he was making his ninth exposure, the

Sun popped out from behind the Moon just before the cap could be replaced.

While the awesome spectacle of a total eclipse is notorious for making people lose their heads, Barnard avoided being thrown off his program by simply not looking up at the Sun. Nevertheless, he was at least semi-conscious of the events taking place around him, and after the eclipse was over sat down and recorded his impressions:

> So impressive was the magnificent spectacle upon the crowd that had gathered just outside our inclosure, that not a murmur was heard. The frightened, half-whining bark of a dog, and the click-click of the driving clock, alone were audible. When the Sun suddenly burst forth, an almost instant and highly surprised cackling of the chickens, that had hastily sought their roosts at the beginning of totality, would have been amusing, could [I] have shaken off the dazed feeling that possessed even [myself] at the sudden and unexpectedly rapid termination of the semi-darkness. My own feelings were those of excessive disappointment and depression – so intent was I in watching the cameras and making the exposures, that I did not look up to the Sun during totality, and therefore saw nothing of the Corona . . . One glance that I cast towards the clouds in the north, showed that they appeared of a sickly, greenish hue.[13]

After the eclipse, Keeler went on to Clear Lake to see Floyd and Floyd's niece, Cora Matthews, in whom he had by then formed a romantic interest, but Barnard was 'extremely anxious' to return to Mt. Hamilton in order to develop his plates. He packed them, still in their plateholders, in his valise. The next 'stage' was due to depart early on the morning of January 3, but Barnard found that two other passengers with a trunk had already reserved all the space, and there was no seat for him. He could not buy them off nor persuade them to leave their trunk. Finally, however,

> After a great deal of persuasion – the roads being one mass of mud and the wagon frail – I secured the high privilege of riding on the trunk and holding my valise in my hands, with the clear understanding that I was to walk up all the muddy hills. It was uphill all the way to Sites!
>
> We started in a heavy rain. It rained hard and persistently all that day. The stage had no cover and was open to the weather, and I had no covering but my overcoat. The trunk dashed wildly forward in going down each hill, and tried to escape backwards in ascending every elevation. The road was through *adobe* soil which, from the long continued rains, had become of the consistency and stickiness of molasses. For a good part of the way our wagon was up to the hubs in mud and water, and every few hundred feet we had to get out and free the wheels from the accumulated mire . . . At last, tired out, drenched with rain, and bedaubed with mud, just as the night closed in, we reached Sites, and from thence the rest of the journey as far as San Jose was by rail.[14]

He was back on Mt. Hamilton by Saturday evening, January 5 – just in time to show visitors the young Moon through the 12-inch refractor.[15]

2

Barnard drew upon all his meticulous darkroom skills to develop his precious eclipse plates. The results left no doubt that the camera had supplanted the sketchbook in the role of historiographer of eclipses. Swift had written of the efforts of a party of fifteen or twenty to sketch the corona at the total eclipse of 1878: 'A comparison of these drawings with the utter absence of resemblance to each other or to the object itself, would astonish the gods. No two were alike in any particular, and the picture least resembling the reality was, strange to say, made by a skilled delineator.'[16] By this standard, the photographs taken by Barnard were a tremendous improvement. As a professional photographer before he was an astronomer, Barnard was thoroughly familiar with the best methods of developing his plates in order to bring out all of their latent details, and they showed beautifully the streamers near the poles of the Sun and the enormous coronal 'wings' spreading far into space on either side of the Sun's equator, thus setting aside 'forever . . . as worthless the crude and wholly unreliable free-hand sketches and drawings previously depended upon,' as Barnard himself pronounced.[17] The winged pattern, incidentally, is characteristic of sunspot minima; a more symmetric, halo-like corona prevails during sunspot maxima, as had first been noted by the English astronomer Arthur Cowper Ranyard in 1879.

Throughout his career, Barnard would have to struggle with the intractable problems of adequately reproducing his photographs. For the Lick Observatory's handsomely published *Reports on the Observations of the Total Eclipse of the Sun of January 1, 1889*, which was issued in an edition of one thousand copies, there were insufficient funds to reproduce any photographs by the photogravure process. Instead, Holden delegated to Barnard the rather monumental task of preparing and mounting an actual silver print in each individual copy as the frontispiece. As a labor-saving device, Holden suggested that Barnard make on each 8 × 10 plate a series of small negatives from his best glass positive, then cut them down to size. The method for mounting and burnishing was, however, Barnard's own, and was arrived at, as he explained, only after considerable experimentation:

> The prints, after being trimmed, were soaked for a few minutes in clear water; each print was then laid face down upon the smooth surface of a sheet of ferrotype plate, the water being pressed from it and a little paste rubbed on; the stiff paper upon which they are mounted was then pressed down firmly on it – a guide being used so that the print should be symmetrically mounted. The prints were then laid aside for half an hour, under slight pressure, until partially dry, and then set upon edge in a warm room, where in from ten to fifteen minutes each print would drop from the plate, mounted and beautifully burnished. With a thousand prints this was a slow process, but the superior finish seems to warrant the time spent upon it.[18]

Barnard's photographs were thought, by the Lick astronomers at any rate, to be the best of the eclipse, and Holden suggested to A. C. Hirst, the president of the University of the Pacific, in Stockton, California, that Barnard's achievement deserved to be

Barnard's photograph of the total solar eclipse of January 1, 1889. Yerkes Observatory

recognized through the conferment of an M.A. degree. Hirst agreed, and Barnard was awarded the degree on November 4, 1889.[19] The man who as a child had had no more than two months of formal education was now a Master of Arts, and henceforth he took considerable pride in signing his name to the articles he authored as 'E. E. Barnard, M.A.'

Good photographs of the 1889 eclipse were also obtained by the Harvard College Observatory expedition to Willows, near Sacramento, which was led by William H. Pickering, the younger brother of Harvard's director, E. C. Pickering. Many of Burckhalter's amateurs also did themselves proud – especially William Ireland, who observed at Norman, California, and William Lowden at Cloverdale. They obtained plates which showed the wings of the corona extending out to $2\frac{1}{2}°$ from the Sun, or a distance of more than four and a quarter million miles. For his pictures, Ireland had used a large portrait lens of the type known to opticians as a 'Petzval doublet' – a two-lens system of which each lens was itself an achromatic doublet.[20] Ireland had borrowed this lens, of six inches aperture and thirty-one inches focal length, from William Shew, the owner of a portrait studio located on Montgomery Street in San

Francisco. Holden was so impressed with Ireland's results that he later purchased this lens, 'at a very low price,' with funds provided by Colonel Charles F. Crocker, one of the wealthiest men in California and a member of the Lick Observatory Committee of the Board of Regents. Inscribed with the date 1859 and the name Willard, it became known as the 'Willard' lens, though Willard was not its maker but a photographic stock dealer; the lenses bearing his name were actually made by Charles F. Usner, of New York City. Large aperture lenses of this type had been used to take fashionable portraits in the days of slow wet-process photography, their large apertures and correspondingly great light grasp helping to shorten exposure times during portrait sittings. When new, such a lens would have fetched several hundred dollars. As the more sensitive dry-plates became available beginning in the early 1870s, they fell into general disuse, but their enormous light grasp and wide fields – in the case of the 6-inch Willard lens, some 20° in diameter – gave them tremendous potential for celestial photography. We will soon have more to say about the 'Willard' lens, which was to become world-famous in Barnard's hands.

3

The eclipse brought many astronomers from other parts of the country to California, and afterwards many of them naturally wanted to visit the famed Lick Observatory. Pickering stopped by before heading on to Wilson's Peak, near Pasadena, where he planned to test the seeing conditions. Ever since the spring of 1888 Harvard had been interested in establishing a temporary observing station there, and the citizens of Los Angeles, motivated by rivalry with those of San Francisco, hoped eventually to set up a telescope four inches larger than the Lick. Among others who visited the observatory were Barnard's old friend Lewis Swift, who had made visual observations of the eclipse from Nelson, California, and Henry S. Pritchett and Father C. M. Charropin of Washington University, who had photographed the eclipse from Norman, near Colusa.

Near the end of January, after all the visitors except Swift had left, several members of the California Legislative Committee on Education visited Mt. Hamilton from Sacramento. Holden later maintained that he had never been informed of their plans to visit. Regardless, when the Committee arrived they received 'a cold greeting,' as the Alta *Daily* announced. 'The summit was reached and we hastened to the Observatory residence,' Senator C. M. Crawford, the committee chairman, was quoted as saying, 'and while we heard footsteps within, even the rapping with a cane could not invoke a host.' Crawford recounted in withering detail that after entering the Observatory by the rear door, the members were received 'by one of the faculty, I suppose,' and after 'being given a peep at Orion and a hasty description of the machinery, were left to amuse ourselves until the Moon rose to light our way back down the mountain. We were chilled through by the cold air, and waited in vain for an invitation to some cosy fireside. We reached the city late at night, in a hungry and generally forlorn condition.'[21] Though Holden later apologized profusely and invited the committee members back for another visit, the newspapers had a field day with the embarrassing incident. The San Francisco *Examiner* published a spoof on the '*Holden Astericidae*, or lonely star fish,' noting that 'The habits of the crustacean are but little known in California, as it has a way of drawing modestly into its hole when approached. Visitors who inspect the tank will find the specimen very rarely in sight.' And the *Examiner* ended by referring the reader to 'recent essays by the late Professor Proctor, who claimed to have made a careful study of the subject.'[22] An article in the Santa Clara *County Official Paper* began 'Hold On, Holden,' and after rejecting Holden's apologies, poured out a litany of complaints:

> We speak by the card when we say that the people of Santa Clara valley and hundreds who have visited the Observatory during the past few months will be pleased to learn that Mr. Holden really admits that this is a 'public institution,' for his pompous airs in the past have served to impress visitors with the feeling that he had set up a little empire of his own on the crest of Mount Hamilton, and that all visitors were intruders.
>
> A notable instance of his lordly style was found on the occasion of the recent eclipse of the sun, when he gave out word in advance that no visitors would be received on that day – that not even reporters would be allowed, but that he himself would prepare press reports. The result was, as was expected, that while full and interesting accounts of observations taken at Dutch Flat, Truckee and other obscure places were telegraphed all over the world, the baldest statement, a half dozen lines in length, was all that was obtainable from the world-renowned Lick Observatory, the one place from which the people of the world were waiting anxiously to hear.[23]

Though eventually Holden did offer the newspapers full reports, they did not appear until several months after the event, and presumably the public's interest in it, had passed.

Yet another nasty row erupted soon afterward, and was the first sign of a split between Holden and Burnham. Article Six of the Observatory's regulations, written by Holden and adopted by the Regents on January 10, 1888, had stated explicitly that all communications from the staff were first to be approved formally by the director. However, Burnham, a crusty individualist, had never cared for the policy and eventually said so. Holden replied that the policy at Lick was no more oppressive than that at 'the Observatories of Greenwich, Rome, Königsberg, Harvard College, Washington, Madrid, Palermo, and many others.' Nevertheless, he was willing to compromise – as long as a paper was first submitted to him for review, the writer could transmit it with his own letter to the journals mutually agreed upon.[24] Though Holden's letter seems reasonable enough on the face of it, Burnham apparently perceived some fault with its tone, for in his reply he fulminated that the Lick Observatory

> is not a military institution, and it is not necessary or desirable to have military or naval discipline. We are all supposed to be doing our best, for the love of the work we have chosen, and it is not conceivable that any abuse of entire personal freedom

in the matter of writing and forwarding astronomical articles could follow on the part of any astronomer here.[25]

The matter was allowed to drop for the time being, but it was the first sign in the official records of the growing rift between the director and his increasingly restive staff.

4

At night the seeing conditions on Mt. Hamilton often proved to be superlative for planetary work. For such work, steadiness of the atmosphere is of the utmost importance, and Mt. Hamilton was fortunate in lying 'above an atmospheric inversion layer, in air that is in laminar flow (no turbulence) from a relatively cool, nearby ocean.'[26]

In 1888, Keeler had made several beautiful drawings of Jupiter with the 36-inch refractor – Barnard had praised them warmly, commenting on his colleague's 'real artistic ability such as very few observers possess.'[27] Keeler had also discovered a new gap in the outer ring of Saturn, as already described. He saw it on the very first night of observations with the great telescope, but not again until March 2, 1889. All that day the sky had been hazy, but at dark, when the observatory was open to visitors, it began to clear. Barnard jotted in his observing book: 'The sky covered with fine veiling of mist and clear in places. The most remarkably steady night I have seen on the mountain.' Keeler was equally impressed with the quality of the seeing – the night was the 'finest . . . for months,' he wrote, and with the 36-inch he 'saw the division in the outer ring of Saturn very distinctly, just as it was a year ago.'[28] After the visitors left, Keeler called Holden, Schaeberle and Barnard to the dome of the 36-inch for a look at Saturn. Holden and Schaeberle left after a brief look, but Keeler and Barnard remained in the dome until well after midnight. As the night wore on, the sky became thick and Santa Clara valley and the canyons filled with fog. Barnard remarked that it was 'a very poor night for transparency – but a marvellous night for steadiness. The planet with the power over 1000 was near the zenith and without a tremour!' The exceptional seeing allowed the wide-eyed observers to boost the magnification up to 1500, and with this enormous magnification, 'Mr. Keeler's division,' as Barnard called it, near the outer edge of the A ring, appeared to him as a hair-thin line. Barnard remarked that it seemed 'as distinctly a division as Cassini's, but probably 1/50th as thick.'[29] This was by far the finest view Barnard would ever have of Saturn – despite many years of observations with the world's largest telescopes, he would never again see 'Mr. Keeler's division.'[30]

5

Though seeing conditions on Mt. Hamilton were often exceptionally fine at night, they were usually awful in the daytime. In 1888, the keen-eyed astronomer Giovanni Schiaparelli, of the Brera Observatory in Milan, had urged Holden and the other Lick

Observatory astronomers to take up the study of Mercury and Venus. Schiaparelli himself had been a pioneer of daylight observations of these planets, realizing the advantages of studying them when they were higher in the sky than during the twilight periods. From a long series of these observations he was on the verge of announcing his conclusion that the rotations of both planets were captured with respect to the Sun. Holden, Schaeberle and Barnard tried to oblige the Italian astronomer by observing these planets during the daytime with the 12-inch refractor, but with little success – the heating of the rocks on the southern side of the mountain, in contrast with the cooler northern slopes, produced an unsteadiness of the air which was almost always present and spoiled the observation of delicate planetary details. Later, Barnard summed up his experience in noting that 'a clear, transparent atmosphere almost always proved unsteady, and this seemed to hold at night also, . . . so that a dark-blue sky really became synonymous with poor seeing.'[31] On the other hand, on some occasions when the air was thick with dust and smoke, the daytime seeing became tolerably good. In particular, on the morning of May 29, 1889, when smoke and dust lay so thick near the mountain that the canyon was impenetrable with haze, Barnard found upon turning the 12-inch toward Venus that 'the seeing was remarkably perfect – I have never seen the image near so steady up here before.'[32] He did not publish his results until 1897, when he wrote:

> I was struck with the remarkably perfect definition. There was not the slightest tremor of the image. The markings on the surface of the planet were distinctly seen, though they were difficult and very delicate . . . A careful drawing was made of these details, which were distinct enough to be drawn with perfect satisfaction. This perfection of definition did not last long enough to show any motion in the spots – the ordinary day-seeing soon taking its place . . . The circumstances under which [my] drawing was made are memorable with me, for I never afterwards had such perfect conditions to observe Venus . . . The planet was watched through many years, but indifferent seeing always baffled me and no satisfaction could be gotten out of it.[33]

The lack of satisfaction was certainly not for want of effort. Barnard made many attempts to improve the image – 'contracting the aperture of the object-glass; using a small diaphragm over the eyepiece; using colored glasses, etc.'[34] What seemed to work best was simply to cover his head and the eye-end of the telescope with a large dark cloth in order to cut out the extraneous light. But all was to little avail. Henceforth the markings on Venus, on the rare occasions when they were seen at all, were always 'too elusive to sketch,'[35] and Barnard never again made out anything on Venus as well-defined as what he saw on that hazy May morning, while his results on Mercury during his seven years on Mt. Hamilton were if anything even less satisfactory.

6

Except when he was away on the mountain, which during his first two years was seldom, Barnard observed every available hour of every night, seven nights a week. If only the sky was clear, he was willing to endure any other kind of weather – and he lost at least one hat on the mountain searching for comets in gale-like winds. A good part of his time was, of course, taken up in comet-seeking, and in 1889 he discovered two more from Mt. Hamilton. The first, 1889 II, was a faint comet which he picked up in Taurus with the 12-inch on March 31. It moves in a very elongated elliptical orbit and will not return again for 1 400 000 years. He found the other, 1889 III, in Andromeda with the $6\frac{1}{2}$-inch equatorial on June 24. It was also faint, but proved to have a short period of only 130 years, so that it is expected to return again in 2019. It is, incidentally, the only one of Barnard's many comets which is likely to be observed again – all the rest have been lost or will not return for thousands of years.

Barnard also made extensive observations of every other comet that was visible from the Lick Observatory, and in the summer of 1889 he began another important line of work – photography with the Willard portrait lens.

Unlike his close friend Burnham, Barnard had from his Vanderbilt days been a firm believer in the promise of photography in astronomy.[36] Holden shared his faith, though his own interest was mainly in obtaining large-scale images of the Moon and planets with the great refractor. Nevertheless, he was enthusiastic about Barnard's efforts to photograph the Milky Way. Of this assignment, Barnard later commented, 'in a photographic experience of twenty five years, I have never seen anything more deceptive to photograph.'[37] His long photographic experience made Barnard realize at once that a portrait lens, by taking in a wide-angle view, would show structures of the Milky Way that were invisible in the telescope because of its small field. However, because of the Milky Way's faintness, a very long exposure would be needed. This meant that there was no choice but to painstakingly guide the camera on a star for hours.

As early as April 29, 1889, a night of good transparency as shown by the visibility of the gegenschein that night, Barnard attempted a one and a half hour exposure on the Milky Way in Scorpio with the 1-inch Voigtländer lens that he had earlier used for the eclipse at Bartlett Springs.[38] Unfortunately, this lens proved to be too small for Milky Way photography – though it registered a number of faint stars on the plate, it failed to give even the feeblest impression of the great Milky Way star clouds. A larger lens – one with greater light grasp – was obviously needed. When Holden purchased the Willard lens, Barnard had the equipment he needed – its 6-inch aperture gave it thirty-six times the light grasp of the Voigtländer. At the end of July 1889, Barnard mounted the Willard lens in a wooden box camera and strapped this apparatus onto the $6\frac{1}{2}$-inch equatorial, which served as his guiding telescope. Though the equatorial was equipped with a clock drive, in order to get sharp images Barnard needed to make manual adjustments as well. Locating a guide star in the center of the field, he kept it exactly centered throughout the exposure by using slow-motion rods to make fine adjustments in right ascension and declination. He later explained his method:

> There was no means of illuminating any spider threads. Fine iron wires were, therefore, inserted between the lenses of a negative eyepiece. These were coarse enough to be just visible, in black relief, on the dark sky. A star in focus would be hidden behind these wires. To render it visible, therefore, the image of the guiding

star was thrown slightly out of focus. The intersection of the wires placed over this small luminous disc for guiding, produced four small segments of light. During the exposure the illuminated quadrants were kept perfectly equal – the slightest deviation from equality could be detected. This method permits great accuracy in guiding, even with a small telescope, but it requires a brighter star to guide on than usual.[39]

It was hard, tedious work, but at least Barnard was used to it from his earlier experience guiding the great *Jupiter* camera for Van Stavoren.

Barnard's first trial with the Willard lens was made on July 28, 1889, when he exposed a plate for 1 hr 17 min on the Milky Way in Scutum, before clouds interrupted his work. Though the region is one of the Milky Way's richest – Barnard himself would later refer to it as the 'Gem of the Milky Way' – this exposure was too short to show it well. Two nights later he made an exposure of 1 hr 30 min on Comet Davidson, a naked-eye comet which had been discovered at perihelion a week earlier from Queensland, Australia. [40]

Then, on August 1, Barnard renewed his attack on the Milky Way. This time he chose a region in Sagittarius which, as he later explained, 'I specially selected as possessing the most intricate and complex structure of any portion of the Milky Way above our horizon.'[41] It was here that he had discovered his 'small black hole' while comet-sweeping in Nashville in 1883. Placing the 'black hole' near the center of the field, he guided the exposure for 3 hr 7 min by keeping a following star constantly bisected by the cross wires in the eyepiece. The plate that emerged from the bath of chemicals in Barnard's developing tray was breathtaking. As he later described it:

> This remarkable picture shows the cloud-like forms like waves of spray. A . . . curving lane [of darkness] runs from the lower left-hand portion of the picture . . . and curves gracefully upwards to the place of *Jupiter*. It is singularly like the stem of a great leaf. At the middle of the picture it is seen to pass behind some of the clouds of stars and emerge beyond, showing us clearly which part of the Milky Way at that point is nearest to us. Imagination may aid one, but it looks as if the lines of the cloud-forms, and of the stars and vacancies, all run more or less concentric with this extensive lane . . . The black hole is seen slightly to the left of the center, with the small cluster [NGC 6520] as a white spot close to the right of it.[42]

The next night, Barnard made several more exposures on the Milky Way, centering on the open cluster M 11, but they were not as spectacular as his plate of August 1. Thereafter he became preoccupied with the remarkable Comet Brooks (see below) and did not make another exposure with the Willard lens until August 21. That night he recorded in his observing notebook: 'I made an exposure on the great neb. of Andr[omeda]. With the Willard camera strapped on $6\frac{1}{2}$ in tube – following by hand (aiding clock). Result. – remarkable. The spiral form shown beautifully and myriads of stars.' Two nights later he exposed a plate on the Pleiades: 'Driving for this not so good, as slow motion in [right ascension] not working good. Sky fine – wind having died down. Developed this at once and, result fairly good. Shows the neb. attached to Merope and that to Electra, etc.'[43]

Barnard's first photograph of the Milky Way, August 1, 1889. UCO-Lick Observatory

Thereafter the $6\frac{1}{2}$-inch equatorial was packed up and sent to San Francisco, from whence it traveled with Burnham and Schaeberle to Cayenne, in South America, for the December 21, 1889 total eclipse of the Sun. In its absence, Barnard tried to guide by mounting the Willard lens on the 36-inch,[44] and he made one or two trial exposures in this way, 'trusting to the clock.' However, by the beginning of October he found to his dismay that the lens had become 'changed some how, so that it gives a penumbra to [stars].'[45] There was no choice but to send the lens away for refiguring to John Brashear's optical shop in Pittsburgh, the funds for this purpose being furnished by Colonel Crocker. Crocker also provided additional funds for the Willard lens to be eventually remounted on its own equatorial and placed in its own dome. However, there would be almost three years' delay – not until June 1892 was the Crocker Photographic Telescope, built around the refurbished Willard lens, ready to use again for the Milky Way photography that Barnard had begun so promisingly in the summer of 1889. Then, Barnard hinted darkly that the malice of the director was responsible for the long interruption of his important work. 'He deliberately attempted to take from E.E.B. his Milky Way photographic work by taking the Willard lens away . . . and [planning to] utilize the same for his own credit,' the young Tennessean later charged. 'And in retaliation for not succeeding took the work out of his hands and [threw] him behind for three years.'[46]

7

On the same fateful night on which he first succeeded in recording the Milky Way star clouds in Sagittarius, August 1, 1889, Barnard went to the 12-inch refractor, which had just been returned to service after a month of repairs. He turned it toward the latest Comet Brooks (1889 V), which had been discovered in the early morning hours of July 7 by Barnard's arch-rival of cometary discovery, who since 1888 had been in charge of the observatory of William Smith, a wealthy nurseryman, at Geneva, New York. The new comet proved to be moving around the Sun in an elliptical orbit, with a period of about seven years, but apart from its short period, it did not seem to be a very interesting object – it was rather small, and never came close to reaching naked-eye visibility. Barnard, however, made a point of observing it, just as he had observed every other comet which had been visible above his horizon since the Lick Observatory opened.

Barnard logged into his observing notebook: 'After closing photographic work on the milky way, observed the comet with the 12-inch – the first observation of it that was possible with this instrument since July 10 . . . A faint nebulosity seen close following and north of the nucleus; angle and distance measured. It is small and has a very faint nucleus. Another nebulous object [in the line between the other two] is small, round, and faint.'[47] The next morning Barnard returned to the telescope and satisfied himself that the two nebulous bodies were companion comets, travelling with the main one. His heart racing with excitement, Barnard woke up Holden to verify his discovery. When the director took his turn at the eyepiece, he glimpsed yet a third nebulous object nearby, and before the night was over Barnard had found three more. Unfortunately, on succeeding nights the sky was 'never sufficiently pure' to decide if any of these were true companions or only nebulae in the same field, as the comet was then passing through 'a region of faint and unknown nebulae.'[48]

Ever since 1882, when he had discovered the nest of companion comets southwest of the Great Comet of that year, Barnard had systematically examined the immediate neighborhood of every comet to determine whether it was accompanied by such companions. At last he had his reward.[49] News of his companion comets was telegraphed around the world. The first to confirm them were Spitaler and Palisa at the Vienna Observatory, and before long they were also seen at the observatories of Pulkowa, Nice, Algiers, Marseilles, and Strasbourg.

On August 3, Barnard observed the main comet (which he referred to as 'A') and its two companions ('B' and 'C') with the 36-inch refractor, while Burnham, who had generously given him observing time on that instrument, recorded the observations. As seen in the great refractor, 'The two brighter companions were perfect miniatures of the larger comet, each having a small, fairly well-defined head and nucleus, with a faint, hazy tail . . . The three comets were in a straight line . . . their tails lying along this line.'[49] On August 4, Barnard discovered two new companions, D and E, with the 36-inch; they disappeared, however, after only a few observations.

During succeeding weeks, Barnard followed the remarkable convoy as it pursued its journey through space. At first, the two companions B and C appeared to be separating rapidly from the main comet, but toward the end of August, the nearer companion B became stationary with respect to the main comet, and at the same time underwent a remarkable change: 'It enlarged rapidly, becoming extremely diffused, and losing all appearance of central condensation.' Soon afterward it dissolved completely – by September 5, Barnard recorded, it had 'disappeared as absolutely from the face of the heavens as did Biela's comet,' a reference to the celebrated comet which had split into two in 1845 and, after one return of the pair in 1852, had disappeared forever.[51]

Meanwhile, the companion C was still receding from the main comet, and as it did so increased steadily in brightness and size, until by August 31 Barnard found it 'perceptibly brighter than the larger comet.' Then it too underwent a process of involution. By the end of September, its tail had disappeared, and its head was large, diffuse, and faint. Barnard entered his last observation of C, made with the 36-inch, on November 25 – at that time he found it 'excessively difficult, [at] the utmost limit of visibility.' By the next night it too had vanished without a trace.

Barnard continued to follow the main comet with the 12-inch until March 20, 1890, shortly before it disappeared into conjunction with the Sun. With the help of an ephemeris prepared by the noted orbit computer Anton Berberich, he recaptured it with 36-inch on November 21, 1890 – it was at the time, Barnard wrote, 'the faintest and most difficult object that I have ever seen in the heavens,' and he added that 'from its present appearance no other telescope in existence can possibly show a trace of it.'[52] It remained just visible with the 36-inch until January 13, 1891.

Barnard disapproved of the term 'fragments,' which was popularly used to describe the cometary companions he had discovered, as this term seemed to him to imply 'knowledge of the method of their formation which we certainly do not possess.' The companions were, he insisted, 'in every respect . . . as perfect in form and individuality as the main comet itself.'[53] Barnard's series of measures seemed to indicate, moreover, that the companions had not recently separated from the main body – he himself had no doubt that if the 12-inch had been available in early July, he would have seen them then. Calculations by Seth Chandler confirmed Barnard's suspicion that the companions had existed for some time. Tracing backward Comet Brooks's orbit, Chandler found that it had swept within only 900 000 miles of Jupiter in May 1886, just inside the orbit of satellite Io. The comet's orbit was drastically changed, its period being shortened from 29.2 to 7.1 years, and Chandler regarded it as all but certain that tidal disruption of the cometary matter had then produced the companions. It was the first such case of a comet shown to be shattered by too close an approach to Jupiter, though the similar break up of Comet Shoemaker-Levy 9 on passing near Jupiter on July 8,1993 showed that its fate is not unique.

The later history of Comet Brooks is of some interest. It (though none of its retinue of companions) was reobserved at the next return in 1896, and again in 1903 and 1911. In 1921, it made another close passage by Jupiter, but though its orbit was again changed, the period was shortened by only 0.2 years. The comet was recovered in 1925 and has been observed at every subsequent return. However, it seems to be gradually fading – it is quite possible that before long it too will vanish from the face of the heavens, like Biela's Comet and its own evanescent companions.[54]

8

Barnard's 'year of wonders' closed with one last observation that was destined to form an immortal page in the annals of science. This was his careful monitoring of the eclipse of Saturn's satellite Iapetus on the night of November 1–2, 1889, by means of which he was able to show that the innermost of the three classical rings (Ring C or the 'crape' ring) is translucent, a condition that had been suspected previously but never proved. This observation has long been recognized as furnishing an outstanding example of Barnard's remarkable intuition at the telescope, but there is also a fascinating but little known sidelight to the story revealing the personality clash and growing interpersonal strains between Barnard and Holden.

With the exception of Iapetus, all of the larger Saturnian satellites lie nearly in the plane of the rings, and so can never be eclipsed. Iapetus's orbit, however, is inclined to the ring plane by some 15°, and once every fifteen years there is a several week period during which Iapetus may pass through one or the other of its nodes at the same time that its orbit is edgewise to the Sun and enter the rings' shadow. Such an event was predicted for the night of November 1–2, 1889 by an alert English astronomer, Arthur Marth, who published a short notice about it in the June 1889 *Monthly Notices of the Royal Astronomical Society*[55]. Oddly enough, Marth's challenge seems to have been taken up only at Mt. Hamilton.

An eclipse of Iapetus, involving its passage through the shadow cast by the entire ring system as well as the globe, takes nineteen hours from start to finish, and so is visible over much of the Earth's surface, but only a part of it can be seen at any one place. From Mt. Hamilton, only the emergence of the satellite from the shadow of the globe and its subsequent passage into the shadows of the inner crape ring and the bright middle Ring B were visible on November 1–2. The night happened to be one of Holden's on the 36-inch. However, despite the rare goings-on in the heavens, he did not change his usual practice of shutting up the dome early and going to bed – he was no doubt already snoring when Saturn, then in the constellation Leo, rose at 12:50 a.m. Mt. Hamilton time.

Barnard, however, was wide awake, and in the 12-inch dome quivered with excitement. As soon as Saturn emerged from the mists near the horizon, he captured it in the 12-inch – a glance showing him that Iapetus was invisible, still deep in the shadow of the globe. However, he sketched in for later reference the positions of the other satellites – Titan, Tethys, Enceladus, Dione, and Rhea, the last two on the following side of the planet. At 1:45 a.m., he recorded in his observing book that Tethys and Enceladus were 'quite near each other, close preceding the end of the ring,' and he began keeping a 'sharp watch' on the approximate point where Iapetus's reappearance was expected. His next few notes, made at intervals of several minutes, record that the satellite remained invisible. Then, just before 2:38 a.m., his keen eye began to detect something – the merest glimmer, quite close to the positions of Tethys and Enceladus but still 'so faint and uncertain that it was not recorded.' Another half minute sufficed to convince him that what he was seeing was not a figment of his imagination. The satellite had reappeared from the shadow of the globe, and in the opening between the shadow of the globe and that of the crape ring was shining much more brightly than Enceladus, indeed rivaling Tethys.

At this point, Barnard wrote afterwards, 'the idea . . . occurred to me that it would be an excellent plan to test the effect of the shadow of the crape ring on the visibility of the satellite, by frequent comparisons of the light of Iapetus with that of Tethys and Enceladus.'[56] He was hoping to do eyeball photometry, using as the standard of comparison the difference in brightness between Tethys (10th magnitude) and Enceladus (magnitude $11\frac{1}{2}$). Barnard 'mentally divided' this quantity into ten equal parts, and according to this scale began recording comparisons in his observing book.

At 3:47 a.m. Mt. Hamilton time, with Saturn at an elevation of approximately 30° above the horizon, Iapetus began fading into the crape ring's shadow. Between then and 5:11 a.m. he documented that 'the absorption of the sunlight became more and more pronounced, until finally the satellite struck the shadow of the inner bright ring, which it rapidly entered and within which it disappeared.' When the event was over, he made a graph of his brightness estimates – eighty in all – which clearly showed the effect of the shadow of different parts of the ring on the satellite's visibility. His main conclusion was that 'the Crape Ring is truly transparent – the sunlight sifting through it; that the particles composing the Crape Ring cut off an appreciable quantity of sunlight; that these particles cluster more and more thickly – or, in other words, the Crape Ring is denser as it approaches the bright rings.'[57] The bright middle ring, on the other hand, was 'so far as the penetration of the solar rays is concerned, . . . fully as opaque as the globe of Saturn itself.'[58]

A few days later, Barnard wrote up a brief account of the eclipse, which no one else on Earth had witnessed, for the *Publications of the Astronomical Society of the Pacific*.[59] He made no mention there of the telescope he had used, but in a fuller account for the *Monthly Notices* he commented: 'Professor Holden requested me to observe this phenomenon with the 12-inch, as it would be out of the reach of the great telescope.'[60] This explanation for why he used the 12-inch has generally been accepted ever since – A. F. O'D. Alexander, for example, in his classic book *The Planet Saturn*, writes: 'He used the 12-inch aperture refractor as Saturn was unsuitably placed for using the 36-inch.'[61]

In fact, it was all a fraud. The true situation is revealed in a Barnard manuscript draft in the Vanderbilt University archives which contains, among numerous 'Charges against Holden,' the following entry:

Instance eclipse of Iapetus as showing how he neglects important work – After 10 [o'clock] the big telescope lay idle the rest of the night while one of the most

important obs[servations] on record was passing unobserved except with the 12 in.
[He] insisted that I state that it was out of reach of 36 in.[62]

In 1889, Holden was still firmly exercising the director's prerogative of passing all
papers before publication, and in order to avoid embarrassment to himself had
suggested that Barnard insert this plausible, though wholly specious, explanation for
the event's having passed unobserved in the 36-inch!

In 1891, Barnard submitted a fuller account of his observations to the journal
Astronomy and Astro-Physics, on whose editorial board he himself served, which at last
gave the lie publicly to the earlier pronouncement in the *Monthly Notices*:

> For various reasons, no other observer in the world saw the eclipse of Iapetus . . .
> The night was fine and clear and especially favorable. The planet rose at 12h 50m.
> At first the seeing was only ordinary, but it increased in excellence until in the latter
> stages of the observations it was superb. By the time the satellite had entered the
> shadow of the crape ring, the planet had attained a high altitude and was excellently
> placed for observing.[63]

Saturn had not, in fact, been 'out of reach' of the 36-inch; Holden's attempted cover-up
was exposed.

Barnard's classic observation of the Iapetus eclipse was at once recognized as of great
importance, and because of the rarity of these events was long in being duplicated. A
lone observer, Rev. W. F. A. Ellison, director of the Armagh Observatory in Northern
Ireland, observed part of Iapetus's passage through the A Ring on February 28, 1919.
However, not until almost ninety years after Barnard's observation – on October 20–21,
1977 – did anyone again see Iapetus pass through the shadow of the crape ring. That
night's eclipse was monitored photoelectrically by professional astronomers at Lowell,
Mt. Wilson, Palomar, Mauna Kea, and Tokyo observatories, and by a team of
observers aboard NASA's Kuiper Airborne Observatory.

Barnard was lucky in 1889. Iapetus was conveniently placed relative to Tethys and
Enceladus, so as to make accurate visual estimates of its brightness possible. Moreover,
he was favored with a view of the most dramatic phase of such an eclipse – the passage
through the crape ring and behind Ring B. But then great discoveries and observations
are often due to a combination of luck and skill. At the same time, such results fall only
to the mind which is prepared and can make the most of its opportunities.

The real story of November 1–2, 1889 is that the glory that night went not to the high
and mighty – the great telescope and the powerful director – but to the humble 12-inch
and the junior astronomer faithful at his post. That night Barnard and his telescope did
well. According to his own frank assessment: 'The observations of that eclipse with the
12-inch equatorial have given us more information about the crape ring of Saturn,
perhaps, than could possibly have been obtained by a hundred years of ordinary
observing.'[64]

1 EEB, unpublished lecture notes from ca 1901; VUA
2 ibid.
3 F. J. Neubauer, 'A Short History of the Lick Observatory, Part 3,' *PA*, **58** (1950), 372
4 E. E. Barnard, 'Report of the Photographic Operations,' in *Reports on the Observations of the Total Eclipse of the Sun of January 1, 1889*, (Sacramento, 1889), 56–73: 57
5 EEB to ESH, December 23, 1888; SLO
6 J. E. Keeler in *Reports*, p. 31
7 JEK to ESH, December 26, 1888; SLO
8 Barnard, *Reports*, p. 58
9 EEB to ESH, December 28, 1888; SLO
10 ibid.
11 EEB to ESH, December 31, 1888; SLO
12 Barnard, *Reports*, p. 58
13 ibid., pp. 69–70
14 ibid., p. 60
15 EEB, observing notebook; LO
16 Lewis Swift, 'The Nebulae,' *SM*, **4** (1884), 2–3. Barnard echoed these sentiments. In 'The Development of Photography in Astronomy-I,' *Science*, **8** (1898), 341–53:345, he wrote: 'The method of free-hand drawing of the corona made under the attending conditions of a total eclipse received its death-blow at that time, for it showed the utter inability of the average astronomer to sketch or draw under such circumstances what he really saw.'
17 Barnard, 'The Development of Photography,' p. 346
18 Barnard, *Reports*, pp. 71–2
19 ESH to A. C. Hirst, April 19, 1889; ESH to TGP, May 24, 1889; SLO. The degree was awarded on November 4, 1889.
20 The optical system had been worked out by a Viennese mathematician, Josef M. Petzval, for the early daguerrotype portrait cameras of Peter F. Voigtländer. See D. E. Osterbrock, 'Getting the Picture: Wide-Field Astronomical Photography from Barnard to the Achromatic Schmidt, 1888-1992,' *JHA*, **25** (1994), 1–14.
21 The Alta *Daily*, January 23, 1889
22 San Francisco *Examiner*, undated; clippings book of SLO
23 *Santa Clara County Official Paper*, January 26, 1889; clippings book of SLO. Swift later attempted to defend the observatory against what he called 'unfounded strictures.' In an article for the Rochester *Morning Herald*, he pointed out that 'Astronomers have much work to do, and the furnishing to the public of a daily bulletin of the labor of the night previous is not for a moment to be thought of. At proper intervals, volumes of observations, including, of course, discoveries, will be published. Meanwhile let the fault-finders keep silence and cultivate patience.'
24 ESH to SWB, February 18, 1889; SLO
25 SWB to ESH, February 19, 1889; SLO
26 Leon E. Salanave, Letter, *Sky and Telescope*, **73** (1987), 5
27 E. E. Barnard, 'Observations of the Planet Jupiter and his Satellites during 1890 with the 12-inch Equatoreal of the Lick Observatory,' *MNRAS*, **51** (1891), 543–9: 546 fn
28 EEB and JEK, observing notebooks; LO
29 EEB, observing notebook, LO; also J. E. Keeler, 'The outer ring of Saturn,' *AJ*, 8

(1889), 175

30   It did, however, appear in the Pioneer and Voyager spacecraft images as a gap about
     200 miles wide, which would correspond to an angular width of only $0''.05$ with Saturn
     at mean opposition. See D. E. Osterbrock and D. P. Cruikshank, 'J.E. Keeler's
     Discovery of a Gap in the Outer Part of the A Ring,' *Icarus*, **53** (1983), 165–73

31   E. E. Barnard, 'Physical and Micrometrical Observations of the Planet Venus, Made at
     the Lick Observatory with the 12-inch and 36-inch Refractors,' *Ap J*, **5** (1897),
     299–304

32   EEB, observing notebook; LO

33   Barnard, 'Physical and Micrometrical Observations of the Planet Venus,' p. 301

34   ibid.

35   EEB, observing notebook; LO. The entry is dated June 12, 1889.

36   See, for example: E. E. Barnard, 'Recent Stellar Photography,' Nashville *Daily
     American*, January 2, 1887

37   E. E. Barnard, 'On the Photographs of the Milky Way made at the Lick Observatory in
     1889,' *PASP*, **2** (1890), 240–44:241

38   EEB, observing notebook; LO

39   E. E. Barnard, 'Photographs of the Milky Way and Comets,' *PLO*, **11** (1913), 13–14

40   E. S. Holden, 'Photograph of the Davidson Comet,' *PASP*, **1** (1889), 34–35

41   Barnard, 'On the Photographs of the Milky Way,' p. 243

42   E. E. Barnard, 'On some Celestial Photographs made with a large Portrait Lens at the
     Lick Observatory,' *MNRAS*, **50** (1890), 310–14

43   EEB, observing notebook; LO

44   E. S. Holden, 'Photographing the Milky Way,' *PASP*, **1** (1889), 74–5

45   EEB, observing notebook; LO

46   EEB, 'notes for Young's Letter,' July 5, 1892; also EEB to WWC, February 7, 1914;
     both SLO

47   EEB, observing notebook; LO

48   E. E. Barnard, 'Physical and Micrometrical Observations of the Companions to Comet
     1889 V (Brooks),' *AN*, **125** (1890), 177–94:178

49   E. E. Barnard, 'Discovery and Observations of Companions to Comet 1889 (Brooks
     July 6),' *AN*, **122** (1889), 267–8: 268

50   E. E. Barnard, 'A Very Remarkable Comet,' *PASP*, **1** (1889), 72–4:72

51   ibid., 73

52   E. E. Barnard, 'On the Reobservations of Comet 1889 V,' *AJ*, **10** (1891), 111

53   Barnard, 'Physical and Micrometrical Observations of the Companions,' p. 177

54   See Gary W. Kronk, *Comets: A Descriptive Catalog* (Hillside, New Jersey, Enslow
     Publishers, 1984), pp. 227–8

55   A. Marth, 'On the Eclipse of Iapetus by Saturn and Its Ring-System, on November
     1–2, 1889,' *MNRAS*, **49** (1889), 427–9. Marth noted: 'The inclination of the orbit of
     Japetus to the plane of the ring being nearly $14°$, while the orbit of the other satellites
     have inclinations of less than $1°$, the rare eclipses of Japetus by the ring system offer the
     only chance of deciding several questions which may be settled with the help of
     observed eclipses. No such observation has ever yet been made. Favorably placed
     observers ought, therefore, to take full advantage of the rare chance they may get on
     November 1.'

56   EEB, observing notebook; LO

57   E. E. Barnard, 'Observations of the Eclipse of Iapetus in the Shadows of the Globe, Crape Ring, and Bright Ring of Saturn, 1889 November 1.' *MNRAS*, **50** (1890), 107–10

58   ibid.

59   E. E. Barnard, 'Eclipse of Japetus, the VIII satellite of Saturn, on November 1, 1889,' *PASP*, 1 (1889), 126–7

60   Barnard, 'Observations of the Eclipse,' p. 109

61   A. F. O'D. Alexander, *The Planet Saturn: A History of Observation, Theory and Discovery* (New York, Dover, 1980 reprint of 1962 ed.), p. 234

62   EEB, 'Charges against Holden,' one of several MS drafts written in 1892; VUA

63   E. E. Barnard, 'Transparency of the Crape Ring of Saturn, and other peculiarities as shown by the observations of the eclipse of Japetus on November 1st, 1889,' *AA*, 1 (1892), 119–23

64   ibid.

11

The young rebel

1

The winter of 1889–90 was of exceptional severity. The storms began in mid-November, and by December the snow was already piled so deep that the stage was blocked for days at a time. Supplies and food had to be carried up the mountain on horseback. The months of January and February were even snowier, and from February 16 to 20, 1890, Mt. Hamilton was swept by one of the fiercest blizzards on record, in which for five days the observatory was cut off from all communication with the outside world.

Barnard on February 19 recorded in his observing book that the 'drifts are 10 feet on trail down to cottages. The snow is drifted higher than the eaves of my cottage and one can walk right up on the roof. All the front [is] filled in completely [and I] have to go around back to get in. The Blue Jays are in a bad condition . . . They are snowbound – found one lying by the side of the door, tried to save it but it died.'[1] Though the astronomers managed to get through the winter without starving, there was little chance to observe – from the middle of November, when the stormy season began, until the end of February, there were hardly any clear days. Instead the observatory's meteorological records show a seemingly interminable streak of clouds, rain, fog, and snow.[2]

The bad weather was to prove an unpleasant harbinger for the whole year, during which the interpersonal climate on the mountain became no less stormy. As the first permanent establishment for astronomical work on a mountain, the Lick Observatory had from the first been conceived as an experiment to obtain the most favorable conditions for astronomical observations. At the same time, it became quite unintentionally a sociological experiment in which the observatory's relative isolation, and the close proximity in which the small staff had to rub against one another, exaggerated all personal differences and animosities. This result seems not to have been foreseen. Only afterward was realized 'the effect of the comparative isolation of such a community, which is apt to be provocative of internal dissension,' as Newcomb put it.[3]

Similarly isolated communities had, of course, existed before, generally held together by a strict rule of discipline. The monastery of Monte Cassino (Italy) was governed by a *Regula Monachorum*, and military rule had held sway over the men inhabiting forts and ships for centuries. Yet the staff of the Lick Observatory was made up neither of monks nor of military men. Except for the academically trained Keeler, all

Barnard's cottage after February 1890 snowstorm. Mary Lea Shane Archives of the Lick Observatory

were self-made men, rugged and fiercely independent. Holden, on the other hand, was a different sort of personality – authoritarian to the point of being tyrannical, a man who thrived on control. Thus it was hardly to be expected that the Lick Observatory could go about its business without a good deal of interpersonal friction. What Herman Melville wrote about the situation on board a ship came equally to apply to the mountain observatory which only two years before Captain Floyd had so hopefully launched into the 'ocean of science':

> Now there can exist no irritating juxtaposition of dissimilar personalities comparable to that which is possible aboard a great warship fully manned and at sea. There, every day among all ranks, almost every man comes into more or less contact with almost every other man. Wholly there to avoid even the sight of an aggravating object one must needs give it Jonah's toss or jump overboard himself![4]

After two years of contact with Holden, Barnard had come to regard him as a 'bumbling, ineffective former Army officer, posing as a scientist,'[5] and was in the process of compiling a long list of grievances against him. The one theme that ran most consistently through his litany of stored-up resentments was the fact that Holden did not allow him regular use of the 36-inch, while the director himself 'wasted the energies of the great telescope' in obtaining results which Barnard regarded as 'merely those of an amateur of no value to science.' On Holden's nights, as Barnard never wearied of complaining, the 36-inch dome was shut as early as 9 p.m. and rarely later than midnight; then the director let the great telescope 'lay idle over half the night while his statements in print would lead to the belief that [it] was never idle.'[6] For a dedicated observer like Barnard, who observed seven nights a week all night long whenever it was clear and managed on only a few hours of sleep a night, this kind of behavior was unforgivable, Holden's heavy daytime responsibilities notwithstanding.

Later, in reflecting from afar on the causes of Holden's growing unpopularity with his staff, Sir William Huggins, the noted pioneer spectroscopist of Tulse Hill, put his finger on the problem to George Ellery Hale: 'What do you expect when a few men are shut up together on top of a mountain, and only one big telescope.'[7] But in Barnard's case, the situation of deprivation was especially cruel. In failing to give the enthusiastic junior astronomer a regular night on the great telescope, or at the very least a share of the time on the nights when it 'lay idle,' Holden, one presumes unwittingly, recreated a situation of severe deprivation in a man whose whole childhood had been one of deprivation, thereby reawakening many of the anxieties and insecurities that lay just below the surface and that he had managed to seal over only with massive effort.

For his part Holden, the former West Pointer, a man of military discipline and rules, cast Barnard as the young rebel, a troublemaker who did not know his place. No one else on the staff gave him so much trouble. The young Tennessean was, as Donald E. Osterbrock notes, 'completely self-centered, hungry for praise, threatened by anything less than adulation, prey to all kinds of illnesses whenever his career was subject to pressure . . . a compulsive observaholic . . . his letters were all handwritten, and his style of writing changes within a letter, from . . . a not very well learned attempt at well formed letters when he was expressing conventional banalities to a childish scrawl when he was pointing out his grievances.'[8] Certainly Holden was ill-equipped to deal with neuroticism on such a grand scale, and even when he seems to have honestly tried to be helpful – which seems to have been much of the time – his motives were always suspect, his attempts at conciliation became merely 'Flattery to gain his point.' Barnard regarded him as a 'Pecksniff,' who hid his real selfishness and corruption behind a display of seeming benevolence.[9] He was 'insolent, overbearing, tricky, and thoroughly unreliable in word and deed,' a 'newspaper puffer, a falsifier and a debaser of talent,' who 'instead of being at the head of a great astronomical observatory should have been at the head of some corrupt organization such as Tammany in its flush where he would have shone with a peculiar luster as a wire puller and intriguer worthy of a corrupt and debased institution.'[10] Thus these two personalities, who did not understand one another, became locked in a grim struggle that came close to destroying the one emotionally and was in large part responsible for the other's downfall.

2

During the spring of 1890, Holden launched vigorously into his long-contemplated program of lunar photography with the 36-inch refractor. Ever since he had seen the impressive results of Burnham's trial exposures of August 1888, he had been promoting the project with such unabashed enthusiasm that astronomers who heard him speak at meetings back East quipped, 'Holden has discovered the Moon!' The newspapers, predictably, reported rather uncritically his sanguine expectations. The Springfield *Republican*, for example, announced: 'Prof. Holden . . . proposes to make a special study of the moon, taking up . . . various sections of the moon's surface [and] noticing carefully any indications of the changes that may have occurred since the same sections were examined by other astronomers.'[11] On the same subject the Atlanta *Constitution* published the following, rather curious, report:

> It is announced that the astronomers in charge of the Lick Observatory in California have made some discoveries in regard to matters and things on the moon's surface of such a startling and incredible nature that they do not dare make them public. The promise of a full disclosure of these remarkable discoveries at some future time is tantalizingly held out, but at present nothing definite can be learned from the astronomers . . . The only thing that the public can do under the circumstances is to brace up and get ready for the shock that will come with the expected deluge of information.[12]

The actual revelations from the Lick Observatory Moon photographs were, however, anything but sensational. On one of the photographs taken by Burnham in August 1888, or more precisely, on a positive enlargement of one of these made by Barnard, Holden grandly announced in the *Publications of the Astronomical Society of the Pacific*: 'I may mention a discovery which I have made . . . It is well known that Maedler (and others) have mapped the walls of the *Hyginus* rill crossing the floor of the *Hyginus* crater. So far as I know, this has only been once seen. The observation is a delicate one, and could only be made when the sun is shining nearly in the direction of the preceding branch of the rill. The walls inside the crater are hardly more than 2000 *yards* apart . . . [y]et they are plainly and obviously visible in this enlargement.'[13] Interesting, yes; but hardly the sort of thing that was likely to turn astronomy on its ear. Later finding an enthusiastic ally in Ladislaus Weinek of Vienna, who became the leading student of the Lick Observatory Moon negatives (and through Holden's influence, received one of the first honorary Sc.D. degrees given by the University of California in 1893; the other went to Keeler), Holden regularly published articles in the *Publications of the Astronomical Society of the Pacific* announcing the discovery of the odd new crater or rill by Weinek – but most of these discoveries, such as they were,[14] were made on Burnham's original plates rather than on Holden's later ones.

The reason was that the director, just as he had shown in his US Naval Observatory work on the Orion Nebula, lacked the patience to become a first-rate observational astronomer. Barnard, who was assigned the task of developing many of Holden's plates, was outraged at the shoddy results – and the waste of his own (and the great telescope's) time. He complained that Holden wasted eight months on the 36-inch trying to determine its photographic focus, 'a work that could be easily done in half an hour on any clear night,' and also that the director refused 'to permit SWB or EEB to do anything with that work except to put the plates in the shields and develop the worthless results.'[15] Furthermore, Holden had once taken '25 or 30 10 × 8 plates of Moon out of focus exposed rapidly in an hour or two in first part of night – taking me all the next day to develop them.'[16] In other photographs the Moon was blurred because Holden had shaken the telescope when making the exposures. Holden's photographs of Mars taken during 1890, most of which were also developed by Barnard, were equally defective, and Barnard bitterly accused Holden of 'throwing away' the important opposition of that year, 'in photogr[aphs] – out of focus – jarred – etc. while [W. H.] Pickering at Mt. Wilson with 13 in. got fine results.'[17]

3

Though Barnard had long fumed privately whenever he saw Holden shutting up the dome of the 36-inch prematurely and heading for bed, in the summer of 1890 his true feelings toward the director could no longer be suppressed and finally broke out into the open. It may be that by this time Holden had already gathered from the junior astronomer's attitude that he did not think much of the director's work – what we do not know from the official records are the glances that had been exchanged between the two men as they passed, the unrecorded words they spoke to one another or the tone of voice with which they spoke them. It was, in any case, the old impasse over the question of Barnard's use of the great telescope that brought the tension between the two men to the point of explosion.

On the evening of June 9, 1890, Barnard was working in the photo lab developing Holden's latest Mars plates, and decided to pressure the director once again for time on the 36-inch. Perhaps he did so in a less than tactful fashion, for his mood was testy to judge from what he wrote afterwards in explaining his motives in approaching the director just then:

> At the time that I spoke to him I was developing exposures which he had made on Mars, and which were worthless because he did not have them in focus and had shaken the telescope in making the exposures. It was in the photo House where I was wasting my time – a fine observing night – developing his worthless work, and I thought it was a good opportunity to speak about my getting some use of the great telescope as he was wasting its energies.[18]

As before, Holden refused Barnard's request – 'He had his usual excuses – he was using it for photography, etc., etc.,' Barnard chafed at the time – and following this encounter Barnard left Holden's presence without saying another word. The next morning he received a long letter from the director which, like all the other letters

Holden wrote, was duly entered into the letter book – the official records of the Lick Observatory. It said in part: 'I wish to repeat to you in writing what I have already said in words – & that is that every instrument in the Observatory shall be used with the sole object of advancing science & not for any person's individual benefit or advantage. This is, has been & shall be my rule.' If Barnard had any research that required the 36-inch, all he had to do was ask and the director would, as he had always done, consider it 'my duty and my sincere pleasure to do all in my power to forward your plans, so far as they are consistent with other work in hand.' He went on to suggest that he had always been particularly considerate of Barnard's needs, by not imposing on him any of the routine duties such as the time service, meteorological observations, and so on, which instead were performed by his seniors, Schaeberle, Keeler, and the director himself, and that 'unless you have taken especial pains to find out how much of this sort of work is done by others (specially by Keeler & myself) you would probably be surprised to learn its amount, & to see to what you owe your freedom, which is an extraordinary one to enjoy at your age.'[19]

When he read this, Barnard became furious. No one could argue with Holden's point that everyone needed to pull together for the good of the observatory; however, Holden had expressed himself ambiguously – probably intentionally so – in order to create the impression that Barnard had not done so. Barnard saw it as an underhanded attempt to enter a false and self-serving account into the observatory records, 'with no reference to anything gone before,' in order to impugn the purity of his (Barnard's) motives. 'The letter gives the impression I had been using the telescopes on Mt. Hamilton for my own personal benefit and not for the uses of astronomy,' he erupted. 'That is the impression it conveys and is what he intended it to do . . . That is how he attempted to silence me so that I should not again refer to the use of the great telescope for fear of having a damnable and false letter put on record against me.'[20] At once Barnard stormed over to Holden's office 'to ask the meaning of this letter – as it intentionally conveys a false impression of the conversation we had.'[21] He insisted on a clarification, which he dictated to Holden. Holden backed down and jotted the dictated clarification *verbatim* in the letter's margin: 'This does not in the *least* mean that you or anyone in the Obsy. has ever used the instruments for anything but scientific purposes.'[22]

Even so the whole matter was by no means closed. Barnard henceforth became obsessed with 'fighting the records of the Lick Observatory,' which he considered a 'fraud.'[23] He invested limitless pains and produced innumerable manuscript drafts over the next several years in his attempt to set the record straight. Holden, he claimed, 'falsified the records of the L[ick] O[bservatory] – as a matter of history – in his favor.'[24] A serious charge, but later he made an even stronger claim: 'The Director of the Lick Observatory has repeatedly injured me by writing and copying into the Records . . . letters – for posterity to read – letters that deliberately and falsely record things injurious to me in the eyes of posterity.'[25]

In reality, of course, Barnard's accounts are no less one-sided and self-justifying than the director's. Under the circumstances, it would be naive to assume that they always present a true and accurate view of the director's actions and motives. On the other hand, there is little doubt that the director was a master of intentional ambiguity, and regularly presented the facts in such a way as to invite entirely false and misleading inferences. Moreover, though perhaps no less dishonest than the average politician, he was more than capable of playing fast and free with the facts themselves when it served his purposes, as the Iapetus eclipse episode shows. Thus, if Barnard at times seemed paranoid, it was not without cause; where there was so much smoke, there was fire.

Returning to the letter in question, Barnard's reaction was to some extent justified, but it also bears witness to his distraught condition at the time. He was then, as always, under enormous strain from overwork, and had hardly spent any time away from the mountain since the opening of the observatory. He worked himself unsparingly all night and then during the daytime found it difficult to sleep. 'He would retire at dawn, almost exhausted with his long hours of work, and fall asleep quickly. But any noise was likely to waken him, and once awake he could not get back to sleep. Everyone on the mountain understood this and special efforts were made to establish what we now call "a zone of quiet" about his house.'[26] Especially each summer and autumn, as Osterbrock has pointed out, 'when nearly every night was clear, the weather hot and dry, and the mountaintop parched, he would become particularly irritable.'[27] Under these circumstances of strain and accumulated fatigue, the skewed aspects of his tortured personality were inevitably brought into high relief.

As acute as his vision was when it came to faint stars, comets, or nebulae, he would become even more keen at perceiving examples of what Burnham later referred to as the director's 'system of petty persecution'.[28] The happy days he had enjoyed on first arriving on Mt. Hamilton were over, and he became depressed and miserable. His distressed state of mind would also be reflected in his work, for the year 1890 would prove, of all his years on Mt. Hamilton, the most barren of results.

4

In July 1890, Barnard first broached the subject of his growing disenchantment with Holden to Charles B. Hill, who by now had left his position as observatory secretary for a more lucrative position as an insurance agent in San Francisco. Hill delighted in gossip, and in turn passed the information on to George Davidson, who also disliked Holden intensely.[29] Evidence of animosity between the director and the members of his staff – and Barnard was not alone, though he suffered the most acutely – now begins to be found with increasing frequency in the observatory's correspondence. At the bottom of what looks on the face of it like a routine order from Holden to Barnard – 'The dec[lination] clamp of the 6-inch telescope is not right yet – and will you please adjust the finder' – Barnard scrawled angrily: 'a cowardly revenge by Prof. H.'[30] And in another letter, Burnham and Schaeberle wrote indignantly to the director: 'We insist at all times that whatever difference of opinion there may be between our views upon scientific subjects your conduct, both in oral and written communications, shall be gentlemanly. Among other things in your communication of yesterday the use of the word *sneer* is unworthy of your official position.'[31]

In August, another major row broke out between Holden and Barnard. This time the subject was Barnard's Milky Way photographs. In March 1890, Barnard had published a first report on these photographs in the *Monthly Notices of the Royal Astronomical Society*, and Holden had taken up the subject briefly in an article on 'Astronomical Photography at the Lick Observatory,' which appeared in the 1890 edition of the *International Annual of Anthony's Photographic Bulletin* and was reprinted in the August 1890 *Publications of the Astronomical Society of the Pacific*. Though Holden praised Burnham and Barnard's 'photographic skill and experience,' he referred to the Milky Way photographs as only 'experiments . . . which promise the most satisfactory results.'[32] The implication was that they had not yet *delivered* satisfactory results. (He had made much the same comment to Agnes Clerke the previous August on sending her two of Barnard's positives on glass: 'They are very interesting as experiments, tho' they have their faults.')[33] Holden also suggested that Barnard's photographs of the Andromeda nebula and the Merope nebula in the Pleiades did not measure up to those obtained at observatories elsewhere. Faint praise indeed from the man who had waxed enthusiastically over each new crater revealed on the Lick Observatory Moon pictures. Barnard considered Holden's comments nothing less than a 'slur . . . on the Milky-Way pictures that I have made here' and on his other work with the Willard lens. Thus provoked, he at once sat down and wrote a long article 'to defend the photographic work of the Lick Observatory which had been done by myself.'[34]

'It seems desirable to give a brief description of the photographs of the Milky Way made by me at the Lick Observatory in 1889,' he began, 'and to call attention to their special and important points which might otherwise be overlooked by those not familiar with celestial photography, and thus their value be under-estimated for the purpose for which they were made':

> One very important feature, and one which must not be overlooked, is that these are the only photographs ever made, here or elsewhere, which show at all the true Milky Way . . . In [these photographs], besides myriads of stars there are shown, for the first time, the vast and wonderful cloud forms, with all their remarkable structure of lanes, holes and black gaps and sprays of stars. They present to us these forms in all their delicacy and beauty, as no eye or telescope can ever hope to see them . . .[35]

Barnard's account was 'immodest but basically correct,'[36] and he defended vigorously, and in much the same terms, his other work, arguing that when due allowance was made for the difference in the size of the instruments, 'my photographs of the *Andromeda* Nebula, etc. compare most favorably with anything made' by astronomers elsewhere. Barnard left this article on Holden's desk with the request that it be published in the next issue of the *Publications of the Astronomical Society of the Pacific*. On reading it over, Holden agreed to publish it, but suggested doing so in his own name, as if he had written it. Presumably, as Osterbrock suggests, he made the proposal partly because he realized that the article 'was too self-glorifying to go out over Barnard's signature,' partly because it would be 'understood by everyone who read it as an attack by a subordinate on his chief, exposing the tensions at Lick.'[37] Holden's suggestion was not at all unreasonable, but as usual Barnard could discern only darker motives behind the director's actions. When the director came to the 12-inch dome to make his proposal, Barnard rejected it angrily, seeing it only as an attempt to defraud him of his labor. The two men quarreled that night in the dome and again the next morning in Holden's office. According to a contemporaneous Barnard memorandum, Holden 'in the most brazen manner began to accuse me of things about the Willard lens that were utterly and absolutely false on the face of them. These fraudulent charges gotten up for the occasion so enraged me that, to avoid any further words with him, I took my manuscript and left him.'[38] Eventually Holden capitulated; the article was published in Barnard's name.

5

At the end of October 1890, Keeler left Mt. Hamilton for a two-month vacation back East, and in his absence Holden assigned Barnard the performance of Keeler's regular duties, including the running of the observatory time service. ('Revenge is Sweet!' Barnard wrote at the bottom of Holden's memo, seeing the director's action as a petty persecution for his having opposed him the previous June).[39] For two months, Barnard faithfully discharged his duty. Then, the day before Keeler's return, on Friday, January 2, 1891, an earthquake shook the mountain eighteen seconds after Barnard had sent out the noon signal. The shock stopped the clock which transmitted the time signals and also all but one of the other clocks in the observatory. As such an event had been completely unforeseen when Barnard had received his instructions from Keeler, Barnard, through an unfamiliarity with all the details of the clock, did not know precisely how to reset it and made a mistake. As a result, though the hands of the clock read correctly, the time signals were sent out exactly one minute in error. This gross error (by astronomical standards) in the Lick Observatory time signals was detected all the way up and down the Southern Pacific railroad's telegraph lines long before it was found and corrected at the observatory. Indeed, Holden first learned of it that same evening from the railroad superintendent.

Considering Barnard's well-known sensitivity to criticism, Holden would have been prudent to let the matter drop. Instead he goaded the young astronomer by passing the letter from the railroad superintendent along with the biting comment: 'You see that the error of 1 m[in] which I pointed out to you was discovered all along the R.R. lines before it was corrected here – which is not to our credit.'[40] Barnard replied defensively that Holden had not notified him of the error until the following evening, after the time signal for the day had already been sent out, and that he had corrected it at his first opportunity to do so the very next morning. He then immersed himself for several days writing various drafts of his own account of the incident – his final version ran to six pages – so that the facts as he saw them could be entered into the observatory record. He protested violently: 'The remarkable and unforseen cause of this error should in itself have been explanatory enough to have prevented the necessity of this letter which I write with much pain.'[41] Barnard's rather hysterical reaction seems to have taken

Holden by surprise, for in a letter to J. H. C. Bonté, one of the University of California Regents, the director seemed to be groping for an explanation of the increasing difficulty he was having with his junior astronomer:

> This is a rather queer place for labor. It is very isolated and not much fun going – and the men get weary, not having either interests or variety. Then too, we are all more or less like the horses of these mountains! We get 'loco,' as they say.[42]

The further parries between the two men – about when Holden first told Barnard about the error, and about why it took so long for it to be corrected – do not make very edifying reading. 'Holden's letters are smooth and reasonable, but self-serving, while Barnard's become increasingly wild and almost hysterical,' notes Osterbrock.[43] At one point Barnard exclaims: 'You seem to take a special delight in goading me with the discredit of this occurrence, and yet you would cover it with the cloak of Charity! No one questions the credit of it, but why harp on it as if it had been done with a desire to injure the Lick Observatory?' He felt that Holden had unjustly 'seized upon this unfortunate occurrence to further depress one whose sole aim has been to aid in the building up the reputation of this great observatory.'[44]

Holden countered that at the time of the incident he had thought of making a written memorandum of the facts that Barnard was now disputing, and only 'out of mistaken consideration' had informed Barnard verbally of the error in the time signal instead; henceforth he would have to make all such communications in writing, 'for my own protection.'[45] Barnard, who was in the habit of making 'translations' of what he considered to be the true meaning of Holden's various statements to him, gave the following interpretation to this one: 'That is, I shall annoy and hound you for the rest of your existence at this observatory, for telling me the truth.'[46] Nevertheless, this time it was Barnard who refused to let the matter drop. He wrote yet a further reply to this reply, which added nothing to the reams of verbiage already written on the subject and seems to have been intended for the sole purpose of allowing him to get the last word in. He insisted that he had done nothing more than defend himself from unjust attacks.[47] Obsessed with having all of his letters carefully copied into the records, Barnard later learned through Burnham's son Augustus, who had replaced Hill as observatory secretary, that this last letter had not been so copied. Infuriated, he marched over to the director's office to confront him about it – and in his own view, at least, did so 'politely':

> [Holden] looked at me in the most insolent manner and said 'Why do you ask me that question. Do you think I ate it?' In return I asked him why he should reply thus to me as I only wished to know that it had been copied . . . He then said it was so copied . . . He was very angry and looked as if he would like to strike me down. I simply said, without loosing my temper, that that was all I wished to know. Still from my knowledge of the man I believed he was telling me a falsehood . . . not the first one by many hundreds.[48]

At Barnard's insistence this letter was finally copied into the observatory records, and only then, after nearly a month, was the case finally – and mercifully – closed. However, the fate of Barnard and Holden's relationship was sealed. Henceforth they distrusted one another so much that at times they communicated only by means of terse written notes. Barnard remained convinced that one of Holden's main objects in life was to persecute him and make him miserable. Though undoubtedly perplexed and exasperated by his junior astronomer, Holden's private views about Barnard were, to his credit, rather more generous. He once confided to the English astronomer E. B. Knobel:

> It is astonishing how excellent an observer he is. He is like Sir W. Herschel for *seeing* and noting what is new. But his reasoning on what he sees is very apt to be quite wrong. This is not only from a defective education, it is in his nature. He is like Sir W. Herschel cut into two parts, and only the observing faculty left . . . His education is unfortunately very limited; incredibly so one might say. He does not seem able to perfect it – or even desirous to do so. There is not a single one of his comet observations (all of which are most carefully made at the telescope) which is entirely correct in the reductions. Something is always wrong – parrallax, refraction, red[uctio]n to apparent date – something.
>
> At first, I had these reductions supervised but he was so dreadfully touchy about them that finally we let them go as they were, especially as the computers learned to check his reductions before using them. I confess this to be an entirely wrong thing to allow – on my part – but it had to be done . . . The curious thing is his indisposition to improve on that side. His refraction (to 2 & 3 places) are all computed with six-place logs! I tried at first to get him to use 4 & 5 place, but after trying a month he gave it up and all his work is done with 6-place . . . All these things are part of his history – but strange as these and many other really serious shortcomings are they do not interfere with one thing – which is that he is the most acute observer and the most tireless of any one I have ever known. Witness his discovery of a double star at occultation, . . . of faint comets etc. A good illustration of the poor way in which he reasons on his splendid observations might be for one example [that] he seriously holds that he can tell whether a comet is periodic or not by its appearance! He has made one or two good guesses, aided by the knowledge . . . that the objects were near the ecliptic, and now he is satisfied that he can predict the fact of periodicity before an orbit is worked out for him! He forgets the cases when his predictions haven't been verified. As in this so in many other cases. And the combination makes a very strange mind. It is precisely as I say. He is one-half of Sir W. Herschel and no more and no less.[49]

Not surprising that the man of detail should react thus when confronted by the man of genius, and wonder at his 'strange mind.' But though anyone with ordinary talent might master the log table and carry out 'reductions,' Barnard's was a much rarer quality – the power of 'right brain seeing,' an intuitive ability when at the eyepiece to see to the heart of the matter, to see what only one in a million could see. If he was 'one half of Sir W. Herschel,' he was the more important half – something to which his irregular education may even have owed something in a curious and half-fathomed way.

6

Barnard's bitter struggle with the director was undoubtedly a great distraction from his work, and though he still 'spent a vast deal of time in comet seeking, using large and small telescopes,'[50] he was not enjoying his former success. The glory for the new comets discovered in 1890 went elsewhere – to Borrelly and Coggia at Marseilles, Brooks at Geneva, Zona at Palermo, and Denning at Bristol.

Barnard did, however, recover the one periodic comet of the year. Ever since April 1890, he had been searching carefully, with the aid of an ephemeris published in the *Astronomische Nachrichten* by Leveau, for periodic Comet d'Arrest, which had been observed at its return in 1877 but missed in 1884. He searched both with the 12-inch refractor and, through the courtesy of Burnham, on occasion with the 36-inch, utilizing every favorable opportunity until September 4, 'after which date the search was given up as hopeless, as the comet was decreasing in light, and was getting into a poor position for finding. From that time the comet dropped entirely from my memory.'[51] Then, on October 6, during the course of his regular comet-seeking, he swept up a 'faint diffused nebulosity' near the southern horizon, which he at once recognized as a comet and telegraphed as new. From three observations made by Barnard on successive nights, Schaeberle computed the comet's orbit. A comparison being made with the orbits of previously observed comets, it became clear that this was none other than the elusive d'Arrest. However, as Barnard noted: 'The discovery was entirely in the line of original search, and without any recollection of the position or even of the existence of d'Arrest's comet.'[52]

In July 1889, Joseph A. Donahue of San Francisco had established a medal to be awarded for the discovery of comets through the Astronomical Society of the Pacific. The terms of the gift did not distinguish between the discovery of new comets and the rediscovery of periodic comets. Thus Barnard was entitled to the medal. However, he declined it for reasons that he explained to the comet medal committee (consisting of Holden, Schaeberle, and also Charles Burckhalter of Oakland's Chabot Observatory):

> I have contended from the first that no medal should be awarded for the finding of any one of the well known periodic comets whose places can be closely predicted . . . There is no merit worthy of special notice in an observer finding, say Encke's Comet, when he has only to take the ephemeris, computed by another person, and set a powerful telescope upon the predicted place. The equally zealous and industrious, but less fortunate, astronomer, who has only a small instrument, has no chance of anticipating the other. Whatever credit there is in a case of this kind belongs rather to the man who computed the ephemeris.[53]

Even though in this particular case the discovery had not been made simply by pointing a large telescope to the place indicated by the ephemeris, Barnard's strong sense of fair play would not allow him to accept the medal for finding a periodic comet whose place had been accurately predicted by an ephemeris – and as he realized after the comet had been discovered, throughout the period in which he had been carrying out 'the most thorough and exhaustive' search for it, it had indeed been close to the ephemeris positions given by Leveau.[54]

Holden, nevertheless, continued to favor awarding the medal in the cases of periodic comets where 'an accurate observation made *early* in the return is of especial use to science and should be rewarded.'[55] Eventually a compromise was reached, in which the medal was to be awarded only for new comets *and for periodic comets observed at their first return*.[56] For his part, however, Barnard made it a point of honor to accept the medal only for the *new* comets which he discovered.

7

With tensions running high on Mt. Hamilton and Barnard's comet-seeking fortunes at a low ebb for more than a year, the San Francisco *Examiner* chose a particularly inauspicious moment to spring a hoax, which Barnard was 'astonished and horrified'[57] to read in the March 8, 1891 edition. The headline over two closely printed columns read:

ALMOST HUMAN INTELLECT
An Astronomical Machine That Dis-
covers Comets All by Itself

The article, supposedly by one Collis H. Barton, began with accurate background information about comet-seeking: 'The comet-seeker has a hard time of it, exposed, as he needs must be, to all sorts of weather, in the open air, with no protection from the wind, at all hours of the night, sweeping over . . . the sky with all his powers of acuteness at the highest pitch, lest the faint object should be overlooked.' So far Barnard himself might have penned these words, for he had often made a point about how much hardship there was in the comet-seeker's lot. The time-intensiveness of the job was also stressed in the *Examiner* article – 'There is nothing to show for an unsuccessful night. In any other branch of astronomical work the observer would have obtained in this time several volumes of results, possibly all of value. No other calls for such an expenditure of useless labor.' Again Barnard could have written these words – and they would have rung especially true just then, given how many hours of 'useless labor' he had personally expended in searching for comets during the previous year. He was always sensitive to the attitude of others who did not fully appreciate the value of the comet-seeker's labor – including, unfortunately, even his close friend Burnham, who had told Barnard on several occasions that 'the one great fear of his life was that sometime he might disgrace himself by accidently finding a comet!'[58]

Given the drudgery of the comet-seeker's lot, Barnard may at times, as University of Michigan astronomer Heber D. Curtis later suggested, 'have beguiled the monotony . . . by dreams of some method of automatic guiding, where the selenium cell or some similar device should relieve [him] of the eye-strain and close attention incident to guiding [his] instruments.' The basis of the *Examiner*'s hoax was its claim that Barnard, with the assistance of Keeler and Schaeberle, had actually invented an automatic selenium cell comet-seeker. The article provided numerous details supposedly obtained through an interview with Barnard himself, and the plausibility of the story was further enhanced by the inclusion of detailed drawings of the comet-seeker, the prism it supposedly used to obtain the comet's spectrum, and even the electrical circuit of its Wheatstone bridge.

The hoax had been written not by Collis H. Barton, a mere *nom de plume*, but by Charles B. Hill, the Lick Observatory's former secretary. It was a clever hoax, but Barnard saw little humor in it and indeed immediately 'wrote hot letters of denial to the *Examiner* and to other San Francisco papers.'[59] Even ten years later, when he related the story to Curtis on the *USS General Alava* in the Indian Ocean as it carried the US Naval Observatory eclipse expedition to Sumatra, he was, on recollecting it, 'able to summon only a rather wan and rueful smile.'[60] But the newspaper was having too good a time of it to publish Barnard's disclaimers. Hill urged patience. The editor of the *Examiner*, he told Barnard, 'asked me to write to you, for him, explaining that they want to wait until the Eastern exchanges come in . . . to find out how many of them have been "taken in." Then they will . . . pub[lish] your letter.'[61]

Among those taken in was Barnard's old friend and fellow comet-seeker Lewis Swift, who wrote to him: 'I have received the *San Francisco Examiner* containing an article regarding your invention to search for comets while you are asleep or using the 12-inch or playing poker. It takes my breath away and makes my hair stand straight towards the zenith to think of it. Although the article appears somewhat fishy I am inclined to think it is still another of the marvelous inventions of the 19th century. I can hardly sleep till I hear more about it . . .'[62] Even two years later Barnard still found himself receiving an occasional letter regarding the automatic comet-finder from others who were desirous of further details. Not until February 5, 1893, after Barnard had become famous for his discovery of the fifth satellite of Jupiter, did the *Examiner* belatedly published its long-promised retraction and apology:

> The *Examiner* seizes the opportunity to express contrition for the annoyance which it caused this eminent scientist by printing some time ago an account of a highly ingenious, but non-existent, machine for scanning the skies and catching wandering comets . . . The inventor was Mr. Charles B. Hill, himself an astronomer of repute, who in order to give a greater appearance of substance to his fancy attributed the origin of the device to Professor Barnard, never thinking that the story would be taken seriously by experts. In this he was mistaken, for the scientific journals of America and Europe were all sold . . . Of course the reputation of a man like Barnard is not to be affected by such a gentle hoax, but nevertheless the *Examiner* extends its apologies and wishes him all the new moons and comets that may be necessary to his happiness.[63]

Barnard certainly was much affected by the 'gentle' hoax. On the other hand, much as he was annoyed by it, it may have acted as a spur to his comet-seeking efforts, for soon after it appeared Barnard began making discoveries almost with the mechanical efficiency of the alleged invention. He discovered or had a share in the discovery of every comet of 1891 – no less than five between March 30 and October 2, including two new ones and the three periodic comets, Encke, Wolf, and Tempel–Swift.[64]

8

Despite the *Examiner*'s own identification of Hill as the culprit behind the comet-seeker hoax, Barnard himself always believed that Keeler was also involved. This is possible, though no firm evidence has ever linked him to it. Moreover, Keeler was at the moment preoccupied with other things. He had become engaged to Cora Matthews, and after five years on Mt. Hamilton was looking seriously at other positions which would offer more of life's comforts to a wife – and a very wealthy one at that – than the modest lodgings he occupied at Lick. Indeed, during his recent trip back East, Keeler had visited Allegheny Observatory in Pittsburgh, where he had earlier worked as Samuel P. Langley's assistant, partly to reestablish his contacts there. He had made it clear that under the right circumstances he was willing to return. Langley, though still nominally the director, was spending most of his time at the Smithsonian Institution in Washington, of which he had been head since 1887. The result of all this was that during spring 1891, Keeler was offered the directorship at Allegheny. He accepted, and tendered his resignation from the Lick staff effective as of June 1. But though Keeler had made his true reasons for leaving perfectly clear, the press blamed the the departure of the brilliant young spectroscopist on Holden.

Indeed, the director – by now a lonely, unpopular, and embattled figure – had not only the constant infighting with his staff to contend with, but also the incessant attacks and savage abuse of the press. In January, shortly after the books were closed on the infamous time signal affair, he had been embarrassed by seismometer observations he had made at Mt. Hamilton of a dynamite blast exploded by a contractor in San Francisco. As the *Examiner* reported gleefully:

> His instrument at the appointed hour appeared to him to yield remarkable and highly satisfactory results. At once he prepared a long telegram descriptive of the same and dispatched it to San Jose by messenger to be wired to the press. The evening newspapers, which arrived on the mountain shortly after the departure of the messenger, disclosed the amazing fact that the blast had been postponed, and had not shaken the globe until three-quarters of an hour *subsequent* to the phenomena observed by Mr. Holden and described by him in the telegram. A mounted man was sent tearing down Mount Hamilton after the messenger, and, providentially for Director Holden's scientific reputation, such as it is, the bearer of the ridiculous dispatch was overhauled before he reached the telegraph office.[66]

News of the 'Social War on Mt. Hamilton' was also leaked by Holden's disgruntled staff, and the first accounts began to appear soon after Keeler's departure, for which the blame was placed squarely on the shoulders of the arrogant director. The San Francisco *Examiner* noted: 'On Mount Hamilton stands the biggest telescope, and also E. S.

Holden, Director of the Lick Observatory. Professor Holden is the distance of his own stature higher than the mountain top.' Referring with heavy sarcasm to the troubles on the mountain, the same editorial continued, 'Twas just a purely social disturbance, provoked, so the star-gazers say, by the masterful way in which Director Holden plants a rude foot on any one with whom he happens to come in contact,' and concluded with the accusation that 'the cavalier way in which he treated Mr. Keeler and Mr. Hill caused them to resign.'[67] In Keeler's case, at least, this was untrue, and he eventually said so explicitly in print.[68]

As soon as Keeler had announced his plans to leave for Allegheny, Holden hired as his replacement William Wallace Campbell, a twenty-nine year old spectroscopist who had trained under Schaeberle at the University of Michigan. He had spent the summer of 1890 on Mt. Hamilton as a volunteer assistant to Holden and Keeler on the 36-inch refractor, and even though the newest and youngest member of the Lick staff, Campbell received two nights on the 36-inch to continue the important spectroscopic work begun by Keeler. Two other nights he was to assist Holden on the telescope with lunar and planetary photography. Meanwhile, the routine duties that Keeler had tended also had to be reassigned, and the director now entertained the sweet revenge of visiting upon Barnard the responsibility for the time service.[69] On being informed of the plan, Barnard, of course, objected vehemently. He agreed to take it over temporarily but insisted that it was impossible for him to do so permanently. His main argument was that he already had enough to do. His results in the cometary department with which he was charged were achieved only 'by the utmost diligence and alertness on my part. It implies nights of sleepless labor. It means days of constant work. My whole time except a small portion for sleep and meals has been devoted to the maintenance and advancement of the reputation of my department of this observatory to the neglect of health and other personal interests.' He pleaded, moreover, that as Holden had in February assigned to him the photography of the entire Milky Way with the (still uncompleted) Crocker photographic telescope, this would eventually make even greater inroads on his time:

> The charge of that observatory does not mean as only so in name. It means that I shall do unassisted every portion of the work of exposing and developing the plates and of reproducing them. This will require long and physically exhaustive work both night and day. Indeed the exposures and developments of the Milky Way pictures are the most trying that can be made. The importance of this work is clearly seen when the photographs which I made in 1889 with the rudest equipment stand to day as the only photographs in existence showing the true Milky Way, notwithstanding the efforts to duplicate that work at Lord Rosse's Observatory and at Sydney, where they have failed entirely to show the clouded structures. The work therefore is of the highest importance and the success of the pictures of 1889 is guarantee that it will be well done.[70]

Holden was unmoved, and after itemizing all of the work required to keep the observatory running in addition to the 'astronomical work of observation & discovery,' he chided Barnard by claiming: 'You have performed no such routine duties . . . for 3 years.' Campbell, with four nights on the great telescope, could not take them all over, as they had proved too much for Keeler 'even with only two observing nights with the telescope, per week.' And yet of all of Keeler's duties Holden had given to Barnard only 'the *easiest*, *simplest* & *pleasantest* – namely the Time Service.' Writing for the record, Holden concluded: 'I am sure that you do not intend it, but the effect of your letter is to ask that others here shall do certain work that you do not like, in order that you may have leisure to do what you do like. It seems to me to be just, that each one of us here should take his share of the less agreeable duties, along with the others. I hope that you will see the force of this & I am sure that if you systematize your time you will not find the care of the Time-Service a burden.'[71]

Needless to say, Barnard did not see the force of this at all. 'It is not for me to question whether Mr. Keeler had too much work to do or not,' he insisted. '. . . It is *my* work alone that enters into the question. I maintain that I have as much as I can do and that the additional work of the Photographic Observatory in which I am specially interested, will take every possible moment of my time':

> I am simply surprised that you should refer to [the number of nights of observation]. Two nights – or four, indeed, in a week, are little enough. I think it will be generally conceded that I observe *seven* nights out of the week, and these do not close at 10 or 12 o'clock but continue throughout the entire night when the conditions are favorable.[72]

This was a line of argument that he could not lose. There could be no doubting that he was a workaholic and always driving himself unsparingly. On Mt. Hamilton there was no Bishop McTyeire to counsel him, 'Rest, rest.' His typical schedule at this time was to arrive at his office in the observatory by 9 a.m. each day, notwithstanding the fact that much of the time he observed steadily from dark until daylight. In a draft of his letter to Holden which he later rejected, he noted: 'I am at the observatory as a rule at least an hour or two earlier than you are. It is a very moderate estimate to say that on an average I am actually in the observatory at least sixteen hours a day and this has been the case for the past three years. The other astronomers here are aware of this fact and have repeatedly protested against it on the ground that it was injuring my health.'[73] He dropped the reference to Holden's hours in the draft actually submitted, but still claimed that, given how hard he was already pushing himself, it was 'physically impossible' for him to do the work of the time service. 'My health is already endangered with the overwork and exertion of the past three years – I have been repeatedly warned by the astronomers here that I was doing too much – and that I should break down physically.'[74]

A few days later this particular struggle between the director and the junior astronomer was resolved when the Regents authorized the hiring of another permanent staff member. Holden cooly notified Barnard: 'This will make it possible to relieve you from the duties of the time-service.'[75] Though Holden's first choice for this additional post was the English astronomer E. W. Maunder, a brilliant spectroscopist at the Greenwich Observatory, Maunder declined when he found out the salary, and instead

Holden hired young Henry Crew, who had studied under Charles A. Young at Princeton and Henry Rowland at Johns Hopkins and had gone on to take charge of the Physics Department of Haverford College, near Philadelphia. He was the first Lick astronomer who had earned a Ph.D. Crew arrived late that summer. However, unlike the diplomatic Keeler, who during his tenure had managed to stay clear of controversy with the director, and also Campbell, who for a long time defended him against detractors, Crew was soon disillusioned and joined the battle on Barnard and Burnham's side. They were, he wrote to Young, 'the two men here whose moral sensibilities are not blunted,' while he considered the director a 'Dictator' and a 'charlatan.'[76] His animosity toward Holden naturally endeared him to Barnard, and the two men became friends and confidants.

9

Barnard's scientific interests still largely centered, as they had since the opening of the observatory, on comets − not only in looking for them, but also in making careful positional and physical observations of them once found. His thoroughness in this regard is well attested by the fact that throughout 1890, he continued to pursue doggedly comets 1889 I and II out to unprecedented distances from the Sun. This success led him to propose that some of the periodic comets might possibly be followed all the way around their orbits.[77]

He also gave much attention to the planets. He was, for instance, the only member of the Lick staff to carefully observe Saturn during the Earth's passage through the ring-plane in October 1891. Most of his observations of the event, which occurs only at intervals of fifteen years, were made with the 12-inch refractor, but on two occasions, through the courtesy of Burnham, he was able to use the 36-inch. Even with the larger instrument, however, he found that the rings were invisible when they were exactly edgewise to the Earth − he was able to convince himself that what seemed to be occasional glimpses of them were only after-images of the dark shadow of the rings on the globe.[78] These observations provided compelling evidence of the rings' exquisite thinness.

No less important was his careful study of Jupiter and its satellites. In 1890 he observed the planet on forty-nine nights with the 12-inch refractor, and on at least as many nights in 1891.[79] The planet, which had been the chief object of his early interest in Nashville, was just then entering a highly active phase. One feature which attracted a good deal of attention was what Barnard described as 'a new oblong red spot' − actually, a prominent dark section of the South Temperate Belt located just to the south of the Great Red Spot. This feature was first observed by him in August 1890, and by October 1891 had become 'by far the most conspicuous and striking object on the planet,' strongly reminding him, in color and intensity, of the Great Red Spot in 1880. Soon after it reached its peak, it began to fade rapidly, and on December 14, 1891, no trace of it could be made out. During the same period, the Great Red Spot was changing from an inconspicuous feature to one 'strongly marked and quite red.' Also in 1891, Barnard observed an outbreak of rapidly moving dark spots in the belt just north of the North Equatorial Belt, which appeared to be an 'exact repetition' of those that had appeared in the same latitude in 1880.[80]

No one can criticize Barnard's observations, and even today they are of great value, but it must be admitted that some of his ideas about the nature of the Jovian surface were rather peculiar. He rejected Keeler's correct view that the Jovian features were simply cloud-forms in the planet's atmosphere. He also disagreed with the then leading authority on the planet, Professor G. W. Hough of the Dearborn Observatory, who maintained that the surface was in the liquid state. Instead Barnard firmly believed that the Jovian surface was in 'a plastic or pasty condition, the belts and markings being merely discolorations in them, due to internal eruptions.'[81] Though Barnard himself found this idea a perfectly 'natural one,' no one else seems to have so regarded it – this particular theory was all but ignored by other astronomers. Nevertheless, Barnard apparently held to it for the rest of his life.[82]

Soon after the great telescope had gone into operation, Holden had noted that 'the satellites of Jupiter, which ordinarily appear as bright points in the telescope, were seen to be bright discs and gave promise of good material for investigation in the future.'[83] The first observations to show anything of real interest, however, were made by Barnard, and not with the great telescope but with the modest 12-inch. On September 8, 1890, he found the first and innermost satellite, Io, nearing mid-transit across the planet's disk. (Barnard himself always referred only to 'first satellite,' 'second satellite,' 'third satellite,' 'fourth satellite,' but to avoid confusion, we will wherever possible use the now-official proper names – thus Io, Europa, Ganymede, Callisto.) With 'as perfect seeing as we have ever had on the mountain' and applying high magnifications of × 500 and × 700 on the 12-inch, he recorded in his observing notebook that Io appeared as a 'pale dusky spot . . . decidedly elongated north and south.' Another look and he convinced himself that 'it is double, I think I distinctly see a line of separation of two equal pale dusky Spots.'[84] He then went to get Burnham to come and have a look. The renowned double-star observer, Barnard noted, 'clearly saw the phenomenon of apparent duplicity and had no hesitation as to the appearance.' Unfortunately, as never quite seemed to fail whenever something interesting turned up, it was a Saturday night, and the two friends were interrupted after only about fifteen minutes by 'a big party of visitors,' who were escorted into the dome by Holden. By the time they had all been shown 'Jupiter and the Milky Way,' Io had glided off the disk.

That night Barnard jotted into his observing-book two possible hypotheses to explain the strange double appearance he had witnessed: 'The satellite is either double or [has] a white belt on it parallel to Jupiter's belts.'[85] In his first published account, in *Astronomische Nachrichten*, he gave equal weight to these alternatives,[86] but privately he leaned toward the double satellite theory. His curiosity was so piqued that he even asked the formidable Simon Newcomb for his opinion – the first time he had dared to consult the master celestial mechanician since they met in Nashville in 1877.[87] Newcomb replied, 'I have to say that I do not see any impossibility in the [double satellite] hypothesis,'[88] but suggested that more observations were needed.

Heedless of the consequences, Barnard fired off a letter to *Astronomische Nachrichten* requesting observations of Io in transit by astronomers with 'powerful telescopes.'[89] As soon as Holden read it, he was irate – after all, the Lick Observatory itself had *the* most powerful telescope in the world, and in a memo he pointed out to Barnard that of course such results could easily be obtained at Lick itself, with the 36-inch. For the records, he reiterated (as he had on so many previous occasions) that he had always supported Barnard's requests for time on the great telescope whenever he had special observations to make, provided only that the photographic corrector lens or spectroscope were not mounted on the telescope at the time it was needed.[90]

Barnard, rather disingenuously, suggested that his letter had not been meant as an attack on Holden's administration but merely as an attempt to mobilize European observers to pay attention to transits which would be invisible in the United States. He had fully expected to take care of 'this end' of the observations himself. 'It was my intention should one of these transits occur on anyone of Mr. Burnham's nights with the 36 in. to observe the transit with his permission and with him, since he has always shown the utmost kindness in permitting me to observe any important event with that telescope on any one of his nights.' But since Holden had mentioned it, 'I would suggest that the observation of Jupiter's first satellite is not the only important work that I should be able to do with the great telescope . . . It would, it appears to me, be simply a matter of justice that one night per week with that instrument be given to me.'[91] Holden remained intransigent. As director he alone had the responsibility for assigning observing time according to his best judgement of what was important, and he told Barnard that he was quite satisfied to leave things as they were. He chided Barnard, moreover, with the fact that in refusing to take over the time service, he had insisted that he had more than enough to do already. That being the case, he would certainly not have time to carry out additional programs with the great refractor.[92] As Osterbrock has pointed out,[93] Holden knew that Barnard simply wanted to transfer part of his work from the 12-inch to the 36-inch and to utilize the larger instrument for investigations that were beyond the reach of the 12-inch, and Barnard remonstrated: 'This would advance . . . my work – the Time Service would be fatal to it.'[94]

Since Barnard could not get the great telescope on any other nights 'without begging Prof. Holden for it,'[95] which he refused to do, he observed only those transits of Io that happened to fall on Burnham's nights. The first of these was on August 3, 1891. (That same night, incidentally, Io was being studied with the 12-inch refractor by the brilliant physicist Albert A. Michelson, of the Case Institute of Cleveland, who at Holden's invitation had come to Mt. Hamilton in order to test his new and ingenious interferometer by using it to measure the diameters of the Galilean satellites' tiny disks.)[96] A single view of Io with the 36-inch and Barnard began to grasp the true explanation of the strange appearance he had seen the previous September. Whereas his first observation had been made when Io had appeared dark in contrast with a bright zone of Jovian clouds across which it was then transiting, he now caught the satellite crossing the planet's dark South Equatorial Belt and it appeared as an 'elongated white

spot.'[97] As soon as he saw this he realized that if his surmise that the satellite had a
bright equatorial belt was indeed correct,

> when the satellite crossed a bright portion of the planet [as in September 1890] the
> white belt would cut it apparently in two, as it would be equal in brightness to the
> surface of Jupiter, and thus leave the two dark polar caps as two separate spots . . .
> and would thus give the observed appearance of duplicity. If, however, the satellite
> should happen to be projected on a dark belt, then the dark poles would merge into
> the surface of Jupiter, and the white equatorial belt alone would be visible as an
> elongated white spot . . .[98]

Though Barnard had had a flash of insight, he did not make up his mind completely
until the fall of 1893. Only then, he said, did an opportunity occur to settle the question
once and for all. Using the 36-inch with magnifications of up to × 1000, he found that
Io 'presented a beautiful appearance,' standing out in 'bold relief like a little globe. The
polar caps were heavily marked and quite dark, while the bright belt was very
conspicuous. The observation was perfectly satisfactory, and the second theory had
become a fact.'[99] He later made the point even more emphatically: 'There is absolutely
no question,' he wrote in 1897, 'as to the correctness of this explanation . . . This matter
I consider settled for all time.'[100] Parenthetically, Barnard's observations have indeed
been shown, by the Voyager spacecraft cameras, to be essentially correct: Io's mottled
disk does indeed, if viewed somewhat out of focus, show a bright equatorial belt and
dusky polar caps, just as Barnard claimed.[101]

Barnard's work on Io stimulated fresh interest in the Galilean satellites. Over the
next several years astronomers competed with one another in their observations of
them, in the attempt to prove the superiority of their vision, telescopes, or atmospheric
conditions. As we shall see, some, such as Barnard himself, were able to make out only
vague shadings; others found the satellites egg-shaped or crisscrossed with strange
linear markings – in the process straining not only their eyesight but credulity to the
limit.

10

As Barnard worked on, productive but unhappy, at Mt. Hamilton, his thoughts in the
dim dome with only the stars shining on him through the narrow slit must often have
wandered back to happier days in Nashville, when he had enjoyed the adulation of
friends and supporters who had cheered each of his discoveries. How different Holden
had proved to be from the kindly McTyeire, the noble Chancellor Garland, the easy
Poole – indeed from his own expectations of him. But how far away all of that seemed
now – like a dream!

Late one clear night, when he was in the dome as usual, he heard music coming
apparently from the sky. He was at a loss to account for this but soon found that the
music was coming up the mountainside and was reflected by the dome. A villager from

the valley had gotten drunk, and had brought his organette up to serenade the dwellers on the peak. For some time he played such classics as 'Annie Rooney' and 'Sweet Marie' to their drowsy ears. When he tired of this and was starting back down the mountain, Barnard said to him, 'Have you got "Dixie" '? 'Oh yes,' he had 'Dixie.' 'Will you play it for me as you go down?' And so, as Barnard afterward recalled, 'I stood there on the peak, listening to "Dixie" as it floated up the mountainside growing fainter and fainter, and when I went back into the dome there was "Dixie" coming down to me from the stars.'[102]

1   EEB, observing notebook; LO

2   E. S. Holden, 'Address of the retiring President of the Society, at the Second Annual Meeting (March 29, 1890),' *PASP*, **2** (1890), 50–68:54–7

3   Simon Newcomb, *The Reminiscences of an Astronomer* (Boston and New York, Houghton, Mifflin & Co., 1903), p. 194

4   Herman Melville, 'Billy Budd, Sailor,' in S. Bradley, R. C. Beatty, and E. H. Long, eds., *The American Tradition in Literature*, (New York, Grosset & Dunlap, Inc., 3rd ed., 1967), vol. 1, pp. 1015–16

5   D. E. Osterbrock, 'The Rise and Fall of Edward S. Holden,' *JHA*, **15** (1984), 81–127:95

6   EEB, 'charges against Holden'; undated MS in VUA

7   Sir William Huggins to GEH, December 10, 1892; YOA

8   D. E. Osterbrock, quoted in Gerrit L. Verschuur, *Interstellar Matters: Essays on Curiosity and Astronomical Discovery*, (New York, Springer-Verlag, 1989), p. 46

9   The reference to Holden as a 'Pecksniff' is found in EEB, 'Charges against Holden,' but Barnard was probably echoing Burnham. Seth Pecksniff was a character with the characteristics described in Dickens's *Martin Chuzzlewit*, and Burnham was notorious for spicing up his conversation and letters with Dickensian phrases. As Barnard wrote in his biographical essay 'Sherburne Wesley Burnham,' *PA*, **29** (1921), 316: 'He was a lover of Dickens and sometimes, when the humor seized him, liked to refer to Mr. Pecksniff or Mrs. Sairy Gamp.'

10 EEB, 'notes for Young's letter,' July 5, 1892; SLO

11 Springfield *Republican*, November 1, 1888

12 Atlanta *Constitution*; undated clipping in SLO. In pasting this item into the observatory's clippings book, Hill had written: 'May the Gods of our fathers defend us from any more like this.'

13 E. S. Holden, 'The Lunar Crater and Rill – *Hyginus*,' *PASP*, **2** (1890), 14–15

14 Many of the craterlets and rills which Weinek 'discovered' were in reality nothing more than irregularities in the gelatin of the negatives. See E. E. Both, *A History of Lunar Studies* (Buffalo, New York, Buffalo Museum of Science, 1961), 25–26

15 EEB, 'notes for Young's letter,' July 5, 1892; SLO

16 EEB, 'charges against Holden,' op. cit.

17 ibid.

18 EEB, 'Translation of E.S.H.'s letter and interpretation of its meaning,' with reference to ESH to EEB, June 10, 1890; VUA

19   ESH to EEB, June 10, 1890; VUA
20   ibid.
21   Note on ESH to EEB., June 10, 1890, in Barnard's handwriting; VUA
22   ibid.
23   EEB to TGP, October 31, 1891; VUA
24   EEB, 'charges against Holden,' op. cit.
25   EEB to the Lick Board of Trust, undated manuscript from *ca* 1894; VUA
26   R. G. Aitken, 'Barnard at the Lick Observatory,' *JTAS*, 3:1 (1928), 20.
27   Osterbrock, 'The Rise and Fall of Edward S. Holden,' p. 95
28   SWB to CAY, July 12, 1891; SLO
29   C. B. Hill to GD, July 20, 1890; BL
30   ESH to EEB, August 1, 1890; SLO
31   SWB and JMS to ESH, September 16, 1890; SLO
32   E. S. Holden, 'Astronomical Photography at the Lick Observatory,' *PASP*, 2 (1890),
      152–9
33   ESH to A. M. Clerke, August 8, 1889; SLO
34   EEB, notes on memo to ESH, August 16, 1890; VUA
35   E. E. Barnard, 'On the Photographs of the Milky Way made at the Lick Observatory in
      1889,' *PASP*, 2 (1890), 240–4
36   Osterbrock, 'The Rise and Fall of Edward S. Holden,' p. 97
37   ibid.
38   EEB, notes on memo to ESH, August 16, 1890; VUA
39   ESH to EEB, September 25, 1890; VUA
40   ESH to EEB, note on a letter from F.L. Vandenburgh to ESH January 6, 1891; VUA
41   EEB to ESH, January 7 to 13, 1891; SLO
42   ESH to J. H. C. Bonté, January 10, 1891; SLO
43   Osterbrock, 'The Rise and Fall of Edward S. Holden,' p. 98
44   EEB to ESH, January 16, 1891; SLO
45   ESH to EEB, January 17, 1891; VUA
46   Note in EEB's handwriting on letter from ESH to EEB, January 17, 1891; VUA
47   EEB to ESH, January 17, 1891; SLO
48   EEB to ESH, January 28, 1891; VUA
49   ESH to E. B. Knobel, June 16, 1893; YOA. This letter was sent by Knobel to Ernest D.
      Burton, president of the University of Chicago, on July 14, 1923. As pointed out to me
      by Donald E. Osterbrock, who called these letters to my attention, though Holden
      thought Barnard's claim that he could tell whether comets were periodic or not by
      appearance alone farfetched, it is not as crazy as Holden thought. Comets with
      eccentricity near 1 – so-called 'new comets' – are brighter and decay more rapidly in
      brightness than 'old' ones, presumably an effect of first passage close to the Sun. D. E.
      Osterbrock to W. Sheehan, personal communication, October 8, 1993.
50   ESH, memorandum, December 10, 1890; SLO
51   E. E. Barnard, 'On the Rediscovery of D'Arrest's Comet,' *AJ*, **10** (1890), 92
52   ibid.
53   EEB, draft of letter to the Donahue Comet Committee, November 14, 1890; VUA
54   Barnard, 'On the Rediscovery of D'Arrest's Comet,' p. 92. Under these circumstances,
      he was led to consider 'the important question of the condition of the comet's light'

during the period in question. A byproduct of his search, incidentally, was that he discovered 'quite a number of new nebulas.'

55   ESH, memorandum, December 10, 1890; SLO

56   C. Burckhalter to ESH, February 13, 1891; SLO

57   H. D. Curtis, 'The Comet-Seeker Hoax,' *PA*, 46 (1938), 71–5 is a valuable account of the whole affair by an astronomer who discussed it with Barnard.

58   E. E. Barnard, 'Sherburne Wesley Burnham,' *PA*, **29** (1921), 309–24:323. There is more to the story, however, for Barnard adds: 'In his work he frequently ran upon new nebulae near some double star or while hunting for new double stars. Notwithstanding this expressed fear he always measured the nebula and went back to it again to see if it really after all was not a comet!' Mary Lea Shane recalled the 'tale' – and unfortunately, it seems to have been only that – 'that one morning at the mess Burnham mentioned that while he had been observing double stars the previous night, he had seen a comet in the field. He hadn't been sufficiently interested even to note the field. Barnard was said to have checked over the whole list of stars Burnham had observed in a fruitless effort to locate it the next night.' Mary Lea Shane, notes for Nicholas U. Mayall's talk, May 13, 1957, to Barnard Astronomical Club at Dyer Observatory in Nashville; SLO

59   Curtis, 'Comet-Seeker Hoax,' p. 74

60   ibid., p. 75

61   C. B. Hill to EEB, March 18, 1891; SLO

62   LS to EEB, March 22, 1891; VUA

63   San Francisco *Examiner*, February 5, 1893. Together with this apology, the *Examiner* also ran an article by Barnard himself, 'How to Find Comets,' on the true methods of searching for them.

64   Barnard and R. Spitaler of Vienna independently recovered Comet Wolf on its first return since its discovery in 1884. The new comet Barnard found on March 30 was independently found by W. F. Denning at Bristol the following morning and became known as Comet Barnard–Denning. Denning also made an independent recovery of Comet Tempel–Swift two days after Barnard first saw it on September 28.

65   Summarizing Donald E. Osterbrock, *James E. Keeler: Pioneer American astrophysicist* (Cambridge, Cambridge University Press, 1987), 107–16

66   'The Trouble on Mt. Hamilton,' San Francisco *Examiner*, August 9, 1892, and also various earlier accounts.

67   San Francisco *Examiner*, August 27, 1891

68   J. E. Keeler, 'Note from Professor Keeler,' *AA*, 11 (1892), 840

69   ESH to EEB, May 30, 1891; SLO

70   EEB to ESH, June 4, 1891; SLO

71   ESH to EEB, June 6, 1891; SLO

72   EEB to ESH, June 10, 1891; SLO

73   EEB to ESH, June 10, 1891; draft of letter, VUA

74   EEB to ESH, June 10, 1891; SLO

75   ESH to EEB, June 12, 1891; SLO. Barnard, however, was still not satisfied until Holden had agreed to copy all of the long self-justifying letters he had written on the subject into the observatory records, and he also made Holden promise that if at any time he sent his side of the correspondence to the Regents he would submit Barnard's along with it. EEB to ESH, June 13, 1891; SLO

76   HC to CAY, December 10, 1891; DCL

77   E. E. Barnard, 'Comets 1889 I and II, and some suggestions as to the possibility of seeing the short-period comets at aphelion,' *AJ*, **10** (1890), 67–8. Not until 1913 would Barnard see this expectation realized. That year, at his suggestion, an ephemeris for Encke's comet was prepared by Frank E. Seagrave, which led to its being successfully photographed at aphelion with the 60-inch reflector on Mt. Wilson. See Gary W. Kronk, *Comets: A Descriptive Catalog* (Hillside, NJ, Enslow Publishers, 1984), p. 245.

78   EEB, observing notebook; LO. His entry for October 26 reads: 'No trace of ring, though I thought once or twice I could catch glimpses of it. This however I am sure was simply persistence of vision and only a prolongation of the shadow and a reversal, by looking steadily at the shadow a few seconds before looking at the ring.' See also E. E. Barnard, 'Observations of the Reappearance of the Rings of Saturn; Observations of the Position-Angles of the Rings, and Observations of the Satellites,' *MNRAS*, **52** (1892), 419–23.

79   E. E. Barnard, 'Observations of the Planet Jupiter and his Satellites during 1890 with the 12-inch Equatoreal of the Lick Observatory,' *MNRAS*, **51** (1891) 543–55; EEB, observing notebook, LO

80   E. E. Barnard, 'Disappearance of the New Red Spot on Jupiter; the Great Red Spot and Other Jovian Phenomena,' *AA*, **1** (1892), 93

81   E. E. Barnard, 'Observations of the Spots and Markings on the Planet Jupiter, Made with the 12-inch Equatoreal of the Lick Observatory,' *MNRAS*, **52** (1892) 7–16:16

82   See, for example, E. E. Barnard, 'The Planet Jupiter,' *The Monthly Evening Sky Map*, **13** (1919): 'Personally, the writer does not believe that the surface we see is exactly a cloud surface. Under the very finest conditions, with the most powerful telescopes, it does not look or act like clouds. The appearance is more that of a pasty nature. One who is only familiar with Jupiter in a moderate telescope is more apt to hold to the cloud theory than he who has studied the planet under the finest conditions with one of the great telescopes of today. To the latter the "pasty" theory seems to be the true one.'

83   *Common Advertiser* (New York), November 1888; clippings book, SLO

84   EEB, observing notebook; LO

85   ibid.

86   E. E. Barnard, 'Apparent duplicity of the first satellite of Jupiter,' *AN*, **125** (1890), 317. Barnard's brief note reads: 'On the night of Sept. 8 while observing Jupiter with the 12 inch, a remarkable phenomenon was witnessed. Satellite I, which frequently transits as a dusky spot, was seen on the disk crossing the bright equatorial region as a pale gray spot. Upon applying high powers (500 and 700 diameters), and with as perfect seeing as we have ever had on the mountain, the satellite distinctly appeared double, the apparent components being in a line nearly vertical to the belts of Jupiter. A line of light was occasionally distinctly seen separating the satellite into two nearly equal parts.

       'A white belt on the satellite parallel to the belts of Jupiter would perhaps satisfactorily explain the phenomenon. If this is not the true explanation, there is no alternative but to consider the satellite actually double.'

87   EEB to SN, April 8, 1891; VUA

88   SN to EEB, April 17, 1891; VUA. See also Barnard's follow-up letter, April 27, 1891, also in VUA: 'I feel more confidently that the first satellite is double.' Newcomb had

asked whether Barnard was the same fellow he remembered having met in Nashville in 1877, and Barnard took great satisfaction in explaining that he had taken the advice Newcomb offered on that occasion to heart and had fulfilled it to the best of his ability: 'I have endeavored all along to do the best and most accurate work that my circumstances would allow. Through following your suggestion I have succeeded in being the first discoverer of fourteen new comets, the second independent discoverer of one (by Brooks Dec. 26, 1885 and by me Dec. 27 before its announcement by Brooks,) and the rediscovery of d'Arrest's comet last year, which happened to be discovered while searching for new comets after the search for it was abandoned.' And he took particular satisfaction in adding that 'the very man whom you held up to me as an example – Mr. Burnham – has become the very best friend that I have.'

89   E. E. Barnard, 'Request for Observations of the First Satellite of Jupiter at its Transit during 1891,' *AN*, **128** (1891). 45. Barnard's letter to *Astronomische Nachrichten* is dated July 9, 1891

90   ESH to EEB, July 11, 1891; SLO

91   EEB to ESH, June [really July] 12, 1891; SLO. He insisted, however, that the one night not be a Saturday, since that night was 'at best but a fragment because of the visitors, and . . . I would scarcely be able to utilize [it] to any advantage, through sheer exhaustion with the entertainment of the regular visitors in the first half of the night.'

92   ESH to EEB, July 13, 1891; SLO

93   Osterbrock, 'The Rise and Fall of Edward S. Holden,' p. 99

94   EEB to ESH, July 13, 1891; SLO

95   EEB, memorandum, September 28, 1891; VUA

96   A. A. Michelson, 'Measurement of *Jupiter's* Satellites by Interference,' *PASP*, 3 (1891), 274–8

97   E. E. Barnard, 'Note on the First Satellite of Jupiter,' *MNRAS*, **51** (1891)

98   E. E. Barnard, 'On the Dark Poles and Bright Equatorial Belt of the First Satellite of Jupiter,' *MNRAS*, **54** (1894), 134–6: 134

99   ibid., p. 135

100 E. E. Barnard, 'On the Third and Fourth Satellites of Jupiter,' *AN*, **144** (1897), 321–30

101 As noted in Roger F. Griffin, 'Barnard and his Observations of Io,' *Sky and Telescope*, **64** (1982), 428–9

102 B. E. Young, *Vanderbilt Observer* (March 1899)

12

'I am tired here'

1

From the opening of the Lick Observatory in June 1888 until January 1892, Barnard had not had – nor asked for – a regular vacation. Exhausted from his strenuous observing schedule and from the draining struggle with Holden, he realized that it was high time, and applied to then acting president of the University of California, Martin Kellogg, for a two months leave of absence. Though he desperately needed the escape, there were also domestic affairs to attend to. He was by now receiving a salary of $1800 a year, and by practicing the usual economy, he and Rhoda had managed to scrape together enough money to purchase twelve acres on the Mt. Hamilton Road, 'five miles this side of San Jose.' The land was near what later became Alum Rock Avenue. There, he told Kellogg, he was eager to plant fruit trees, and for the sake of economy planned to personally supervise the work and do a portion of it himself.[1] For one who throughout his boyhood had known only poverty and a kind of unrooted, vagabond existence, his purchase of this property is touching and recalls his statement on another occasion: 'I had always longed for such a home where one could plant trees and watch them grow up and call them our own.'[2] Rhoda, incidentally, would henceforth spend much of her time tending the 'ranch,' especially in the long, hot, dry summers, when conditions in the small cottage on Mt. Hamilton where the Barnards were still living became very close and uncomfortable.

In taking his first leave from the observatory, Barnard told Kellogg that he did not wish to spend the entire two months away from his observing. 'There are long intervals of cloudy or poor observing weather in the winter when no observing can be done. I wish to be at liberty to go and come where necessity requires it [and] by this means I can continue my regular observing without as much loss to it as if I were absent during the two months.' The notes in his observing-book show that, though gone the greater part of the time, he did make a few sporadic observations between then and his return to his full-time duties in mid-March.

One of the first objects of intensive study following his return was a new comet (1892 I) just discovered by his old friend Lewis Swift. The weather conditions did not cooperate until April 4, by which time Swift's comet had a 3rd magnitude head and had developed an impressive tail, 'fully twenty degrees long and straight and slender.' On examination with the 12-inch refractor, Barnard found that the tail consisted of 'two branches, well defined on their outside edges.'[3]

On the morning of April 5, Barnard used the Willard lens – not yet mounted in its dome on Huygens Peak, but as in the early days strapped onto the $6\frac{1}{2}$-inch equatorial – to photograph the comet. Though for most of the hour-long exposure the sky was covered with haze and clouds, the resulting image 'showed a remarkable state of affairs,' Barnard wrote. 'There were now three main branches to the tail – a new one having sprung out between the two which were seen on the previous morning.' A photograph taken on the morning of April 6, in a break in the clouds, showed that the short northern branch had disappeared, and the two remaining branches had 'blended together more or less to form a single flat train very narrow where it joined the head.'[4]

At this time, Barnard was going down one day a week to San Jose for a series of lectures he was giving to the University Extension Club. Unlike Keeler, who had always resented the interruption of his research caused by visitors' nights on Mt. Hamilton, the young Tennessean had always made it 'my pride to exert myself to the utmost to entertain the visitors.'[5] The same attitude extended to his public lectures, for which he was always in great demand – not least, perhaps, because he scrupulously refused to accept any compensation. As a lecturer he was, as the Sacramento *Daily Union* described him after an earlier lecture, 'a young man, of quiet presence, sensitive almost to the point of timidity [who spoke] deliberately, in a conversational manner, with deep earnestness and without any effort at elocutionary effects. He has a vein of humor and it crops out . . . here and there.'[6] On April 6, Barnard went down at noon to San Jose to give his evening lecture. For a hardbitten observer who at the same time took his public service responsibilities very seriously, the situation created for him an agonizing dilemma:

> Every morning's picture [of Swift's Comet] increased the interest and importance of the work . . . I did not want to disappoint the people, and I certainly could not let the comet go by unphotographed. San Jose was nearly a mile below us in vertical height, and twenty-seven miles distant by stage road. The only possible way for me to secure the photograph and not disappoint my audience was to return to Mount Hamilton that night after the lecture. At ten o'clock I hired a horse and buggy in San Jose and drove up that lonely mountain road, the journey taking five hours, and arrived at the summit at three o'clock in the morning, in time to make the photograph of the comet . . . I must say that a good many thrills passed over me during that lonely mountain ride in the dead of night – some for the chance that I might drive over into a canyon to death, and others for the possible interruption of my terrestrial existence through an encounter with some hungry, roaming mountain lion. In the main the journey was a most impressive one – alone in the mountains, with only the horse in front and my friends, the stars, above me. I doubt if my courage had not failed me entirely if the friendly stars had not encouraged me with their presence.[7]

The comet was indeed showing the most extraordinary changes. The photograph taken on April 7, which Barnard called 'the most successful of the series,' showed an abrupt bend at one point in the tail that had not been there the previous morning, 'as if its current was deflected by some obstacle.' In other places the tail reminded him of

Comet Swift (1892 I) photographed by Barnard on April 7, 1892. UCO-Lick Observatory

'crumpled silk.'[8] The photographs Barnard obtained of this object, together with others by W. H. Pickering at Arequipa, Peru, were historic in being the first to show the extremely rapid transformations that take place in the tails of comets, which henceforth became one of Barnard's most important areas of research.

Soon after Barnard finished photographing Swift's comet, the Willard lens, refigured by Brashear, was placed on its own mounting, and set up permanently in its own dome on Huygens Peak. The improvements, which were paid for by Colonel Crocker, were modest enough. The lens, in its wooden box, was fastened to the declination axis of an ordinary equatorial by means of a flat iron plate, and Barnard personally volunteered the small telescope he had built long before with Braid for the guide telescope. He triumphantly entered in his observing notebook on June 18, after making a four-hour exposure centered (as his first successful image of three years earlier had been) on the small 'black hole' in Sagittarius: 'Result fine . . . More details shown than in 1889.'[9]

2

That same month, Holden was faced with yet another crisis when Burnham resigned.[10] Actually, the wiry double star observer had been considering leaving for sometime, but as he later explained to C. A. Young: 'The only means I [had] for enduring the life on Mt. H. so long [was] that I could not afford to quit without something definite in view . . . I hesitated about leaving because I knew that such a step meant the abandonment of all my astronomical work so far as the telescope is concerned and as I was more interested in it than I was before, it was pretty hard for me to think of such a thing, but it had to be. Life is not long enough to be abridged as it would be . . . by living in such a situation.'[11] Something definite had finally come into view when he had received the unexpected offer of a position of clerk of the United States District and Circuit Courts of Chicago, which had become vacant through the illness of the man that had been occupying the position, and Burnham accepted at once.

Burnham reported to Barnard what transpired when he turned in his resignation to the director: 'I . . . went into his room to have a talk – that I regretted leaving very much, but that things were getting from bad to worse with no hope . . . of any change for the better . . . He positively refused . . . to talk about the matter at all in any way . . . I retired, and that is the end of it. Of course you understand that his game will be as it has always been so many times . . . finally to get some comparatively unknown person here who can be made a tool of.' Burnham predicted that Holden would soon succeed in driving Barnard out as well and that the Lick Observatory would before long become a 'fourth class place':

> It is too bad and it is no less serious now that I have no personal interest in it. I am afraid there is no help for it. The Regents cannot see these things as they are . . . If they could once see how much worse it is than anyone has represented – that one half has not been told, they would save the obs[ervator]y by putting an honest man in charge of it.[12]

Though with his large family, the salary of $7000 a year that Burnham was offered in the clerk's position – twice what he was making at Lick – would have been difficult to pass up under any circumstances, Burnham made it clear to the Regents that even so he would not have left but for his extreme dissatisfaction with the state of affairs on Mt. Hamilton. Barnard at the same time wrote to Colonel Crocker: 'The same cause has embittered my life here and it is simply a matter of time when I too must throw up every interest at this point. This state of affairs is sincerely to be regretted for we are not men of the world. We are prone to submit to many indignities rather than to disturb our minds with the contentions that have existed here, and it is only when the burden is too great that we rise to defend ourselves.'[13]

Burnham's departure was, in retrospect, probably inevitable. From the first he had chafed under Holden's controlling and autocratic regime. Keeler later summed up the situation perfectly in a letter to Campbell: 'Burnham is a poor ship-of-the-line, but a splendid independent cruiser.'[14] His first conflict with Holden had been over the submission of articles in his own name, and later he had quarreled bitterly with the director about the form in which he had wished to publish his double star observations. An exasperated Holden had written to Keeler: 'Vol. II [of the Lick Observatory Publications] in its original shape was abandoned because Burnham did not wish *his* observations to be printed in the same volume with anyone else's . . . For the sake of peace (which I did not get, by the way!) I consented to the change.'[15]

Naturally the resignation of the celebrated observer did not escape comment in the local newspapers. An editorial, signed only 'Astronomer' but most likely penned by Hill, ran in the San Francisco *Examiner* under the headline 'Misery on Mount Hamilton.' It asked plaintively: 'What is the matter with the Lick Observatory? Why is it that peace does not dwell upon the mountain in that eyrie of science, but only discord continually? How comes it that no man of first-rate ability can long continue to serve there, and that one after another they depart, declaring that life is made intolerable there?' The editorial went on to predict that Burnham's resignation would soon be followed by those of other valuable members of the staff, since 'it appears to be quite impossible for a man of independent mind and high reputation to devote his whole mind to his scientific work on Mount Hamilton.'[16] It was generally expected that Barnard would be next to go.

3

Barnard confided to Crocker that he was 'almost heartbroken' by the departure of his closest friend. But he was not paralyzed. He immediately seized the opportunity to apply for Burnham's two nights on the great telescope. In a letter designed for the record, he wrote to Holden:

> It is impossible for me to do anything like justice to myself and to the Lick Observatory without the use of the 36-inch equatorial. You will remember that I have been refused the use of this instrument except upon special occasions and the work then done stands as among the finest and most important ever performed with the great telescope.
>
> The most important and valuable of my observations are those which demand the most powerful telescopic aid. The lack of this has been a serious drawback to my labors here . . . I respectfully request that these two nights be assigned to me . . .'[17]

As on countless prior occasions, Holden refused Barnard's request, telling the young Tennessean that Burnham's nights had already been reassigned to Crew and

Schaeberle.[18] He thereby set up a showdown with Barnard before the Regents, to whom Barnard immediately appealed the decision.

In making his appeal, Barnard first drafted an emotionally charged letter to Timothy Guy Phelps, chairman of the Regents' Lick Observatory Committee, which he may or may not have sent (it survives only in draft form): 'I wish to say to you personally that the way I have been treated with reference to the great telescope is disgraceful and shameful. If I had had 1/10 of the time that has absolutely been wasted with it through the instrumentality of the Director, I should have had valuable work to show for it and in all probability discoveries of great importance. So it is now, since Mr. Burnham's resignation, I shall have essentially no chance to work with the telescope, for when Mr. Burnham was here I could, by breaking in on his time, get the use of it on certain important occasions without the mortification of a possible refusal from the Director, who has not failed to mortify me upon every possible occasion.' Though Crew had expressed a willingness to relinquish his additional night in Barnard's favor, even this would not do. 'I want both nights,' he insisted. 'It is time that the great telescope should do something to justify its existence. In the hands of Mr. Burnham it has done noble work . . . In the hands of the Director it has been a failure, and his pure jealousy and spite will not permit me to have a chance.' He hinted that, as he had thrown up 'as bright prospects as ever a young man had to come here simply in the hopes of bettering myself by more powerful means of research,' he might, if he continued to be denied the opportunity to use the great telescope, be forced to seek his fortune elsewhere, and concluded almost hysterically: 'If you hesitate for one moment over the justice of my request and of my ability, I will go with you to the University of California and show you in the astronomical journals of the world, to be found there, that since the opening of this institution the reputation of the Lick Observatory has been maintained before the world mainly by the work and discoveries of Mr. Burnham and myself. I am ready any time to go with you and prove this assertion. I am sick of the continual efforts of the Director to crush me because I have opposed him in the cause of right and Justice. I am sick of the petty jealousy that promp[t]s him to disregard the interests of this observatory and to prostitute honor and fair dealing to the level of a corrupt politician.'[19]

The same day he wrote this draft, Barnard also penned a smoother, more restrained letter of appeal to the full Board of Regents, which was submitted to them through Holden as required by the observatory rules.[20] Burnham, from Chicago, also did his best to shore up Barnard's cause with a long letter of his own to the Regents. 'Don't let Barnard be driven away from the Observatory,' he urged Phelps.[21]

The result of all this activity was that Phelps and Crocker personally came to Mt. Hamilton to interview the divided staff. They decided that Barnard deserved regular observing time on the great telescope, and also recommended that he be given a raise (to $2200 a year; at its meeting a month later the Board approved an even more generous raise, to $2400). Holden, informed of the Regents' decision, sent Barnard a terse memo on June 23: 'Mr. Barnard will please take charge of observations with the 36-inch telescope on Friday nights, for the present.'[22] This brief communication from the director officially acknowledged that, after four years on Mt. Hamilton, Barnard had finally won the opportunity that he had craved for so long.

Despite his partial vindication, Barnard was disappointed. He had wanted both of Burnham's nights on the 36-inch, not one only, and he had also hoped that once the Regents came to Mt. Hamilton they would see for themselves the corruption of Holden's administration and demand his dismissal. However, the inscrutable Schaeberle, though critical of the director and his methods at times, at the crucial moment rallied behind him, as Barnard bitterly recounted to Davidson in the high sounding biblical prose that he often fell into whenever he was working himself up into a self-righteous tirade: 'The very man who has opposed the Director all along saved him at the last moment by indorsing his actions and methods up here. It is a mighty strong soul that can resist the seduction of flattery. I am sick of every thing and every body up here. I am going away. I am too sick to even wag my pen. Fraud is dominant. Falsehood flappeth his wings and croweth in triumph. I have fought the good fight and have lost it.'[23]

Meanwhile, Burnham urged him to 'Fight it through till this fall . . . and then you can see your way clear to something else if worst comes to worst.'[24] But privately he worried about his highstrung friend, confiding to C. A. Young: 'It will not be possible for Barnard to endure much more of things as they have been. He has been breaking down and aging very rapidly. The last three years have done more than 15 years of hard work would in any decent place. He cannot stand such things. It worries and bears much more on him than it would on me. He must get out of there soon or it will be too late.'[25] Crew, also writing to Young, added that the tension on Mt. Hamilton was so great that Barnard's hair had turned prematurely white – a slight exaggeration, inasmuch as, though Barnard's hair would indeed become completely white by the age of forty, at the time Crew wrote this he showed no more than a touch of gray around the temples.[26]

4

Barnard's first regular observing night with the great telescope was Friday, July 1. Burnham urged him: 'Now you have a chance with the 36-inch, I would give Mars some little attention.'[27] But Barnard needed little coaxing. Not only had the red planet been one of his first interests as a budding astronomer, but Mars fever was running epidemic during the summer of 1892, when the planet approached closer to the Earth than it had for years.

Already, at the two previous oppositions, in 1888 and 1890, interest in Mars had been at a fever pitch, and a steady stream of articles had appeared in the newspapers of the day to feed a hungry public the latest news about the planet and its possible inhabitants. Above all, curiosity was piqued by the so-called 'canals' of Mars, linear markings criss-crossing the planet's surface, which had first been described in detail by the Italian astronomer Giovanni Schiaparelli in 1877, though hints of them had already appeared in a few earlier drawings of Rev. W. R. Dawes, Father Angelo Secchi, and even Holden himself. Schiaparelli had also recorded, from 1882 onwards, a curious doubling of some of his canals, so that where one had run its course, there were now two, in parallel courses. This, needless to say, had added even more to their strangeness.

Naturally it was hoped that when the great Lick refractor finally went into operation it would shed some light on these phenomena, but it was not ready in time to take full advantage of the opposition in 1888, when most of the observations were not made until the planet was already very far from the Earth, while in 1890 the results were also disappointing. Holden blamed the severe weather which lasted unusually late into the spring, so that the fine seeing usually found on Mt. Hamilton during the summer did not begin until late July or August, by which time Mars was already receding from the Earth and too low in the west to be well observed. (Barnard, of course, had another explanation for the lost opportunity – Holden's bungling attempts at photography.) At the end of the 1890 opposition, Holden, in a widely reprinted interview with the New York *Herald*, expressed no definite views as to the 'theory that the strange parallel lines on the planet . . . represent great canals which have been constructed by intelligent beings.' To the contrary, he was quoted as saying: 'The markings of reddish yellow have been supposed to be land; the dark have been supposed to be water, and the whitish ones to be snow. It is entirely uncertain.' But this waffling pronouncement from the director of the observatory with the world's largest telescope was at once attacked by the excitable French astronomer Camille Flammarion, who was himself unrestrainedly enthusiastic about the idea of Martian life. Flammarion wrote sarcastically:

> 'I do not know.' That is very simple. One could make the same reply to every imaginable question.
> 'What is the human skull?' 'I do not know.'
> 'What is the nerve system?' 'I do not know.'
> 'What is the origin of man?' 'I do not know.'
> 'How is it that the sun heats us?' 'I do not know.'
> It seems to me that if we were always content with this answer humanity would still be in the age of carved stone and caverns inhabited by rhinoceroses and bears.[30]

Holden nevertheless stood his ground, and in a dignified reply pointed out to the French astronomer that though he and the other Lick astronomers had mapped the canals, 'I am by no means ready to say that [they] are veritably filled with water, nor that they even probably are the work of human hands. If you ask what these dark markings are really, I am obliged to answer that I do not know. If this answer seems inadequate I submit that it is a scientific answer and the best possible under the circumstances.'[31] What Holden was saying was very similar to what Barnard had said in his unpublished 1880 Mars manuscript ('It is quite possible we may have formed entirely erroneous ideas of what we actually see. The greenish patches may not be seas at all, nor the ruddy continents, solid land.') Yet however justified Holden's position, it failed to satisfy anyone, and Burnham later referred to 'the Mt. H. fulminations with regard to Mars' and the director's 'making the most of the occasion for a blow. So far . . . it amounts to nothing, as one would expect.'[32]

As disappointing as the oppositions of 1888 and 1890 had been, the very favorable approach of Mars in August 1892 offered the hope of something more definite from the world's largest telescope. But as in 1890, the planet's far southerly declination made it difficult to observe under optimal conditions. Barnard found a magnification of only 260 on the 36-inch was best for showing the details, and only 175 on the 12-inch. As Holden pointed out, 'How unfavorable the circumstances have been can be estimated when it is remembered that powers of 1000 and even more have been employed on Jupiter and Saturn with good result.'[33]

Under these circumstances, Barnard made only a few drawings of Mars. In publishing them he noted, 'I have carefully avoided putting anything on record that was not certainly seen, and this may account in the main for any lack of detail in my drawings.' Nevertheless, he quickly added: 'What is shown . . . can be relied on.'[34] He thus expressed a philosophy which he was careful to follow over the years and which came to distinguish his planetary observations from those of others who chased wildly after fugitive details, and crowded their maps with them. Despite Barnard's personal vulnerabilities and neurotic tendencies, somehow at the telescope all of this seemed to fall away. There he was aloof, cool, and supremely confident, all of which qualities separated him as a planetary observer from others of the day and would make him a figure of great, almost Olympian, stature.

On the front cover of his observing book, Barnard pasted a copy of Schiaparelli's 1877 Mars map, which he had clipped from one of the newspapers. As he began observing the planet, he was at once struck with the fact that if this and other earlier charts 'are at all correct – and they doubtless were – important changes are at work on the planet.' He found a small dark spot following Schiaparelli's Solis Lacus (Lake of the Sun), which was not shown on the Italian observer's celebrated chart, and there were other changes as well – indeed, Barnard noted that the entire region around Solis Lacus appeared much different from Schiaparelli's representation, while the 'lake' itself was larger than he had shown it. This made him wonder

> whether what we see before us in the heavens is really another world like our own,
> with relatively fixed oceans and continents, or whether it is not a world like our own
> in its younger days when continents were shifting and oceans shifting, before the
> surface of the earth became firm and fixed by the process of cooling. If the latter is
> the case we can quite readily decide that Mars is not inhabited by the higher orders
> of life. [35]

Though Barnard saw at least a few of the so-called 'canals' of Schiaparelli, he was unable to verify their supposed doubling. However, he regarded the whole question as still open, as the low altitude of the planet had interfered with the most sensitive observations. Some idea of what he was dealing with can be gleaned from the entries he made in his observing book. Thus on July 29 he wrote, 'Seeing has been so poor tonight, while observing Mars, that scarcely any trace of detail could be seen on the planet. No drawings possible.' On August 14 he noted, 'There is a great deal of detail showing now but it is confused by the low altitude of the planet.'[36] In short, the Lick observations of 1892, like those at the two previous oppositions, proved on balance to be a keen disappointment, and the most sensational reports were obtained from elsewhere, especially from W. H. Pickering at Arequipa in the Peruvian Andes, about which we shall have more to say later.

5

In August 1892, Mars had to share newspaper headlines with the latest round of scandals on Mt. Hamilton. With the observatory still reeling from Burnham's resignation, Crew, at the end of July, tendered his resignation to Kellogg, effective as of September 1, 1892.[37] This led once more to violent incriminations against Holden in the press. In addition to commenting on the director's uncanny ability to drive away the most talented members of his staff, the San Jose *Mercury* printed the accusation that the director had attempted to establish on Mt. Hamilton a literary monopoly by allowing nothing to be sent out for publication unless it had first passed his editorial censorship, while charging exorbitant rates for his own articles (Burnham and Barnard were exacting their bitter revenge). 'It is only the pretender, the charlatan, who seeks to hide his littleness with the high-heeled shoes of pomposity,' the San Francisco *Examiner* chimed in, 'and to swell his proportions with the robe of impressive remoteness. Greatness does not think much of itself; littleness does not think of much else. A Herschel is incapable of the Holdenesque delusion that astronomy was invented to give him personal importance and to furnish him with an income.'[38] The embattled director attempted to defend himself by pointing out that he had 'steadily refused' to accept payment for any articles since 1888,[39] but this only drew the *Examiner*'s further retort:

> There can be no question at all that up to that time [1888] Mr. Holden was a very voluminous and highly paid contributor to California newspapers. The monopoly of news from the heavens, which his position on Mount Hamilton gave him, was abused by him in the extortion of rates for his productions far beyond those demanded by other scientific writers of infinitely greater ability.[40]

That same day, August 9, the *Examiner* unleashed another editorial under the headline 'The Trouble on Mount Hamilton,' which contained the most systematic and damaging attack on Holden's administration to date. Though signed only 'Observer,' it bears, on the basis of internal evidence, every indication of being Burnham's work. Barnard, at any rate, was convinced that it was.[41] The writer attempted to gather in one place every episode to Holden's discredit as a scientist, including his failure to observe the eclipse of Iapetus with the 36-inch refractor:

> Here was an opportunity for studying the construction of the ring system which, as Barnard subsequently remarked, 'gave more information about the crape ring of Saturn, perhaps, than could possibly have been obtained by a hundred years of ordinary observing.' Barnard was the only astronomer in the world to make this observation. I find from his writings to the scientific journals that 'Professor Holden requested [him] to observe it with the twelve-inch equatorial, as it would be out of

the reach of the great telescope.' Yet it appears that the most interesting phases of the eclipse took place when the planet was at an altitude of more than two hours above the horizon, and for two hours later Professor Barnard made records with the means at his command . . . If Lick's giant refractor cannot be used on an object three hours and more above the horizon there is something radically wrong in the mounting, which has been so much praised. But perhaps a sufficient excuse is again to be found in the unseemly hour at which the phenomenon occurred.[42]

The *Examiner* had, in fact, resolved on an all-out effort to bring Holden down, and for this purpose sent a correspondent to interview Burnham in Chicago. 'Very little is necessary to secure [Holden's] dismissal,' Burnham was notified in a telegram seeking his cooperation. 'Examiner agrees with public sentiment that such course is best for observatory.'[43] Burnham, of course, was more than agreeable, and wrote to Barnard that he gave the correspondent 'very fully . . . the condition of things at the L.O.' And he added: 'I hope it will do some good, but I am afraid it won't amount to much. It is worth trying. If this fails, you may as well give up the ship as soon as other plans can be perfected, but don't be in a hurry. You will have a lonesome time when Crew leaves, but you can stand it for awhile.'[44]

In the printed interview, 'Burnham Versus Holden,' Burnham enumerated his complaints against Holden. 'The desire of one man to pose as the whole embodiment of the Lick Observatory, all other persons connected with the institution to be known as mere useless apendages, except as moved by him – that is one of the chief reasons for the resignations and dissatisfaction at Mount Hamilton . . . It is a pity that such a magnificent institution as the Lick is dominated by an individual whose nature is so thoroughly selfish that the petty difficulties resulting at nearly every turn make life there almost intolerable to a man of spirit':

> Director Holden seems possessed of the martinet idea and a determination to run people at the heads of departments as though they were boys having no independence, intelligence or self-respect of their own. It is true that an appeal may be had to the regents. That resource has been utilized on occasions, but it is out of the question for the vexatious matters arising perhaps ten or a dozen times a day, and each of itself something which if standing alone might be perhaps ignored, altogether.[45]

Holden, however, weathered this barrage of attacks just as he had previous storms. Presumably the onslaught from the newspapers 'made the Czar feel a little uneasy,' as Burnham confided to Barnard.[46] But the *Examiner*'s expectation that it would take very little to bring the director down was in error. Holden still enjoyed the confidence of the Regents, and though embattled he remained secure in the directorship. As a result, there did not appear to be any immediate end in sight to Barnard's misery, and with Burnham's encouragement the frustrated young astronomer began, rather reluctantly, to cast around for other positions.

6

All that summer Burnham had been talking in earnest to George Ellery Hale, who would figure increasingly in Barnard's future plans. The son of William Ellery Hale, who had made a fortune building the hydraulic elevators used in the skyscrapers that sprang up in Chicago following the great fire of 1871, Hale was a prodigy, interested in science from an early age. When he was only fourteen he had met Burnham, who was then living on Vincennes Avenue, and discovered the $18\frac{1}{2}$-inch refractor of the Dearborn Observatory. 'G. W. Hough . . . was kind enough to let me look through it frequently,' he wrote later. 'I was also permitted to aid the feeble gas-engine in turning the dome, or rather drum, of the observatory':

> Naturally my ambitions were thus stimulated, but neither Hough nor Burnham had the slightest interest in astrophysical research, and I could not have devoted my life to such work as they were doing, valuable as it was to science. The reason lay in the fact that I was born an experimentalist, and I was bound to find the way of combining physics and chemistry with astronomy. Fortunately, it was not far to seek.[47]

Hale found his way in the then infant science of astrophysics.

> My father, always ready to encourage serious efforts, enabled me to buy a small spectrometer. This had a simple prism, and I lost no time in fitting it with a small plane grating. For by this time I had learned of the work of [L. M.] Rutherfurd and [H. A.] Rowland and had acquired some slight conception of the possibilities of high dispersion. Nothing could exceed my enthusiasm in observing the solar spectrum and in measuring the principal lines. I bought Lockyer's *Studies in Spectrum Analysis* and began the observation of flame and spark spectra and their comparison with the spectrum of the Sun. At last I had found my true course, and I have held to it ever since.[48]

In 1886 he entered the Massachusetts Institute of Technology, and in his spare time carried out spectroscopic work under E. C. Pickering with the 15-inch refractor at Harvard. During his summer vacations he returned to Chicago, where his father built for him, next to the family mansion at 4545 Drexel Boulevard in Kenwood, a private observatory with a 12-inch Brashear refractor, as well as his own fully equipped physical laboratory. Following his graduation from MIT in 1890, he married. For their honeymoon the newlyweds went west – the high point was, in Hale's estimation at least, a visit to the Lick Observatory, where they were hosted by Burnham. Holden was so impressed with the young man that then and there he offered him a position as a volunteer observer. However, as Hale's main interest was always the Sun, he declined on the grounds that the daytime 'seeing' at Mt. Hamilton was not good enough. Hale also had one or two other offers, but instead decided to return to his private observatory in Chicago.

At about this time the University of Chicago was being organized in nearby Hyde Park as a great research university, with the financial backing of John D. Rockefeller.

The university's dynamic young president, William Rainey Harper, offered Hale a faculty position on condition that he turn his observatory and physical laboratory over to the University. But Hale was not to be won over on these terms. He protested that if he was not competent enough to be appointed on his own merits, he would just as soon continue his research on his own. Harper eventually managed to smooth things over with the sensitive young man, and approached him again after Hale returned from a trip to Europe in the spring of 1892. This time Hale accepted.

Hoping to lure Barnard to Chicago eventually, Burnham made a point of keeping Barnard apprised of these developments. 'I was at Hale's this evening,' he wrote in mid-July 1892, 'and we were talking about trying to get you in this vicinity at the Chi[cago] University. They want to attach Hale and his outfit to it, and are making efforts generally to get the best men and give the thing a big start from the first . . . They have lots of money and aim to make a big thing of it.'[49] A week later Burnham wrote to Crew, who had accepted a position in the physics department at Northwestern University north of Chicago, 'I am sure you will be delighted with the change. After awhile we will get Barnard here, and then the L.O. may go to the lower region where it belongs.'[50] For Barnard, the prospect of a position where he could be close to his old friends was undoubtedly inviting, but as yet there was nothing firm for him to take hold of. Besides, he was reluctant to leave balmy California, where his fruit trees were just beginning to take hold.

7

Barnard found, as always, a refuge in hard work. During August 1892, he was photographing the Milky Way with the Willard lens (in its new incarnation as the Crocker telescope), observing on every clear night with the 12-inch except on Fridays, when he had the 36-inch. He was now so busy with these researches that he did not have time to spare for comet-seeking, and by the end of the summer resolved to give it up altogether. By then he had, in a little more than a decade, discovered fifteen new comets (and soon afterward would discover a sixteenth; see below). Only Pons himself had discovered more, though the dedicated Brooks, who worked on with single-minded zeal, would soon surpass Barnard to take over second place on the all-time list. Barnard, who did not suffer from false modesty, nevertheless told J. T. McGill shortly after Brooks's death in 1921: 'If I had continued seeking for comets until now, it is probable that I would have been first. But I ceased comet seeking for other work that I thought was more important.'[51]

8

One of Barnard's most puzzling observations was made on Friday August 12. That night he used the 36-inch to make routine measures of the Martian satellites, and also looked at Jupiter, timing the emergence of the second satellite, Europa, from occultation behind the planet and the passage of the Great Red Spot across the central meridian. Finally, a half hour before sunrise on Saturday morning August 13, 1892, he turned the great refractor to Venus. After studying the planet for twenty minutes, his attention was captured by the presence in the field of a star which he estimated 'must have been of at least the 7th magnitude to be seen in strong daylight.' Normally he would have secured a position, but as he jotted in his observing book: 'The position of the tube with reference to the high chair made it impossible to get the measures before daylight killed [the object] out.'[52] He later gave a somewhat more detailed account: 'The position was so low that it was necessary to stand upon the high railing of a tall observing chair. It was not possible to make any measures, as I had to hold on to the telescope with both hands to keep from falling.'[53]

On attempting to identify this star by checking its position against his reference chart, Barnard was surprised to find that there did not appear to be any stars answering to its description, nor were any of the brighter asteroids nearby at the time. For some reason he did not publish the 'unexplained observation' until 1906. His belated account then appeared in the *Astronomische Nachrichten,* and drew critical comment from a German astronomer, Rudolf Pirovano, who suggested that Barnard had made a simple mistake about the date. On the date August 13, 1892, he noted, Venus was a full 26° above the horizon, which seemed to contradict Barnard's recollection that 'the position was so low that it was necessary to stand upon the high railing of a tall observing chair.'[54] Barnard immediately replied that there had been no mistake:

> In reply to Rudolf Pirovano . . . I would say that with the large telescopes of this country 26° would be considered a low altitude for the observation of objects. These telescopes are not expected to be used near the horizon. When it is required to observe near the horizon, it is necessary to have a high observing chair in addition to the elevating floor . . .[55]

Barnard recalled that there had been two observing chairs at the Lick Observatory, one that was almost always used, and the other – 'a great cumbersome affair' – seldom used except for getting at the photographic plate when the 33-inch corrector lens was on the telescope. He recalled that on the night in question he had used the smaller chair, but to make sure that his memory had served him faithfully, he went to the trouble of writing to W. W. Campbell, who by 1906 was director of the Lick Observatory, asking him to set the telescope to the specified position – that is, with the elevating floor at its highest point – and to measure for him the distance from the floor to the eyepiece.[56] Campbell complied and reported that with the telescope in this position, 'the eye piece of the 36-inch is 11 feet above the highest position of the floor. The top step of the smaller observing chair is 8 feet above the floor.'[57] There could no longer be any doubt as to the accuracy of Barnard's recollection, but the question remained: what was the object that Barnard had seen?

Possibly, as Joseph Ashbrook has suggested, it was a nova.[58] Certainly no more plausible explanation has ever been offered, but we may never know for certain. The other question, why Barnard waited fourteen years to publish his observation, is less profound but psychologically interesting. Barnard was always extremely sensitive to criticism and reluctant to publish any except the most ironclad results. As we have seen, his conservatism had led him to publish his 1892 drawings of Mars under the disclaimer, 'I have carefully avoided putting anything on record that was not certainly seen.' Under the circumstances, his hesitation about his 'unexplained observation' is understandable – indeed, when after fourteen years his sense of his obligation to science to make known a potentially valuable observation finally overcame his fear of ridicule, the announcement, as he expected, met with disbelief. The whole episode illustrates the researcher's dilemma in needing to balance a desire for priority, which Barnard felt as strongly as anyone, with the risk of announcing a mistaken result through premature disclosure. Except for the premature announcement of the unconfirmed comet at the beginning of his career, for which he had suffered such keen embarrassment, Barnard's tendency was to hew to the latter course.

9

Within a fortnight of Barnard's 'unexplained observation,' Campbell made a remarkable discovery about Nova Aurigae. Discovered by the Scottish clergyman T. D. Anderson as a 5th magnitude object the previous January, it had since appeared to be steadily on its way to extinction – Burnham had last seen it as a 16th magnitude star near the end of April. Now it had flared up again to magnitude $10\frac{1}{2}$. At once Campbell summoned Holden and Schaeberle to the dome. All three agreed that the star's appearance seemed to differ from that of other stars of the same brightness, and Campbell suspected that 'its disk was larger and its light duller,' though he was uncertain because of interfering moonlight.[59]

When Holden cabled the news of the star's surprising revival to Harvard College Observatory, from which it was picked up by the wire services, Burnham wrote incredulously to Barnard, 'I see by this morning's paper a dispatch from H[olden] that the new star in Aurigae is now visible of $10\frac{1}{2}$ m[agnitude]. This may be so, but it seems to me improbable, and I fancy some other star has been mistaken for it.'[60]

Barnard could hardly wait to examine the nova with the 36-inch, and had his chance two nights later, on Friday August 19. As soon as the nova climbed high enough into the sky, at just past 3 o'clock on Saturday morning, he scrutinized it carefully. In the position where the previous spring a star had stood, he saw 'a small bright nebula with a star-like nucleus of the 10th magnitude.' His measures with the micrometer showed the new nebula to be about 3″ of arc in diameter.[61]

That same night Campbell made spectroscopic observations on the 12-inch, with equally dramatic results. The spectrum showed bright emission lines, and he announced: 'The general character of the spectrum justifies us in calling this . . . object a planetary nebula.'[62] From his observations of the lines in the spectrum of Nova Aurigae, Campbell deduced that the nebula was rushing toward the Earth at a speed of about 175 miles per second – as we now know, the speed at which the shell of gas thrown off from the star at the time of its explosion was then hurtling Earthwards.

Both Barnard and Campbell independently announced to Holden on the morning of August 20 the sensational discovery, from the previous night's work, that the star was nebulous. One had made the discovery visually, the other from spectroscopic observations.[63]

This was a penetrating insight into the origin of those puzzling objects, planetary nebulae. It was also a great triumph for Barnard, who less than two months after the Regents had overruled Holden in giving him his night on the 36-inch had used it to make a major discovery. Even so, Barnard's mood was far from exuberant. Burnham and Crew were gone, and he was feeling discouraged and alone. 'It has been lonely here since you left,' he sulked to Crew in a letter written on September 8. 'I have very little to say to anyone and they have as little to say to me. – Look out for anything in my line. I am tired here.'[64] Thus his mood on the eve of what would prove to be the most fateful day of his life.

1   EEB to Martin Kellogg, January 9, 1892; SLO
2   E. E. Barnard, 'A Few Unscientific Experiences of an Astronomer,' *Vanderbilt Quarterly*, 8 (1908), 273–88:277
3   E. E. Barnard, 'Observations and Photographs of Swift's Comet of March 6, 1892,' *AA*, 11 (1892), 386–8
4   ibid., p. 387
5   EEB, memorandum, September 21, 1891; VUA
6   Sacramento *Daily Union*, November 12, 1891
7   Barnard, 'Experiences,' pp. 283–4
8   Barnard, 'Observations and Photographs of Swift's Comet,' pp. 387–8
9   EEB, observing notebook; LO
10   SWB to Martin Kellogg, June 7, 1892; SLO
11   SWB to CAY, July 12, 1892; SLO
12   SWB to EEB, *ca* June 10, 1892; VUA
13   EEB to Col. C. F. Crocker, June 1892; VUA
14   JEK to WWC, September 20, 1892; SLO
15   ESH to JEK, January 26, 1893; SLO
16   'Misery on Mount Hamilton,' San Francisco *Examiner*, June 16, 1892
17   EEB to ESH, June 13, 1892; SLO
18   ESH to EEB, June 13, 1892; SLO
19   EEB to TGP, June 13, 1892; VUA
20   He again protested that his work was 'among the most important in the realms of astronomy' and that it was a matter of sheer injustice for the director to deny him use of the great telescope which had for four years been employed 'for a large part of the time in far less important work than that which I wish to put it to.' EEB to the Honorable Board of Regents of the University of California, June 13, 1892; SLO
21   SWB to TGP, June 21, 1892; VUA
22   ESH to EEB, June 23, 1892; VUA

23 EEB to GD, June 27, 1892; BL

24 SWB to EEB, June 21, 1892; VUA

25 SWB to CAY, July 12, 1892; SLO

26 HC to CAY, July 5, 1892; DCL

27 SWB to EEB, July 7, 1892; VUA

28 E. S. Holden, 'Note on the Opposition of Mars, 1890,' *PASP*, **2** (1890), 299–300

29 Even Holden had to admit that 'many photographs . . . were made, but none of any real excellence.' ibid., p. 299

30 'A Coolness Between Two Eminent Scientists,' New York *Herald*, October 1890; newspaper clipping in SLO

31 'Astronomers in Dispute – Are There Land, Water and Human Beings on the Planet Mars? – Prof. Holden Answers Camille Flammarion,' San Francisco *Examiner*, October 8, 1890

32 SWB to EEB, August 4, 1892; VUA

33 E. S. Holden, 'Note on the Mount Hamilton Observations of Mars, June–August, 1892,' *AA*, **11** (1892), 663

34 E. E. Barnard, 'Preliminary Remarks on the Observation of Mars 1892, with the 12-in. and 36-in. Refractor of the Lick Observatory,' *AA*, **11** (1892), 680–4: 680

35 ibid., p. 683

36 EEB, observing notebook; LO

37 HC to Martin Kellogg, July 29, 1892; SLO. He informed Kellogg that he had accepted an appointment as Professor of Physics at Northwestern University, near Chicago.

38 San Francisco *Examiner*, August 8, 1892

39 E. S. Holden to the Editor, San Jose *Mercury*, August 6, 1892

40 'The Lick Observatory,' San Francisco *Examiner*, August 9, 1892

41 EEB to GD, Aug. 12, 1892; BL

42 'Trouble on Mount Hamilton,' San Francisco *Examiner*, August 9, 1892

43 San Francisco *Examiner* to S.W. Burnham, telegram received at Chicago August 9, 1892; SLO

44 SWB to EEB, August 9, 1892; VUA

45 'Burnham Versus Holden – The Astronomer Says that the Director of Lick Observatory is Unfit for the Place – His Selfish Ambition and Constant Interference Combine to Make Positions at Mount Hamilton Unbearable.' San Francisco *Examiner*, August 10, 1892

46 SWB to EEB, August 19, 1892; VUA

47 W. S. Adams, 'George Ellery Hale,' *Ap J*, **87** (1938), 369–88:371

48 ibid.

49 SWB to EEB, July 14, 1892; VUA

50 SWB to HC, July 20, 1892; SLO

51 EEB to J. T. McGill, November 18, 1921; VUA

52 EEB, observing notebook; LO

53 E. E. Barnard, 'An unexplained observation,' *AN*, **173** (1906), 25. As nearly as he was able to estimate it from the position of Venus at the time of this observation, the star's position was at about RA 6h 52m 30s, declination 17° 11'.0.

54 Rudolf Pirovano, 'Notiz betr. E. E. Barnard, An unexplained observation,' *AN*, **173** (1906), 207–8

55 E. E. Barnard, 'Reply to Mr. Rudolph Pirovano's remarks . . . concerning "An unexplained observation,"' *AN*, **174** (1907), 315–18

56 EEB to WWC, September 14, 1906; SLO

57 WWC to EEB, September 20, 1906; SLO

58 Joseph Ashbrook, 'Barnard's "Unexplained Observation,"' *Sky and Telescope* **15** (1956), 356. The same explanation is favored by Richard Baum, who treats the subject in detail in *The Planets: Some Myths and Realities* (Newton Abbot: David & Charles, 1973), pp. 84–91.

59 W. W. Campbell, 'Nova Aurigae,' *PASP*, **4** (1892), 192

60 SWB to EEB, August 19, 1892; VUA

61 EEB, entry dated August 19, 1892 in ESH, diary; SLO. See also E. E. Barnard, 'Nova Aurigae a Nebula,' *AA*, **11** (1892), 751

62 W. W. Campbell, 'Nova Aurigae,' p. 192.

63 E. E. Barnard, 'Note on Professor Campbell's Observations of Nova Aurigae,' *Ap J*, **5** (1897), 277–8

64 EEB to HC, September 8, 1892; SLO

13

Immortality

1

In August 1892, a reporter for the New York *Herald* paid a visit to the dome of the 26-inch refractor of the US Naval Observatory in Washington, DC. After passing the two dogs that roamed the front halls of the observatory which, he reported, seemed to be the 'only things alive in the dark and spooky passages leading to the main observatory room,' he found Asaph Hall mounted on the steep ladder peering at Mars through the telescope. This was the same telescope with which Hall had discovered the two tiny moons of Mars fifteen years before – 'the greatest discovery in astronomy of the century,' the *Herald* reporter described it. On this night, when Hall had finished his observations, he invited the reporter to the top of the ladder. 'If you want to see Mars, young man, come up and look at him.' The result was disappointing, for what the young man saw seemed to him 'an awful little thing to make such a big fuss over . . . a ball of butter with blue fringe around it. It did not look like very good butter, either. It had the appearance of the second hand variety, made up of job lots of various shades of yellow.'[1] This is not a bad description of the usual telescopic appearance of Mars.

In 1877, when Hall made his great discovery, the 26-inch refractor had been the largest in the world. Fifteen years later the 36-inch Lick telescope reigned supreme, yet Barnard saw little more on Mars even with that great telescope. The satellites however were, he noted, 'very conspicuous,' even with the dazzling planet in the field, and during July, August and September, Barnard made numerous measures of their positions with a filar micrometer. (In a filar micrometer, the observer sights on a pair of parallel threads; one is movable by means of a screw and the other fixed. As one turns the screw, the number of revolutions, or parts of a revolution, needed for the thread to traverse the space between two objects yields a measure of their angular separation.)

As Barnard pursued these observations, the circumstances in which Hall had made his great discovery must have been very much in his mind. That discovery had been partly due to the powerful glass Hall had used, partly also, however, to his boldness in pushing his search closer in to the planet than had any of his predecessors. Now Barnard wondered whether his keen eye and powerful glass might, if applied with the same boldness to the search, reveal something equally significant. 'Since July 1 of this year,' he explained, 'I have had the use of the 36-inch refractor on one night each week. Previous to this I had no regular use of the instrument, and the observations made with it were of specified objects, the time being limited to the objects. Among other things that I have devoted the instrument to on my nights, was a search for new objects.'[2] Though he did not say so, this undoubtedly meant new satellites.

The possibility that new satellites might be revealed in the great refractor had, of course, been quite in the cards ever since it had first been turned toward the skies. Indeed, such discoveries were almost to be expected – hitherto almost every time a superior instrument had been applied to the quest, one or more new satellites had been captured. Galileo's small telescope, modest as it was, had represented a marked improvement over the eye and had shown him the four satellites of Jupiter in 1610. Similarly, the discoveries of Huygens, Cassini, Herschel, Lassell, and Hall were all made with telescopes that represented a significant advance over their predecessors. Holden, strangely enough, had been guarded about the chances of like success with the Lick refractor, noting in a lecture given shortly before it went into operation: 'Galileo's discovery of the four satellites of Jupiter, Herschel's of the planet Uranus, Le Verrier and Adams's discovery of Neptune, Hall's of the satellites of Mars – perhaps such as these can never be repeated. It may well be that there are no more than eight planets – that all the satellites have already been discovered. So it may be that these glaring, obvious, and popular discoveries, so to say, are come to a natural limit. I do not say this is so. It may well be so.' But he added: '[W]hat we cannot see with our telescope, the best of all, in our elevated situation, the best in the world, need not be searched for with inferior telescopes or less favored situations. We shall be justified in publishing our negative results.'[3]

Neither Holden nor anyone else, during the first four years of the great telescope's existence, seems to have been particularly diligent in searching for satellites. Though occasional sweeps for them must have been made, the observers kept their efforts to themselves. Barnard, also, did not make it common knowledge that he was searching for new satellites during the summer of 1892. At first, he must have been chiefly in pursuit of additional companions of Mars, but by September 9, 1892 he was throwing the net of his scrutiny around Jupiter as well.

2

That Friday evening, Barnard began his work with the 36-inch refractor with a series of measures of the separation and position angle of β^2 Capricorni, the double star he had discovered at Vanderbilt. He then turned to Mars and its satellites. As soon as he put eye to the eyepiece, he noted that 'Phobos and Deimos . . . are so very near [Mars] that it is hard to tell which is the brightest.' Over the next several hours he regularly set the wires of the micrometer on them to obtain their positions. As midnight approached, he swung the great telescope toward Jupiter, which was then in Pisces, near the 5th magnitude star μ Piscium.

He was consciously searching for faint Jovian satellites. As he later admitted, 'It was simply with a desire to satisfy one's own eyes (by a personal search) that no other moon existed in the Jovian family that a new satellite was looked for.'[4] In the hopes of revealing such an object, he slid the overpoweringly bright planet just outside the field of the × 520 eyepiece that he was using. 'At 12 o'clock as near as may be,' he afterward recalled, 'to within a few minutes, I detected a tiny point of light close following the planet and near the 3rd satellite [Ganymede] which was approaching transit. I immediately suspected it was an unknown satellite and at once began measuring its position-angle and distance from the 3rd satellite. On the spur of the moment, this seemed to be the only method of securing a position of the new object, for upon bringing the slightest trace of the planet in the field the little point of light was instantly lost.'[5]

Barnard recorded that when he made his first measure, at about 12:13 a.m., the faint object was about 29″ of arc from Ganymede. Though the object was well seen, it was swamped out by glare as soon as Jupiter was brought back into the field. In his observing book, he made a sketch of its position relative to Ganymede and Jupiter itself, indicating it as a tiny dot below which he wrote: 'This must be a new satellite at elongation. It is apparently smaller than either of the satellites of Mars.'[6]

Barnard was able to get two sets of distances and one set of position-angles of the suspected satellite relative to Ganymede. By 12:48 a.m. he noted that it had moved appreciably, and was now within 22″ of Ganymede. He attempted to measure the position relative to Jupiter itself, but found that one of the wires of the micrometer was turned at a large angle to the other. On looking further into the matter he discovered that one wire had broken out and that the other was loose. Thus his later measures were not as reliable as the first two sets. When he caught his last glimpse of the presumed satellite at 2:33 a.m, it was 'rapidly closing in on Jupiter and possibly moving northward (?) 10 m[in] later it could not be seen.' Thereafter he kept a careful watch for its reappearance at the other limb of the planet, but gave up at 4:50 a.m. when he recorded: 'There is no show for it now. The seeing is v[ery] bad.' By then day was beginning to break.

Though positive that he had a new satellite in his grasp, Barnard was too conservative to make an announcement without getting more measures. The next morning he replaced the wires in his micrometer, and he appealed successfully to Schaeberle, who had Saturday nights on the great telescope (after the visitors left), to let him have it again that night. Then he nervously waited out the rest of a long day.

When evening finally came there was the usual throng of Saturday visitors waiting to file past the telescope for a look. Their numbers had thinned somewhat, however, from the peak numbers of a month earlier, when as many as two hundred of them had swarmed the mountain in the hopes of getting a glimpse at Mars near opposition. The last of the visitors left by 11:30 p.m, when Barnard entered in his notebook: 'Have examined carefully on each side [Jupiter]. Not any object near.' A few minutes later, he spotted the new satellite leaving the planet on the following side, and by setting the wires of the micrometer perpendicular to the belts and then throwing the planet out of the field, he obtained a good set of measures of the satellite at its eastern elongation, which occurred, he estimated, at about 12:53 a.m. This observation would be of critical importance in working out the satellite's orbit. On subsequent nights, he used a strip of

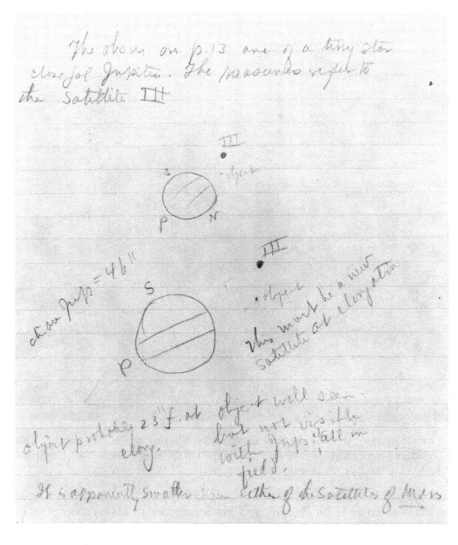

Page from Barnard's observing notebook recording discovery of fifth satellite of Jupiter, September 9, 1892. UCO–Lick Observatory

mica, carefully smoked, in front of the eyepiece to occult the planet, which facilitated his measures by allowing the satellite and the planet to be seen at the same time.[7]

Early the next afternoon, September 11, the electrifying news of the great discovery was telegraphed to the Harvard College Observatory and to the Associated Press and United Press wire services. From there it went out to the rest of the world. Though he did not say so at the time, Barnard later made clear that the discovery was by no means an accident, but 'the direct result of a careful search for just such a body.'[8] The telegraph message Barnard composed contained only such information as the satellite's distance from the center of the planet (112 400 miles), its period of revolution

Lick Observatory at about the time of Barnard's discovery of the fifth satellite, with horsedrawn carriages and visitors assembled in front of the 36-inch dome. Mary Lea Shane Archives of the Lick Observatory

(erroneously given as 12 hr 36 min), and his September 10 observation of its eastern elongation. To this matter of fact summary, Holden added a colorful introduction which did much to capture the public imagination: 'The Lick Observatory desires to announce that Professor Barnard has added a fifth Satellite to the four Satellites of Jupiter discovered by Galileo on January 7th 1610.'

The day after the world first learned of the exciting news, Barnard awoke and found himself famous. Congratulations began pouring in from all quarters. Burnham, who had been informed of the discovery by a personal telegram from Barnard even before it had appeared in the newspapers, at once telegraphed back: 'Your discovery is the greatest of this nineteenth century. Congratulations.'[9] Keeler was also thrilled for his former colleague. 'I have just read of your great find, and hasten to congratulate you, for I suppose this is not a newspaper "fake". I would rather you had made this discovery than anybody I know, not even excluding myself! It is not always the fellow that blows that gets there.'[10] Charles Burckhalter of the Chabot Observatory wrote: 'Just think, Galileo and Barnard – alpha and omega walking arm in arm[.] Nearly three centuries apart – the first & the last to discover Jovian satellites!! Its *almost* too good to be true. I cant get that blambed Satellite out of my head.'[11]

Among all the expressions of good will Barnard received from former colleagues and friends, a note from Joseph S. Carels, the assistant postmaster who had befriended him during the days when he had been a street urchin in Nashville, elicited the most interesting reaction from the discoverer. 'I have been getting telegrams and letters of congratulations for the past week,' he told Carels in reply, 'but of all, yours brings back memories that are sad of struggles that were hard to bear sometimes and of hopes and disappointments that clustered thick about me in that newly built cottage away out on Bellmont Avenue . . . I know you did your part in bringing me to pubic notice – in letting the people know that I was trying and wanted to do something . . . ' He went on to recall the dear friends that had helped him in that difficult, but still hopeful, time of his life – Albert Roberts, editor of the Nashville *American*, Anson and Fanny Nelson, Braid, Peter and Ebenezer Calvert, Rodney Poole, Judge John M. Lea, the kindly instrument-maker Charles Schott, Bishop McTyeire, Chancellor Garland, Landreth, Vaughn, Baskerville. But he paid the highest tribute to Carels himself. 'Clinging to me through life has ever been the memory of [your] kind word and nod and smile of recognition to a poor sick ragged boy on his way to or from work . . . This is not sentiment. It is plain and substantial reality.'[12] Without forgetting those who had lent him a helping hand along the way, that poor, sick and ragged boy had somehow managed to climb from a miserable existence. Now suddenly he stood at the pinnacle of the astronomical world.

3

Nowadays, when we have grown used to satellites being discovered by the score by spacecraft cameras, it is not easy to recapture the mood of a time when the announcement of one small satellite, which Barnard himself estimated could scarcely be more than 100 miles across, could produce such a tremendous torrent of popular interest. In the nineteenth century, however, such discoveries were relatively rare events. Before Barnard found the fifth satellite of Jupiter, there were only twenty moons known in the whole Solar System, and none at all had been added since Hall's discovery of the satellites of Mars in 1877. Barnard's discovery came, moreover, at a transition point in the history of astronomy. It was the last satellite to be discovered visually. Henceforth all discoveries would be made photographically – and there would be many such, beginning with W. H. Pickering's discovery of Saturn's ninth satellite, Phoebe, in 1898. These later moons were seen more as captives of advancing technology than of the skill of the astronomers who found them, and they did not make their discoverers heroes in the same way that Barnard's discovery made him.

The newspapers of the day were eager to soak up all the information about Barnard's satellite and its discoverer that they could, and eagerly consulted every expert who was willing to give them a quote. For public consumption, Burnham expanded on his

Barnard's self-portrait with the 36-inch refractor, taken shortly after discovery of fifth satellite (note stain on trousers). Mary Lea Shane Archives of the Lick Observatory

telegraph message to Barnard: 'E. E. Barnard has earned a niche for himself beside Galileo and Herschel, and the permanency of his fame is secure. The discovery of this satellite is the greatest astronomical achievement of the century and will cause the world of science to ring . . . It is far greater than the finding of the satellites of Mars':

Barnard's work makes an addition to the solar family in a field that was well worn out and which was thought to have been so thoroughly sifted that astronomers gave up the task of looking for fresh facts. Then, too, Jupiter has been the most observed of all the planets for the reason that it is so very large and such a grand body withal that it invites inspection. I can tell you that this will be grand news for the astronomers of Europe.[13]

Barnard's old friend George Davidson, of the Coast Survey, also commented on the discovery. Describing it as Burnham had, as 'one of the discoveries of the century,' he called special attention to the keenness of Barnard's eyesight. 'Barnard doesn't realize he possesses the marvelous keenness of vision that he does . . . The average person with the naked eye can count six stars in the Pleiades group. Mrs. Herschel was able to count ten without a glass. One very clear night I counted eleven stars. When I mentioned this to Barnard he smiled and said that in Tennessee one night he had made out twelve stars with perfect distinctness.'[14]

Among other reactions to the discovery was that of the Irish astronomer Sir Robert Ball, who observed that 'since the invention of the telescope some 280 years ago the great planet Jupiter has never been the object of so much interest as it is at the moment . . . Here was this system which everyone knew, which had occupied so much attention, and now we are told on the best authority that there is something to be seen in it which has eluded all the eyes that ever looked at it before.'[15]

A Mrs. A. E. Barnard, of Los Angeles, named her son – as she then supposed – after the famous astronomer, because as she later explained to him, 'I think when the powers of the intellect begin to develop themselves in a boy in many instances – the foundation of the character is laid – and the disposition formed after the one by whom he is named.' Unfortunately, she for some reason thought that the astronomer's name was Edward Eugene, and this was the name given him.[16] Another unique tribute was paid to the astronomer by three hardy California mountain climbers. On September 24, 1892, W. L. Hunter and his two sons John and William became the first men to reach the summit of an as yet unnamed 14 000 foot peak in California's Sierra Nevada range, located near the state's tallest summit, Mt. Whitney. Able to think of no better name for it than that of the celebrated astronomer, the climbers christened it Mt. Barnard.

4

The fifth satellite was the innermost in order of distance from the giant planet, just over half the distance of Io from the planet's center. Taking into account the satellite's extreme faintness and its tendency, owing to its proximity to the planet, to be swamped out by glare, Barnard suggested that 'it is scarcely probable that this satellite will be seen with anything less than 26 inches, and only with that under first-class conditions.'[17] Barnard himself, on one night when the conditions were highly favorable, took every precaution known to him to try to see it with the 12-inch refractor, but 'no trace of the little moon could be made out, though it was then at its greatest distance from the planet.'[18]

Aside from the 36-inch refractor itself, there were, then, only a few telescopes in the world that had a chance at it. But in the weeks following the announcement of the new satellite's existence, observers elsewhere were frustrated in their attempts to confirm it as much because of the inaccurate information that had initially been given out as because of the satellite's intrinsic faintness. Barnard's first estimate of the satellite's period had been 11 hr 36 min. From further observations he later amended this to 11 hr 49.3 min, and then finally to the correct figure – 11 hr 57 min. Unfortunately, the first telegrams to Harvard College Observatory and the press had incorrectly given the period as 12 hr 36 min, while the telegram sent to European observers contained not only this erroneous period but dropped the reference to Barnard as discoverer – it began only, 'Professor Holden announces that the Lick telescope has revealed the existence of a fifth satellite to Jupiter.' It also omitted the crucial information about the time Barnard had observed the satellite at its eastern elongation. Still later telegrams to Europe made the confusion even worse. Because of an error in the transmission, they gave an even more wildly misleading period – 17 hr 36 min.[19]

Among those who were thrown off the track by such misinformation was the English engineer and amateur astronomer, A. A. Common, who had set up a five-foot reflector at his private observatory at Ealing, near London. The inaccurate telegrams caused him to look for it at times when the satellite was hidden behind the planet. With October quickly passing and without a definite sighting of the satellite from Ealing, Common wrote angrily about the way that the discovery had been announced at the Lick Observatory:

> It is very much to be regretted that this brilliant discovery, for such it most undoubtedly is, has been announced in such a very peculiar way. We have an announcement giving 12h 36m as the period, 112 400 miles as the distance, and 13 as the magnitude, and the Lick telescope as the revealer: this is the first telegram. Then we find that Professor Barnard is the discoverer and that the period is 17h 36m, – still without the time of discovery fully stated, from which with either period some idea of its place in its orbit could be found and for weeks this erroneous time stood uncorrected and no further information was given . . . [H]ere, where an hour or two of good seeing is quite a rarity, the hopelessness of trying to see such an object by chance is too apparent. Had it been some faint comet we should have had all the information necessary for its full observation in the course of hours or a few days at the most; but in the case of this most important discovery after nearly seven weeks we know very little more than the first telegram gave us. It is only a short time since an important discovery was communicated by a diffident amateur by means of a post-card: a resort to this method by the Lick authorities would have been preferable to the very inadequate method they have adopted.[20]

Apart from Barnard himself, the first to succeed in confirming the satellite was Tyler Reed, C. A. Young's assistant at Princeton, who aided by a more accurate position than had so far become available elsewhere captured it with the 23-inch refractor of the

Halstead Observatory. Young himself, who had been hitherto indisposed with a bad case of sciatica, saw it with the same instrument a few nights later,[21] and soon other reports followed, including a sighting by G. W. Hough with the 18½-inch refractor at the Dearborn Observatory, the smallest instrument in which it was reported visible.

Nevertheless, the satellite continued to elude European observers, including searchers equipped with the large refractors at Nice and Vienna. The French astronomer Flammarion wrote to Holden on October 16, 'Fifth satellite of Jupiter doubted by European astronomers. Please give date of the last observation.'[22] At the same time E. W. Maunder of the Greenwich Observatory reported to Barnard:

> I have tried hard again and again to catch a glimpse of your fifth satellite with our new 28-inch telescope, but only succeeded on two occasions in just *fancying* I saw it for a moment. Night after night I could find no trace. It is true Greenwich is a most unhappy site for a great telescope, but my ill-success has given me a very high idea of the skill, patience, & keenness of sight which you must possess to have made the original discovery.[23]

Finally, when better information about the satellite's whereabouts became available, it was seen 'pretty certainly' by Common with his five-foot reflector from Ealing,[24] and by observers with the 25-inch Newall refractor at Cambridge – the only observations of it from Europe.

5

Even as skilled observers using large telecopes struggled to see it, the satellite was claimed by several amateurs with very small instruments. Thomas S. Cogley, a Washington DC lawyer, asserted that he had found the satellite four years earlier with only a 5-inch telescope! In fact Cogley accurately described what he had seen, but misinterpreted his observation. As explained at some length in the California newspapers by both George Davidson and Charles B. Hill, Cogley's observation had been made on June 8, 1888, the evening before Jupiter was due to occult the 6th magnitude star 47 *Librae*. On looking at the planet with his telescope, Cogley had seen what he took to be four satellites on the western side of the planet. Not knowing about the approaching occultation, he had assumed that these four, which actually consisted of three satellites and the star, were the Galilean satellites. As he continued to observe, he saw another satellite emerge from behind the planet on its eastern side – this was Io, but Cogley in his confusion believed it to be a satellite hitherto undiscovered, and wrote to Simon Newcomb about it at the time. The assistant Naval Observatory astronomer who wrote the reply for Newcomb's signature tried to explain the mistake that had been made, but after Barnard's discovery Cogley brought the whole matter up again and continued, beyond all reason, to press his claim, which was given prominence, as the San Francisco *Bulletin* noted, in the 'Eastern journals.'[26] On receiving a clipping advancing Cogley's claim and headed 'Not Barnard's Moon,' Barnard became incensed enough to finally defend his own priority in print. 'Such claims invariably follow in the train of every great discovery,' he noted bitterly, 'and I am therefore not at all surprised that some one else wants California's new moon':

> It is customary for astronomers to pay no heed to such claimants where the observation is so preposterously absurd as in the present case. However, since Californians have taken such a deep interest in the welfare of the new member from Santa Clara County, and since they are themselves not sufficiently familiar with the difficulties attending the discovery of such an object to decide for themselves, it is only just that some statement would be made by me with reference to the matter.[27]

Quoting a remark by Newcomb that 'the chance of finding a new satellite to Jupiter with a small telescope is about equal to that of finding a gold mine under Pennsylvania Avenue,' Barnard showed how far-fetched it was for Cogley to believe that he had made such a discovery:

> I have stated that this new satellite to Jupiter is of the thirteenth magnitude. There is a certain relation between the aperture of a telescope and its penetrating power, or the faintness of a star that can be seen with it. Now a five-inch telescope would not show this satellite if it were isolated on the dark sky away from the planet. Place now this tiny point of light close in the immediate glow of the giant planet, and it becomes one of the most difficult objects in the heavens and can only be seen with special precautions, such as hiding the planet behind a bar in the eye piece to get rid of its dazzling light. The claim, therefore, of Mr. Cogley is of the highest absurdity. One of his five moons was a bright star – perhaps visible to the naked eye – and if he will go back to where Jupiter was on the date of his observation he will find the selfsame star right there – where it has been since the dawn of creation.[28]

Though these arguments ought to have affected Cogley 'as a snuffer affects a candle,'[29] the lawyer continued to press his claim. 'Professor Barnard thinks I could not have discovered the fifth satellite to Jupiter with a 5-inch telescope,' he wrote. 'Galileo discovered four of Jupiter's satellites with a 1-inch instrument.'[30] If the Eastern journals were indeed, as the California newspapers suspected, on Cogley's side, there was no doubting the loyalty of the local papers. Thus an editorial in the San Francisco *Call* demanded that Cogley 'show the 5th moon to anybody else through his precious instrument. But he cannot do that, and unless he wishes to be taken for a blind prevaricator he had better close the controversy.'[31]

No sooner had Cogley been beaten back than the claims of a Detroit amateur, O. E. Cartwright, began making the rounds in the press. Cartwright boldly announced that he had detected a new satellite of Jupiter two years before Barnard, and had written about it to the directors of both the Harvard College Observatory and the Lick Observatory. Barnard found Cartwright's claim that these letters had directly inspired his own discovery 'as detestible as it is contemptible,' and again felt the need to defend his priority in print. Under the heading 'Barnard is Bitter,' the San Francisco *Chronicle* published Barnard's long retort, which said in part:

> I would say that up to the present moment any letter ever sent to the Lick Observatory by Mr. Cartwright has never been seen or even heard of by me, and I

know nothing absolutely of the existence of any such letter. I have no doubt but that every observatory of any consequence in the world has dozens of such letters on its files with claims [such as]: From 'a mythical planet within a couple of degrees of Uranus, and with a satellite up near the north pole of the heavens – as pointed out by a spirit on the star Vega;' to 'the discovery of the double canals on Mars with the naked eye.' These two letters actually exist.

Barnard once again emphasized the failure of observers even with large telescopes to see his moon, and ended defiantly: 'I suppose Mr. Cartwright and Mr. Cogley have a definite understanding between them which of the two is to have the satellite – the fortunate one then settling his difference with me.'[32]

By then still other claimants had appeared. One California newspaper announced that, 'as far as yet known, the satellite had been discovered by a group of local young ladies,' while a minister back East wrote to Barnard claiming that he had discovered the satellite on showing the planet to friends in his small telescope on September 8. Barnard excerpted this letter in the *Examiner* in order to show 'how positive a man can be in a mistaken idea, and how difficult it would be to convince an amateur observer – who had made one of these "discoveries" – that his claim was entirely fallacious.' Among other things, the minister had said: 'I am prior to you in seeing this newly discovered satellite . . . I saw it also on the night of the 9th – the whole five being in plain view – as much earlier than you as California is west of this place . . . May I have the honor of being accredited the first to see this new satellite? It is something I wish to treasure to myself – "the first one in the world to see the five moons of Jupiter."' The minister added that no less than twelve people had seen the fifth satellite before Barnard had seen it, and he was prepared to 'furnish you abundant witness to the truth of this statement, which, considering my profession, you will not hesitate to understand.' Barnard had patiently to explain that it was unfortunate that on the night of the discovery there was a bright star near Jupiter, so that 'every amateur who looked at the planet on that date will now recall to himself that he then also saw the "five moons of Jupiter."'[33]

Perhaps the strangest story of all appeared in the Wilkes Barre *Telephone*, which reported that a Professor Coles, 'the tall scientist of the Wyoming Valley,' discovered with the aid of his 'Electric Eye' that Jupiter has seven moons. The *Telephone* report continued: '[W]e make the claim for him at this time, so that when the other eminent astronomers at the little end of their wonderful instruments shall make the same discovery, as they must in time . . . that Prof. Coles shall receive the credit as their first discoverer. This would be no more than right. We recently had the pleasure of seeing this wonderful photograph of Jupiter, and seven moons in their different phases of full, half, quarter and new moons, were plainly and distinctly seen.'[34]

Barnard did not dignify this story with a response, and indeed had by this time decided that silence was the best policy. 'I certainly can not enter into any correspondence about the matter,' he announced to the San Francisco *Examiner* on declining to reply to a defiant 'Open Letter' from Cartwright. 'I have already foolishly noticed these claimants in the press, and I regret having done so for it only gives more importance to them.'[35]

Though Cogley, Cartwright and the rest soon faded into richly deserved oblivion, one claim that continued to surface from time to time long afterward concerned John Winthrop, Jr., the colonial governor of Connecticut, who had supposedly discovered the fifth satellite with a small telescope as early as 1664. Winthrop had made an observation of what he believed to be a fifth satellite of Jupiter, which he described in a letter to Sir Robert Mobray, then president of the Royal Society of London, 'Having looked upon Jupiter with a telescope upon the 6th of August last I saw 5 Satellites very distinctly about that planet; I observed it with the best curiosity I could, taking very distinct notice of the number of them by severall aspects with some convenient time of intermission.' Needless to say, what he saw was in reality nothing more than an ordinary star, as he himself would have realized had he only kept up his observations for a few more nights. [36]

6

Though these preposterous claims for the discovery of the fifth satellite irritated Barnard, in responding to them he was kicking against the pricks – no one with any real knowledge of science ever took them seriously. As rightful discoverer, the privilege was his and his alone of selecting a suitable name for the tiny moon, though he received a good deal of advice in the matter from colleagues and the press alike.

Holden, in keeping with the traditional system of using mythological names for newly discovered planets, satellites and asteroids, proposed that the new moon be called 'Metis.' This goddess, he explained, was not only the mother of Minerva ('wisdom') but of Porus ('plenty – i.e. there are more satellites to come').[37] Barnard's antipathy to the director no doubt allowed him to eliminate this proposal from consideration without a moment's further reflection. Camille Flammarion wrote to him to suggest Amalthea, for the nurse of Jupiter,[38] while the English astronomer W. T. Lynn, recalling that the moons of Mars had been named for the attendants who had harnessed the war-god's steeds, asked:

> With what should we especially associate Jupiter in a similar way? Surely with the thunderbolt. My suggestion, then, offered in all diffidence, is that the Galilean satellites be still called, as formerly, by their numbers I., II., III., IV. (now used in so many books that they have almost become names in this connexion), and that the new interior one be designated Fulmen, or (if a Greek form be preferred, as in the case of the satellites of Mars) Keraunos.[39]

An editorial in the San Jose *Mercury* agreed that, as 'the other moons have classical names, . . . the new one will probably be named in the same way; otherwise Barnard itself would be a good name for it.'[40] However, a writer in the San Francisco *Call* argued for breaking with the mythological tradition altogether. 'How will the new moon be named?' he asked:

> The four that Galileo discovered have classical names. The fifth might be baptized after Barnard, or James Lick, or Harrison, Cleveland, Edison, Lowell, Longfellow

or any other American celebrity. It is quite time a departure was made in the nomenclature of the stars. A new plant gets the name of the discoverer or some friend. Why not a new satellite?[41]

Though immediately after the discovery Barnard was quoted in the newspapers as saying, 'It devolves upon me to find a name for it. This will, of course, be some mythological name, one connected with Jupiter in some way,'[42] he himself soon began having second thoughts about the mythological scheme, and was urged by Burnham to 'name the satellite to suit yourself. If it was mine I would give it something besides a heathen name. I think we have got beyond that time over here, but whatever you [do], don't let anyone else boss it.'[43]

Barnard had personally disliked the names Io, Europa, Ganymede and Callisto that had been given for the Galilean satellites by Simon Marius, a German astronomer who had disputed with Galileo priority for their discovery. His objection to them was clearly stated in a letter to Simon Newcomb written on October 12:

> I have been thinking of breaking away from the mythological method of naming in this case. First of all, every name that would be suitable had been applied to an asteroid. I would like to give it a name that has some connection with our own country. Secondly, the names that have been applied to the satellites of Jupiter are those of the mistresses of Jupiter and the whole system is so immoral that I see no good in keeping it up.[44]

When he wrote this, Barnard was seriously considering the name *Columbia*, in recognition of the fact that the satellite was found in the year marking the four hundredth anniversary of Columbus's discovery of America. He also liked the name *Eureka*, Archimedes's exclamation of discovery and the state motto of California. Not only did this name make explicit the satellite's association with California, where it had been discovered, but the date on which the discovery had been made, September 9, was Admissions Day, when the state had officially been welcomed into the Union in 1850. But Newcomb did not fully endorse either of these suggestions. 'The name Columbia,' he replied, '. . . would be very appropriate at this time if it were a larger and more important body' (suggesting by implication that it was not appropriate for the body in question). He added, moreover: 'I would not like to speak either for or against the mythological method of naming without considering the subject more fully than I now have any occasion to do. I certainly do not see why any ideas of morality or immorality should be associated with the fables of early humanity.'[45]

Newcomb suggested that Barnard ought to gauge the opinion of European astronomers who, he no doubt realized, were unlikely to look favorably on the abandonment of the mythological scheme. Indeed, the opinion there was no doubt as Lynn had summed it up soon after the discovery, namely, that though 'it is the undoubted right of a discoverer to name the body or place he has discovered . . . this right, like others, appears to have certain recognized limitations, the principal one being that the name selected should be not only appropriate but of a similar kind to those of other bodies similar[ly] circumstanced.'[46] In the end, Barnard continued to procrastinate. After repeated statements of the fact that the satellite 'still is nameless,' his own clear preference finally began to emerge. In an article in *Popular Astronomy* in 1893, he wrote: 'It would seem . . . almost to have found itself a name – "The Fifth Satellite." '[47] A year later, in the *Astronomical Journal*, he added: 'So far no name has been given to the satellite. This is principally because the mythological names long ago assigned to the other four are never used in practice, and promise eventually to be altogether dropped. It is far more convenient to speak of these moons as I, II, III, and IV, than to write or say *Io, Europa, Ganymede,* and *Callisto*':

> The numerals, to me at least, stand as names . . . I think, therefore, that this new moon should continue to be called the 'Fifth Satellite,' or Satellite V, as I have always called it. This will also be correct if we assume the other satellites were numbered in the order of discovery . . . There is certainly nothing to be gained by giving this object a special mythological name. However, if it is the general desire of astronomers that it should bear such a name, I will select one for it.[48]

To the end of his life, Barnard never made such a selection. In the absence of any choice by the discoverer, Flammarion's name, Amalthea, came into regular use among those partial to the mythological scheme – despite the fact that Barnard did not wish to see the satellite named for the nurse of Jupiter. 'The smallness of the satellite would make this name rather inappropriate,' he wrote.[49]

Ironically, it is this name, which Barnard specifically disliked, that has now become official. In the days when the Jovian satellites had to be scanned from afar through telescopes, Barnard could rightly argue for the convenience of the numerical system. But as the spacecraft flybys have now shown, each satellite is a dramatic and highly individual world in its own right, and the adoption of proper names has become irresistible. Amalthea is no exception. A mere speck as seen from Earth in even the largest telescopes, it was imaged by the Voyager spacecraft cameras in 1979, and from a distance of about 300 000 miles it began to show signs of being an interesting little world. It is irregularly shaped, measuring some 170 by 100 miles, and its surface, which shows crater-like features, is intensely reddish. It is also, incidentally, no longer the innermost known satellite to Jupiter – two smaller satellites discovered by Voyager, Metis and Adrastea, lie inside it.

But though no one would any longer claim that the capture of this little moon was the greatest astronomical discovery of the nineteenth century, as Burnham and others in the heady excitement of the moment hailed it, it did fire the imagination of the time as few others did. It was, moreover, a defining moment of Barnard's career. For his brilliant discovery he was awarded, within the year, the prestigious Lalande and Arago prizes of the French Academy of Sciences. The words on the obverse of the Arago medal are especially apt:

Laude Damus: Posteri Gloriam
(We give praise – Posterity gives glory).

1   New York *Herald*, August 3, 1892
2   E. E. Barnard, 'Discovery and Observations of a Fifth Satellite to *Jupiter*,' *AJ*, **12** (1892), 81–5:81
3   ESH, text of lecture to the Society of California Pioneers, May 1887; SLO
4   E. E. Barnard to the editor of the San Francisco *Examiner*, December 25, 1893. Also EEB to SN, January 13, 1893; LC
5   E. E. Barnard, 'An Account of the Discovery of a Fifth Satellite of Jupiter,' *AA*, **11** (1892), 749–50:749. In addition to the accounts given by Barnard himself, the most valuable summary of his discovery of the fifth satellite is: D. P. Cruikshank, 'Barnard's Satellite of Jupiter,' *Sky and Telescope*, **64** (1982), 220–4
6   EEB, observing notebook; LO
7   E. E. Barnard, 'Discovery and Observations of a Fifth Satellite to *Jupiter*,' *AJ*, **12** (1892), 81–5:81
8   E. E. Barnard to the editor of the San Francisco *Examiner*, December 25, 1893
9   SWB to EEB, telegram, September 12, 1892; VUA
10 JEK to EEB, September 13, 1892; VUA
11 Charles Burckhalter to EEB, September 15, 1892; VUA
12 EEB to Joseph S. Carels, September 21, 1892; VUA
13 San Francisco *Examiner*, September 15, 1892
14 San Jose *Mercury*, September 15, 1892
15 San Francisco *Examiner*, November 11, 1892
16 Mrs. A. E. Barnard to EEB, *ca* 1905; VUA
17 Barnard, 'Discovery and Observations,' p. 81
18 San Francisco *Examiner*, September 25, 1892
19 A. A. Common, 'A Fifth Satellite of Jupiter,' *Observatory*, **15** (1892), 357
20 A. A. Common, 'Jupiter's Fifth Satellite,' *Observatory*, **15** (1892), 388. Holden later tried to absolve himself of responsibility in a letter dated October 25, 1892 and published in ibid., pp. 452–3: 'The desired information was not "withheld." I have no knowledge as to why the telegram did not reach England in the form in which it was sent from here. I regret that any failure occurred. We are, however, in no wise responsible.'
21 San Francisco *Chronicle*, October 12, 1892
22 C. Flammarion to ESH, October 16, 1892; SLO
23 E. W. Maunder to EEB, January 2, 1894; VUA
24 A. A. Common, letter to *The Times* of London, December 28, 1892
25 San Francisco *Examiner*, September 17, 1892
26 San Francisco *Bulletin*, September 26, 1892
27 San Francisco *Examiner*, September 25, 1892
28 ibid.
29 San Francisco *Bulletin*, September 26, 1892
30 San Francisco *Bulletin*, September 28, 1892
31 San Francisco *Call*, September 27, 1892
32 San Francisco *Chronicle*, October 3, 1892
33 ibid.
34 Wilkes-Barre *Telephone*, December 17, 1892; newspaper clipping in SLO
35 EEB to San Francisco *Examiner*, December 12, 1892; VUA

36   John W. Streeter, in 'John Winthrop, Junior, and the Fifth Satellite of Jupiter,' *Isis*, **39** (1948), 163, identifies the 6th magnitude star which Winthrop took for a moon as HR 7128. For that matter, Christoph Scheiner, a Jesuit astronomer, had reported a fifth satellite of Jupiter as far back as 1612, but it too was an ordinary star which happened to share the same field with the giant planet, and was soon left behind as the planet and its true retinue of moons continued their motion to the east. A year before his death, Barnard replied to a scholar, Frederick E. Brasch, who called his attention to an article promoting Winthrop's claim to have discovered the fifth satellite: 'I saw this article a good many years ago. It seems to me it hits pretty hard at Wm. Herschel and Lord Rosse and others with powerful telescopes that are nearer to Winthrop's time! I suppose they never looked at Jupiter. It makes me feel specially humiliated just now, for I have only caught the feeblest glimpses of the fifth satellite with the [Yerkes] 40-inch during the past two years. Winthrop's telescope must have been specially good for him to have seen this faint object . . . in 1664. I am comforted, however by the fact that Dean Swift in Gulliver's Travels made the Liliputians discover the two little moons of Mars such a long time before Hall found them with the 26-inch at Washington. The satellite of Venus was also a product of the early telescopes.' EEB to F. E. Brasch, January 18, 1922; VUA

37   ESH, undated manuscript page; SLO

38   C. Flammarion to EEB, February 14, 1893; C. Flammarion to EEB, June 11, 1893; VUA

39   W. T. Lynn, 'Fifth Satellite of Jupiter,' *Observatory*, **15** (1892), 427–9:429

40   San Jose *Mercury*, September 15, 1892

41   San Francisco *Call*, September 13, 1892

42   San Francisco *Examiner*, September 15, 1892

43   SWB to EEB, September 19, 1892; VUA

44   EEB to SN, October 12, 1892; LC

45   SN to EEB, October 24, 1892; VUA

46   Lynn, 'Fifth Satellite of Jupiter,' p. 428

47   E. E. Barnard, 'Jupiter's Fifth Satellite,' *PA*, **1** (1893), 76–82: 81

48   E. E. Barnard, 'Micrometrical Observations of the Fifth Satellite of Jupiter during the opposition of 1893, with measures of diameters of Jupiter, 1892–1894,' *AJ*, **14** (1894), 97–104: 97

49   E. E. Barnard, 'Jupiter's Fifth Satellite,' p. 82

14

Travels and travails

1

During Barnard's darkest days on Mt. Hamilton, George Davidson, of the US Coast Survey and a leading figure in the California Academy of Sciences, had assured him that '[t]he devil will certainly get his due in time.'[1] The time had seemingly come. Though Barnard's troubles with Holden did not end with the discovery of the fifth satellite, he found himself in a greatly strengthened position. Among other things, he was able to turn the director's long resented monopoly of information about the Lick Observatory and its work on its head. Holden knew that Barnard, who had earlier resigned from the Astronomical Society of the Pacific in protest against him and had ever since boycotted its *Publications*, would never agree to cooperate with him by providing fresh material about the fifth satellite. Therefore, as editor of the *Publications*, Holden had no choice but to be satisfied with reprinting a brief excerpt from the *Astronomical Journal* article in which Barnard had first described his discovery (though rather self-servingly, Holden failed to reprint Barnard's comments about not getting use of the 36-inch on a regular basis until July 1). He added a few of his own remarks, though even these were based largely on second-hand information. As a result, the *Publications* were singularly reticent about the most talked about discovery of the day; Holden's account ran to a mere page, prompting Barnard to gloat to Crew: 'I send you a copy of A.S.P. Pub that I have come across. You will notice the brilliant account of the discovery of the V Sat. of Jupiter as given there by the Director! Oh! had it been somebody elses! The A.S.P. Pub would have had to be doubled in size!'[2]

Though Barnard's colleagues, Schaeberle and Campbell, would have been eager for the glory of publishing the first orbit for the new satellite, Barnard regarded them as Holden's allies, and snubbed them by sending his measured positions directly to the *Astronomical Journal*. Holden later complained to E. B. Knobel, 'You noticed his blunders about the distance in miles of his new satellite. They all came from not understanding and nobody looks over his work now[.] I just let it go because half of it is always good - and he is responsible for what he prints. The orbit of his new satellite I offered to have computed for him as he couldn't possibly do it himself - but he preferred to print his observations in full . . .'[3]

2

Burnham grasped immediately the opportunity that Barnard's sensational discovery offered for fresh intrigues against the director. 'I need not say how immensely glad I am,' he enthused to Barnard immediately after the announcement of the fifth satellite. 'This puts you on the top of your crowd there and all others everywhere. This will settle a lot of difficulties at Mt. H. and I hope you will take advantage of it.'[4] A week later he advised Barnard: 'You will insist upon having every chance now with the 36-in, both for measuring your [satellite], and other purposes. I think you may assume that it will not be observed elsewhere, to any extent, if at all, and therefore you want to attend to that yourself . . . I would insist upon the glass three nights per week for the present, and if [Holden] won't consent appeal to the Regents':

> You might as well make a test of this thing now, and if [the Regents] are going to keep things as they have been, you can afford now to throw up your hand and bring about a crisis at once. It would result in [Holden's] removal, and ultimately in your having charge. I fancy that now is the time for action . . . [T]he golden time has come . . . You are a lion now, and you must make the most of it, as any one else would.[5]

Burckhalter shared Burnham's assessment of the situation: '[Barnard] is in a position to *demand* anything almost, and get it,' he assured Burnham. 'There are two good friends of [Barnard's] here [in San Francisco] . . . that think he ought to resign, and they say the Regents would decline his resignation, and fire H[olden].'[6]

At the moment, Barnard contented himself with requesting more telescope time. Thus on September 22, 1892, he sent a brief note to Holden: 'I should like to observe the new satellite tonight . . . The East[ern] elongation is slipping away and it will be a long time before the western is favorable here.'[7] For some reason Holden did not favor this particular request, but a few days later Barnard wrote more boldly: 'I really think the satellite should be observed three nights a week while favorably placed – inasmuch as very soon the bad weather will cut it off altogether.'[8] During October and November, Barnard did indeed manage to obtain an average of one extra night a week on the great telescope expressly for the purpose of continuing his series of measures of the tiny moon.

3

Barnard was well aware of the embarrassed position in which his discovery had left Holden, writing to Crew: 'There is a terrible revulsion of feeling going on here – the Director is in a terrible position. The new discovery has hit him and his followers as nothing . . . ever hit before.'[9] Burnham argued that it would be 'easy enough' for Barnard to take over the director's duties himself – the job appeared complicated, he insisted, only because Holden approached it 'from the martinet and red-tape standpoint.' If Barnard cared to become director, he might easily take care of the business side of the job simply 'by finding a good commensense man to do it who need not be an astronomer at all.'[10] Yet Barnard had little relish for the idea of becoming director or resuming the struggle with Holden. The publicity over the discovery had left him 'all used up' and too 'unwell' to do anything, he confided to C. A. Young, and he added: 'My ill health is produced by worry, and I don't know how to improve it unless I go and jump in the Bay or clear out for the East.'[11]

Though he decided against jumping in the Bay, the idea of 'clearing out for the East' was becoming more and more attractive. 'I haven't had a moments peace since the new satellite turned up,' he complained to Crew. 'Every body is writing to me – I did not know I had so many friends all over the world.' This general adulation did not in the least improve his situation on Mt. Hamilton, which was, he pointed out, 'more strained than ever – and now I certainly have no desire to stay – I have got to get on to something elsewhere now, which I may take hold of.'[12]

He was in touch with Newcomb, who soon after the discovery of the fifth satellite had asked him whether he had done enough 'book studying' of mathematics and practical astronomy to be able to pass a civil service examination.[13] In a follow-up letter, Newcomb explained: 'I would like very much to see you here [in Washington], if a place suitable for your peculiar talents could be found. There is at present, I believe, a vacancy at the Observatory.' Nothing ever came of the offer, but the paternalistic Newcomb also gave the younger astronomer some advice. Never particularly fond of observing himself, he had regarded the night work at the US Naval Observatory as 'a drudgery,' and whenever he and his assistant got tired used to 'vote it cloudy' and go out for a plate of oysters at a neighboring restaurant. Not understanding the psychology of an observaholic like Barnard in the least, he suggested that Barnard ought to 'vote it cloudy' more often. '[R]est on your laurels as far as working after midnight is concerned,' he recommended. '. . . I think, in the long run, you will gain more by doing so.'[14]

Following this advice was, of course, impossible for Barnard. Indeed, in the months after his great discovery he pushed himself harder than ever. Naturally the fifth satellite of Jupiter claimed much of his attention, for he recognized that 'observations of the satellite, for this year at least, would rest almost alone upon my efforts.'[15] But as always, he pursued other lines of research simultaneously. His most important work was his Milky Way photography with the Willard lens (henceforth known as the Crocker telescope), which will be described fully later. On one of his exposures, on the constellation Aquila, taken on October 12, 1892, he accidentally registered the fuzzy trail of a 13th magnitude comet. This comet (1892 V) is memorable for being the first ever to be discovered photographically. Unlike his other comets, this one 'sought me,' he noted.[16] When the comet's orbit was computed, it was found that it had the unusually short period of 6.3 years, but though Barnard himself searched carefully for it at its expected later returns, it has never been seen again.[17]

Though Newcomb's offer of a position at the US Naval Observatory went nowhere, Burnham continued to dangle before Barnard the prospect of a position at the

University of Chicago, where, as earlier mentioned, President Harper had succeeded early that summer in winning over George Ellery Hale to the faculty. As part of their agreement, Hale was willing to allow the university to use his own Kenwood Observatory for students under his direct supervision, and also Hale's father, the elevator magnate, offered to donate to the university the observatory, with its 12-inch refractor and other instruments, on condition that his son still wanted the job after a year and Harper was successful in raising $250 000 or more for a larger telescope within another two years.[18]

An opportunity for the larger telescope arose in August 1892, when the younger Hale met Alvan G. Clark at the meeting of the American Association for the Advancement of Science in Rochester. Clark told Hale that two partly finished glass disks by Mantois of Paris, forty-two inches in diameter, were available to make the largest telescope in the world. They had originally been purchased by a group of Southern California promoters who had hoped to set such a telescope up on Wilson's Peak, near Pasadena, but their plans were based on land speculation, and when this speculative bubble burst the Clark firm, which had contracted to make the lens, were left holding the bag with the unfinished glass disks.

On October 2, 1892, Hale and Harper approached Charles Tyson Yerkes, one of the few men who were even richer than Hale's father, about providing the money to finish the lens. Yerkes was the slippery Chicago streetcar magnate who had introduced the system of cable transmission of power used to develop the surface traction lines of Chicago, and had already carried the cable under the river by a tunnel at LaSalle Street.[19] Harper and Hale caught him in a euphoric mood – he had just married a chorus girl whom Harper described as 'the most gorgeously beautiful woman I have seen for years,'[20] and he was now, according to Harper, 'red hot' to become the 'owner' of the world's largest telescope. Indeed, Harper was amazed to find that Yerkes did not hesitate 'on any particular,' even though the whole enterprise promised to cost at least half a million dollars.[21] What may have sold Yerkes on the idea was the information that he did not have to advance the money for the expensive dome and building, or even for the object glass, for two or three years, and he is said to have remarked to an associate that 'by that time his fortune would be fully made, or else he would be completely "busted." '[22] He may have had another motive as well. At that time his business was expanding rapidly, and he was in need of considerable credit. As a result of his announcement to build a great observatory to house the largest refracting telescope in the world, at a cost of several hundred thousand dollars, he found that his credit was greatly strengthened, and he was able to borrow all that he needed for his business ventures.

Thus, the University of Chicago seemed well on its way to obtaining its own great · refractor, and Burnham wrote to Barnard that Hale was 'very anxious to have you [here] and his idea is that the different observers shall be absolutely independent in their lines – no interference or oversight by any one.'[23] After his bitter experiences with Holden, Barnard could hardly be more receptive to the idea of a less tyrannical administration. But the Chicago offer was also appealing because it meant being close to Burnham once again. As Barnard confided to Crew: 'I envy your being so near to Mr. Burnham. I don't know how it is, but some how my life seems wrapped up in his. Besides my most sincere admiration for him as a man and astronomer there seems to be a love for him that surmounts everything else.'[24] Even Holden fully expected Barnard to leave, telling E. B. Knobel:

> I rather think he will be invited to go to Chicago to do the visual work on the New Yerkes Telescope. We shall be sorry to lose him - very - but I think he will be happier there. I wish him well wherever he goes. In one sense I shall not be sorry to have him go - for in his place we shall get a professional - and I must say that Amateurs are hard to get on with! The Chicago telescope will give him all the field he needs and we shall hear more of him yet.[25]

By the end of September 1892, it was being reported in the San Francisco papers that Barnard had already 'resolved to resign his position under Holden so soon as an opening presented itself elsewhere.' There was also the usual hue and cry for Holden's head that occurred on such occasions. Thus, a San Francisco *Examiner* editorial noted:

> Candidly, the people of California are sick of Mr. Holden and his overbearing methods. Were there any intrinsic worth hidden beneath this thick and unpleasant coating of selfish egotism, we could afford to disregard the storm of popular criticism . . . Even if Holden is incapable of intelligent research himself it is only fair that he should be restrained from antagonizing those at the Observatory who are.[26]

In order to stir up matters even more, Burnham, without Barnard's permission or even, apparently, his knowledge, leaked to the press the rumor of his interest in the Chicago position. The *Examiner* reacted immediately and with outrage: 'Not satisfied with dimming the luster of the Lick Observatory by the promised construction of a telescope . . . greater in diameter than that on Mount Hamilton, Chicago is now endeavoring to deprive that institution of the services of one of the greatest astronomers of the century, if not *the* most eminent.'[27] From far away England, Sir William Huggins agreed with the *Examiner*'s analysis, writing to Hale, 'I presume the recent disruptive forces apparently at work at the Lick have really originated the Chicago Observatory which is to lick the Lick.'[28]

The speculation about Barnard's imminent departure for Chicago caused a good deal of consternation locally, and indeed would continue to circulate in the rumor mill for the next several years. However, it was considerably premature. Even Burnham had to temper his initial enthusiasm about luring Barnard to Chicago, for the simple reason that there was not yet a position to lure him to. 'Now it will probably be at least a year before anything can be done practically,' he admitted to his friend, and at least two years before the 40-inch telescope would be ready (an estimate that itself proved to be overly optimistic by three years). Under these circumstances Burnham could only urge, 'In the mean time you can act with entire independence with this for the future. If you can get the first place at L[ick] O[bservatory], had you not better take it? Leaving the future as to change to be considered when the time comes.'[29]

4

Instead of remaining on Mt. Hamilton and intriguing to bring Holden down, as Burnham and others wished he would do, Barnard decided to parlay an invitation to lecture in Nashville into an extended tour back East. On December 31, 1892, he submitted his request for a two months leave of absence.[30] A few days later he learned that the French Academy of Sciences had voted to award him its prestigious Lalande Prize for his discovery of the fifth satellite, and decided to ask for an extension of four more months in order to travel overseas to receive the prize in person and make a grand tour of the leading European observatories. The Regents, only too eager to gratify their famous but increasingly restive young astronomer, granted his request at once, with full salary, which as the San Francisco *Bulletin* did not fail to remark was 'a perquisite that many Berkeley men have often sighed for, but seldom received.'[31]

Accompanied by Rhoda, Barnard left Oakland by train on the evening of February 11, 1893. After giving lectures accompanied with slide presentations of his comet and Milky Way photographs in Butte, Montana and Northfield, Minnesota (at Carleton College), he arrived in Chicago on February 23, on a 'raw wet cold' afternoon, and stayed a few days with Burnham. He then continued south to Nashville, which neither he nor Rhoda had seen since their departure for the Lick Observatory five years earlier. Just before the now famous native son's lecture to a large and enthusiastic audience at the Vendome theater on March 6, 1893, Chancellor Garland, eighty-three years old, announced the wish of the young astronomer's *alma mater* 'to anticipate the result of the world in bestowing . . . honors' by conferring upon him the honorary degree of Doctor of Science.[32] Barnard's lecture that night, on 'The Astronomer and His World,' was so well received that he was prevailed upon to repeat it to an equally packed house a few nights later.

After leaving Nashville, Barnard lectured in St. Louis and in Chicago again (where he was falsely rumored in the Nashville papers to have received $1000 for one lecture)[33]. Continuing east, he visited Warner and Swasey in Cleveland, Keeler at the Allegheny Observatory in Pittsburgh, and Newcomb in Washington. In Rochester, New York, he was introduced on the lecture platform by his old friend Lewis Swift. There were further lecture stops in Hartford, Connecticut, and New York City. While making preparations to embark from New York to Liverpool, England, to continue the European part of the tour, Rhoda suddenly fell sick, as she so often did on these occasions, but she recovered in time for them to depart as planned on the morning of May 20, on the *Aurania*, a steamer of the Cunard line.

After landing at Liverpool, the Barnards traveled by rail through the beautiful English countryside to Bradford, in Rhoda's native Yorkshire.[34] She would remain there with relatives while Barnard continued the rest of the journey alone. Though one suspects that she would have liked to have been with him when he received his accolades, she had always been intensely uncomfortable in social situations with astronomers, and especially their wives, most of whom came from a much higher social class – it is a well-documented fact that she almost invariably developed vague,

Barnard in 1893. Mary Lea Shane Archives of the Lick Observatory

probably psychosomatic illnesses just before he had to travel somewhere to accept an honor or award.[35] She was no doubt relieved to be able to stay among her own humble people in Yorkshire.

In London, Barnard visited the Tulse Hill observatory of Sir William and Lady Margaret Huggins and met astronomical historian Agnes Clerke and British statesman Lord Randolph Churchill. After visits to the observatories of Greenwich and Cambridge, he set out for the Continent with Arthur Cowper Ranyard, a prominent figure in English astronomy of the day and editor of the illustrated journal of science *Knowledge*. The channel crossing between Folkstone and Boulogne was uneventful, but on the train to Paris, Ranyard was stricken with a very bad case of hayfever, but even so he never lost his extreme politeness, as Barnard marvelled at the time.

In France, Barnard and Ranyard visited the Paris Observatory, which had been founded by Louis XIV, and Meudon, where the 33-inch refractor, the masterpiece of the brothers Paul and Prosper Henry, had just been erected under the supervision of the eighty year old solar astronomer, Jules Janssen. They also made the brief trip to Juvisy, about half an hour by train from Paris in the direction of Fontainbleau, to see Camille Flammarion, the greatest popularizer of astronomy in the nineteenth century and the founder of the Société Astronomique de France. Flammarion entertained them in his beautiful eighteenth century château, which he had received as a gift from a wealthy admirer. Barnard, seated opposite to the French astronomer during *déjeuner*, had an excellent opportunity to study his face, which he thought 'an extremely interesting one – poetical, full of thought and expression. [It] is rather swarthy with dark penetrating eyes.' Some of Flammarion's assistants were also there – in particular, Barnard was taken with one young man who 'also had the face of a poet . . . and soft brown eyes.' This may have been Eugène Michael Antoniadi, a Greek-born astronomer who had just come to Juvisy to work as Flammarion's assistant. Flammarion proposed a toast to the health of each visitor, as well as of Barnard's absent wife, though Barnard himself abstained, to the great surprise of his host. Ranyard, on the other hand, became full of conviviality in fluent French, so that as Barnard commented afterwards, 'So much did he enter into the spirit of these Frenchmen that Flammarion finally complimented him by saying "Mr. Ranyard is a perfect Frenchman."' After being shown Flammarion's library and observatory, which was equipped with a fine 9-inch refractor, the visitors went into the orchard immediately adjoining the château and wandered among the cherry trees eating the succulent cherries. Barnard took some of the pits along with him, telling Flammarion of his plans to plant them in his orchard on the Mt. Hamilton Road, and he promised that if they turned out satisfactorily he would call them *cerises Flammarion*.[36]

After Paris, Ranyard returned to London, while Barnard continued alone on his grand tour to the observatories of Marseilles and Nice. In Milan, he met Giovanni Schiaparelli, the discoverer of the 'canals' of Mars, who though he did not speak English, was able to read and write it perfectly, so that the two astronomers were able to converse by writing.[37] From Milan Barnard crossed the St. Gothard Pass to Lake Lucerne, and traveled on to Munich, Vienna, and Dresden, where he met a fellow traveler abroad, Mark Twain. In Berlin, Barnard saw the telescope that Johann Gottfried Galle had used in 1846 to make the visual discovery of Neptune. He visited the observatories of Potsdam, Bonn, Coblenz, Heidelberg, and Strasbourg before returning, via Brussels, to England. Still he had not had enough, and after rejoining his wife in Yorkshire, the couple visited the observatory at Edinburgh and that at Birr Castle, in Ireland, where Lord Rosse had set up his great reflectors. So long insecure and unsure of himself, Barnard thoroughly enjoyed the adulation of the European astronomers. The poor boy had made good, and now had the princes of the astronomical world sitting at his feet.

The Barnards finally made the transatlantic crossing once again, and on the way back to California stopped in Chicago, which was then holding its Columbian Exposition – 'the greatest thing the world ever saw,' Barnard wrote.[38] No doubt of special interest to him were the pier and mounting of the great Yerkes telescope, ordered from the firm of Warner and Swasey of Cleveland only the previous autumn and – in what was a remarkable engineering feat – already completed that summer in time to be installed as an exhibit in the Exposition's manufacturer's building in Jackson Park near the Lake Michigan shore. However, the telescope's final destination had yet to be determined. Aside from the fact that it was felt to be impracticable to situate the new observatory at a distance not much greater than 100 miles from the city of Chicago, 'without materially affecting its value as one of the departments of the University,'[39] which eliminated Wilson's Peak in California from the running, there were any number of offers of land and other inducements held out by individuals and by towns hoping to lure the prestigious institution their way, including two dozen sites in Illinois and one – Lake Geneva – in Wisconsin.

5

When Barnard returned to Mt. Hamilton on August 24, 1893, he pointedly avoided reporting in person to Holden. The director, who had noted to Bonté, the secretary of the Regents, shortly before Barnard's return that during his absence he had felt almost as if he were on vacation,[40] complained to Phelps and Crocker of this neglect of official courtesies, which included not only his failure to report to him on his return to Mt. Hamilton but his failure subsequently to inform him of his absences from the mountain for the purpose of delivering lectures. The Regents were dragged in yet again to resolve the dispute, and wearily penned yet another letter in which they informed Barnard, in as nonconfrontational a way as possible, that 'whatever the personal relations of men may be, official courtesies should never be ignored or overlooked.'[41] Henceforth Barnard informed the director of the occasions when he needed to be away from Mt. Hamilton by means of terse written notes. It was clear that Barnard's leave of absence from the observatory had resolved nothing at all – relations between the two men resumed on as strained a basis as ever, if not more so.

Though his conflicts with Holden were no secret, having been so well publicized in the newspapers, Barnard had up to this point generally managed to keep his bitterness out of the professional journals. However, this had begun to change. Even before he had left for Europe, he had published in the *Astronomical Journal* his latest set of measures of the fifth satellite, to which he had added the comment: 'The greatest enemy . . . to accurate measures is the wind . . . Very often, in the exposed position that Mt. Hamilton occupies, there are winds prevailing night after night [and] when it is necessary to face these winds, as has been the case in many of the observations of this new satellite . . . the great portion of the telescope exposed to them causes a constant vibration of the tube.' He suggested that in the construction of domes for large refractors, a protecting screen could be introduced in the observing slit to cut off the direct wind from the tube, and for this purpose proposed using a canvas curtain which could be raised from a roll at the base of the tube.[42] In response to these comments, Holden later published a note in *Astronomy and Astro-Physics*, a new journal edited by Hale and W. W. Payne, which explained that, in response to similar complaints by Burnham, the director had hired a machinist three years earlier to put up two rods inside the dome, one on each side of the slit, together with a series of loose sliding rings to each rod. Holden elaborated, 'We never went so far as to attach canvas to these rings, for Mr. Burnham finally decided that he did not wish it to be done on his account. The rods and rings have been in place for several years, and any of the experiments suggested can be tried at any time in half an hour by anyone interested who knows of the existence of the rods, etc.'[43] After wasting countless hours trying to make sensitive measures while the great telescope was shaking like the proverbial reed in the wind, Barnard was furious to learn that such a device had already existed but that he had never been informed of it: 'I was very much surprised in reading the note from the director of Lick Observatory . . . concerning certain experiments that had been commenced . . . to provide means for reducing the effect of the wind upon the 36-inch telescope. I had never heard of any such experiments . . . I had never heard of the existence of the "rods," etc. which had been attached to the slit for the purpose of conducting such experiments.'[44] Burnham also entered a note of sarcasm: 'Certainly Professor Holden in saying that I finally did not care to have this improvement made on my account is in error, and he has probably confused me with some one else in respect to this matter. As a matter of fact I was not aware that the rods referred to had been attached, but as my work in the dome was entirely at night, they might have easily escaped my notice.'[45] Despite the recognized need, it was not until August 1895 that a windscreen was finally put in place, and Barnard wrote sarcastically to Hale: 'I noticed today that a windscreen . . . was being put on the slit of the 36 in. dome!!! "To him who waits, all things will come." '[46]

In these disputes the public always took Barnard's side and wanted Holden ousted, and the California newspapers, especially the San Francisco *Examiner*, were no less relentless in their attacks on the director. Astronomers, however, generally tried to take a more evenhanded view of the controversies. Rather typically, Huggins confided to Hale: 'We at a distance are rather puzzled what to think. The general view seems to be that the Director is a little difficult to get on with.' Even Newcomb felt an obligation to tell Barnard confidentially and at length 'how things look from this side of the continent':

> To the outer world the regime of the Lick Observatory seems of the most liberal kind. You and the rest of the staff seem to make what observations you can, and to publish them at pleasure without supervision, whereas the general rule is, at all observatories of a private character, that the director alone has the right to publish

work done at the institution. The theory is that the work belongs to the obs[ervatory] and not to the worker, so the former alone has the right to it, and in publication the director represents the institution.

Now I would like you to think over another point you mention. Why should H[olden] want to get rid of men like yourself and Burnham? He was the latter's most devoted friend, and supporter, and when the observatory was started, invited him to it because he believed him the best man in the world to do work of a certain kind. Why such a change came over the spirit of his dream? Not jealousy of the work he did or of the reputation he was making, because he himself knew *all* about those before he sent for him . . .

Similar remarks apply to your own case. What has seemed most questionable in his conduct toward you was his failure to publish officially the discovery of the 5th satellite of Jupiter. But it has also been rumored that he left the announcement to you because you wished it. I asked you about this at my house, but you evaded the subject . . .[48]

Barnard wrote back that 'If I ever have the opportunity to see you again I shall expose to you the rascality and humbugery of the man who has made my life here one of misery.'[49] But though Barnard had many such opportunities, Newcomb remained unconvinced – he qualified but never completely withdrew his support from his former protégé.

Keeler, too, refused to join the attacks on Holden. In response to an editorial by W. W. Payne in *Astronomy and Astro-Physics* about the defections of the best and the brightest from the Lick staff, the brilliant spectroscopist drafted a carefully worded defense of his former director: 'I desire to say that I resigned more than a year ago for private reasons which were in no way connected with the administration of observatory affairs.'[50] Barnard felt betrayed, and protested to Davidson, 'I am not so much surprised at the note from Mr. Keeler, after knowing as I do the power that was brought to bear on him when he left and since he left':

It is pretty hard to withstand a severe attack of flattery and adulation. This was piled on for political reasons – and no one knows how to carry a point that way better than the Director of the L.O. I regret that Mr. Keeler has done it because I well knew his feelings and his estimate of the Director during the greater part of the time he was here, what ever may be his feelings and opinions now . . . Well there are two men in this world who will go down to the grave without writing such a note and one is S. W. B[urnham] and the other can be no one else but E. E. B[arnard]. I have a great regard for Mr. Keeler and I cannot believe he has carefully weighed the manner in which his note will be used by one who never had any love for him and who has flattered him only for political reasons. Ah! Me. So goes the world. But they will all be even, and all rights will be straightened out and all wrongs will be distorted and twisted and finally come out on top when we are all dead![51]

Among others expressing an opinion, Lewis Boss, the director of the Dudley Observatory, regarded Barnard's behaviors in an unfavorable light, but suggested that they were partly to be excused as 'due to the former influence of another' (i.e.,

Burnham).[52] Far less charitable was Seth Carlo Chandler, who regarded Barnard as little more than a spoiled *prima donna*: 'If not trenching on too delicate a matter,' he told Holden, 'I have for some time desired to say, with reference to certain public references to the trouble which thin-skinned and big-headed assistants of the Lick Obs. have been stirring up, that my sympathies are with you':

> So far as we outsiders can judge, you appear to have been most magnanimous and liberal . . . I fear that Barnard has been under bad influences. He ought to have developed into a good man. But it looks as if he had been spoiled by adulation, and over-estimate of his own abilities and standing . . . I know too that other astronomers feel much the same way. It would have been far better for him if he had seen the need for instruction and training; but I suspect that he thinks he knows it all, because he has a good eye. He ought to reflect that if old Galileo were alive today, in the present condition of astronomy, he would have been cooking larger fish than satellites; also that Hall's position in the world of science (not in the popular estimation) is entirely disconnected with his discovery of the Mars satellites. Yet I fear he considers that Galileo, Hall and Barnard, occupy or will occupy companion-niches in the future temple of science; and consequently has no need of wise kindly counsel by lesser men. In this I may be doing him a wrong, but all we can judge such things by is his actions and speech, which do not seem, in all things to have been in the best taste.[53]

The vicious tone is surprising coming from one who in earlier days had been one of Barnard's staunchest supporters, but is probably sufficiently to be accounted for by professional jealousy. In any case, its assessment is decidedly unfair. No one was more eager than Barnard himself to 'cook larger fish than satellites.' Indeed, rather than rest on laurels won, Barnard threw himself into varied activity, and the next two years, for all his personal unhappiness with conditions on Mt. Hamilton, were to be arguably the most productive of his career. Among the objects he studied, none more arrested his attention – or that of the astronomical world – than the planet Mars, which in 1894 came to opposition and was better placed for observation than it had been for years.

1   GD to EEB, April 14, 1892; VUA
2   EEB to HC, October 11, 1892; SLO
3   ESH to E. S. Knobel, June 16, 1893; YOA
4   SWB to EEB, September 12, 1892; VUA
5   SWB to EEB, September 19, 1892; VUA
6   Quoted in ibid.
7   EEB to ESH, September 22, 1892; SLO
8   EEB to ESH, September 27, 1892; SLO
9   EEB to HC, September 18, 1892; SLO
10   SWB to EEB, October 6, 1892; VUA
11   EEB to CAY, October 19, November 9, 1892; DCL
12   EEB to HC, October 11, 1892; SLO

13   SN to EEB, September 30, 1892; VUA

14   SN to EEB, October 24, 1892; VUA

15   E. E. Barnard, 'Micrometer Measures of the Fifth Satellite of Jupiter,' *AJ*, 12 (1893), 161–74

16   EEB to J. T. McGill, November 10, 1921; VUA

17   As noted in Gary W. Kronk, *Comets: A Descriptive Catalog* (Hillside, New Jersey, Enslow Publishers, 1984), p. 222, the comet's orbit was carefully recalculated by D.K. Yeomans in 1975. Yeomans' orbit indicated a close approach, within 0.09 AU, of Jupiter in September 1922, which when 'coupled with the uncertainties in the initial conditions, would prevent realistic computation of [the comet's] motion beyond 1922.'

18   William E. Hale to WRH, July 1, 1892; RL

19   Yerkes fascinated novelist Theodore Dreiser, whose trilogy *The Financier*, *The Titan*, and *The Stoic*, was largely based on Yerkes's life.

20   WRH to F. T. Gates, undated; RL

21   WRH to F. T. Gates, October 10, 1892; RL

22   E. B. Frost, *An Astronomer's Life* (Boston and New York, Houghton, Mifflin & Co., 1933), p. 97

23   SWB to EEB, October 14, 1892; VUA

24   EEB to HC, October 11, 1892; SLO

25   ESH to E. B. Knobel, June 16, 1893; YOA

26   'Holden and the Regents,' San Francisco *Examiner*, September 25, 1892

27   'Holden's Notorious Record,' San Francisco *Examiner*, October 14, 1892

28   Sir William Huggins to GEH, December 10, 1892; YOA

29   SWB to EEB, October 14, 1892; VUA

30   EEB to ESH, December 31, 1892; SLO

31   San Francisco *Bulletin*, February 24, 1893

32   Nashville *Daily American*, March 6 and March 11, 1893. The faculty of the Academic Department at Vanderbilt had originally recommended that the honorary degree of Master of Arts be conferred on E. E. Barnard, but at almost the last minute changed in its mind in favor of the Doctor of Science. The question of whether the degree was, despite the University's policy of not conferring honorary degrees, indeed honorary is considered at length in Robert T. Lagemann, 'Has Vanderbilt Granted Honorary Degrees,' Vanderbilt Graduate School *Almanac*, Fall 1972.

33   Nashville *Daily American*, March 11, 1893. In the scrapbook in which he kept these clippings, Barnard wrote next to this entry 'Not so.' VUA

34   EEB, travel notes; VUA

35   D. E. Osterbrock to W. Sheehan, personal communication, March 29, 1990

36   EEB, untitled MS draft describing the trip to Juvisy; VUA

37   The notes are in the Vanderbilt University Library, and were first published by George Van Biesbroeck, 'E. E. Barnard's Visit to G. Schiaparelli,' *PA*, 42 (1934), 553–8.

38   EEB to SN, August 12, 1893, LC

39   G. E. Hale, 'The Yerkes Observatory of the University of Chicago: Selection of the Site,' *ApJ*, 5 (1897), 164–80: 171

40   ESH to J.H.C. Bonté, August 3, 1893; SLO

41   T. G. Phelps, C. F. Crocker and H. S. Foote to EEB, November 21, 1893; SLO

42   E. E. Barnard, 'Micrometer Measures of the Fifth Satellite of *Jupiter*,' *AJ*, **12** (1893), 161–74:161

43   E. S. Holden, 'Screens to Protect Telescopes from Wind Tremors,' *AA*, **12** (1893), 471

44   E. E. Barnard, 'Wind at the Lick Observatory,' *AA*, **12** (1893), 573

45   S. W. Burnham, 'The Lick Telescope Disturbed by Wind,' *AA*, **12** (1893), 572–3

46   EEB to GEH, August 7, 1895; YOA

47   Sir William Huggins to GEH, December 10, 1892; YOA

48   SN to EEB, August 19, 1894; VUA

49   EEB to SN, September 11, 1894; LC

50   J. E. Keeler, 'Note,' *AA*, **11** (1892), 840.

51   EEB to GD, November 17, 1892; BL

52   Lewis Boss to ESH, August 13, 1895; SLO

53   SCC to ESH, April 16, 1893; SLO

15

Barnard and Mars

1

Fed by the gross sensationalism of newspaper reports, the Mars mania of the 1890s grew steadily with each opposition, and the public's expectation of further news about Mars ran very high. As earlier noted, at the favorable opposition of 1892, the planet's low southerly declination led to disappointing results at most northern observatories. Nevertheless, 'projections' on the planet's terminator, of which some had been reported at the 1888 and 1890 oppositions, were again observed, leading to speculations about Martian signal lights being used in the attempt to communicate with Earth. Barnard dealt humorously with such reports, and went so far as to write a story in which by means of paper letters 100 miles long earthlings finally succeeded in sending Mars the message: 'Why do you send us signals?' At last the answer comes back: 'We do not speak to you at all, we are signaling Saturn.'[1]

Easily the most sensational reports of the opposition, however, were received from Harvard's William Henry Pickering, who in 1892 observed the planet with a 13-inch refractor at the Harvard Observatory's southern station at Arequipa, 8100 feet high in the Peruvian Andes – an altitude nearly twice that of the Lick Observatory. Pickering maintained that the atmospheric conditions at Arequipa were nearly perfect, and cabled his observations of Mars to the New York *Herald*:

> September 2. Mars has two mountain ranges near the south pole. Melted snow has collected between them before flowing northward. In the equatorial mountain range, to the north of the gray regions, snow fell on the two summits on August 5 and melted again on August 7.
> October 6. [Have] discovered forty small lakes in Mars.[2]

Noting that 'these and similar telegrams from South America . . . were received by the astronomers at the Lick Observatory with a kind of amazement,' an incredulous Holden demanded: 'How does [Pickering] know the dark markings are lakes? Why does he not simply call them dark spots? And is he sure there are forty?'[3]

Aside from Mars, Pickering also paid close attention to the Galilean satellites of Jupiter. On October 9, while making micrometer measures of the diameter of Io with the 13-inch refractor, his attention was arrested:

> At first glance I noticed to my surprise that its disc was not circular, but very elliptical . . . Observations on the next evening confirmed my first measurements.

Some of the other satellites were also measured, and I then returned to the first one, when to my astonishment, instead of showing an elliptical disc, it showed one that was perfectly circular, precisely like the other satellites. I could scarcely believe my eyes, but as I continued to watch and measure, I saw the disc gradually lengthen again and assume the elliptical form, and I then understood what had really been found. The 1st satellite has the form of a prolate spheroid or ellipsoid, or in popular parlance is 'egg-shaped.' The two minor axes are approximately equal, and the satellite revolves about one of them, or as we may say, it revolves 'end over end.'[4]

A few days later Pickering cabled these latest results to the New York *Herald*, and followed up with a series of articles on the Galilean satellites in *Astronomy and Astro-Physics*, a short-lived journal which, in 1895, was succeeded by *Popular Astronomy* and the *Astrophysical Journal*. Here he described his further observations of Io, from which he announced a refined rotation period of 13 hr 3 min, and also his discovery that the disks of the other Galilean satellites varied between circular and elliptical.[5] His assistant, Andrew Ellicott Douglass, confirmed his results, though the influence of suggestion is strongly indicated; Pickering noted that initially 'Mr. Douglass, while readily confirming the ellipticity of the 1st, declared that the others always appeared to him circular . . . Of late, probably owing to training of his eye, Mr. Douglass has been able to confirm my observations upon the three outer satellites, and we now both see them elliptical at the same times, and our position angles agree with one another within a few degrees.'[6]

Given his own experience with Io's elongated form during its transits, Barnard's interest was naturally piqued by these findings, and as soon as he returned to Mt. Hamilton from Europe he made it a high priority to carefully examine the forms of the Jovian satellites with the 36-inch refractor. He had by now been promoted to two nights on the great equatorial every week, Sundays and Mondays. But though he used magnifications of up to × 1500, not once did he suspect the slightest deviation from the circular form in any of the satellites whenever they were seen against the dark sky. In publishing his conclusions, he assumed the tone of the Olympian, supremely confident of his keen eye and of his powerful telescope. '[T]he 36-inch refractor of the Lick Observatory,' he announced,

is unquestionably better adapted to settle a question of this kind than a 13-inch – even though the 13-inch may have a better atmosphere to work in, and I am not sure it has . . . Understanding this, it must be clear that the Lick telescope should be a criterion in this case and by its verdict, if rightly interpreted by me, the Arequipa observations of the forms of these satellites must stand or fall.[7]

Barnard was emphatic that large apertures were required to decide such questions, and on this occasion his verdict was – as it almost always would be whenever he set out to scrutinize the supposed discoveries of others using smaller apertures – negative.

Parenthetically, we now know that the Galilean satellites, and especially the innermost Io, are alternately squeezed and unsqueezed by tidal forces each time they orbit the planet. But these distortions are much too small to have had anything to do with the effects perceived by Pickering, for which some kind of optical effect, probably slight deformations of disc outlines caused by variations in atmospheric seeing, can only have been responsible. But whatever the exact explanation, Pickering continued diligently to record the varying disc shapes for many years. How he could have spent decades interpreting such manifestly optical phenomena as real is a question that obviously belongs to the psychological realm, and raises, at the very least, serious questions about his critical judgement.[8] But his view of the Jovian satellites was entirely on par with many of his other outrageous claims – for instance, his later conviction that he had found evidence not only of active volcanic activity, but of snow and ice, clouds, vegetation, and even swarms of insects on the Moon! Barnard, who generally regarded the Moon as an object to be 'execrated for spoiling the night with its undesired brilliancy as the Sun had already partly done by cutting out a large portion with which to make the day,' but who did observe the Moon on a few occasions, reacted to these results by calling them 'so startling – bordering, as they do, on the sensational – that one hesitates to accept them, or rather to accept his conclusions.'[9]

2

In 1894, Pickering had not yet alienated serious astronomers because of his extremist views, and his prominence and passion for visual studies of the Moon and planets made him a natural choice as a mentor for Percival Lowell, a bright, energetic and restless Bostonian who had just then developed an all-consuming passion for Mars. Lowell would become, 'of all the men through history who have posed questions and proposed answers about Mars, the most influential and by all odds the most controversial.'[10] He was also to prove in some ways a foil to Barnard, so that his career deserves to be considered in some detail.

Unlike Barnard, who had struggled heroically to overcome the disadvantages of extreme poverty and little formal education, Lowell was the scion of a prominent and wealthy New England family. After graduating with honors from Harvard, he spent several years managing his grandfather's textile business, then abruptly forsook the family business and spent ten years in the Far East.

But astronomy was also among the interests of this self-described 'man of many minds and many moods.' As a boy, Lowell had observed Mars with a small telescope from the flat roof of the family mansion in Boston, and on his last trip to the Far East, in the fall of 1892, he had carried with him a 6-inch Clark refractor. Just before sailing out, he had gone so far as to arrange, through Burnham, a meeting with Barnard at the Palace Hotel in San Francisco, and though unfortunately there is no record of whether they actually met and if so what they discussed, one would guess that Mars would have ranked prominently among the topics.[11]

In January 1894, only a month after returning from this last trip to the Far East, Lowell arranged to meet Pickering and Douglass in Boston. The meeting was probably arranged through his cousin, A. Lawrence Rotch, an amateur meteorologist who had

Percival Lowell. Mary Lea Shane Archives of the Lick Observatory

accompanied W. H. Pickering on the Harvard College Observatory eclipse expedition to California in 1889 and had also joined him at Arequipa for the purpose of setting up a meteorological station there. Pickering, especially, made a strong impression on Lowell. His emphasis on the need for a supremely steady atmosphere in order to make the most sensitive planetary observations struck Lowell as a revelation. Then and there he resolved to set up a new observatory in Arizona, a location where Barnard among others had predicted that very favorable observing conditions might exist, for the express purpose of studying Mars at the forthcoming opposition of October 1894. Though the planet at its closest would be, at 40 000 000 miles, slightly farther away than at the 1892 opposition, it would be much higher in the sky and thus more favorably placed for scrutiny from northern observatories, and Lowell intended to make the most of it.

At first, the expedition seems to have been planned as a joint venture with Harvard, but before long Lowell was chafing at the implication, mooted in the press, that he was simply 'going along' as its well-to-do patron, and the official relationship with Harvard was somewhat acrimoniously dissolved – much to the delight of the gossip-mongering Chandler. Writing to Holden, Chandler noted gleefully that Harvard's director Edward C. Pickering, older brother of William H., had tried unsuccessfully to capture the Lowell Observatory for Harvard, 'but found he had caught a Tartar in Lowell who is a man who can see through a mill-stone, and not one it is safe to play as a sucker.' He added: 'Very curiously [Lowell] has the very lowest possible estimate of Edward's capabilities as a man of science . . . but, astonishing to relate, he is very much struck on William H.' Chandler correctly anticipated some of the dangers of this infatuation:

> Lowell, I am sorry to say, has not selected the right kind of companions for his astronomical picnic. He is handicapped by them. What he needed was some young, well-equipped astro-physicist (I mean with education & the mental qualifications of a scientific man), sound and conservative, to sit on his coat-tails and keep him down to business, and prevent wild flights of fancy . . . [H]e was well-warned before he started . . . as to his unfortunate environment, and professed to know what he had to deal with in that respect, but expressed unbounded confidence in his ability to keep W.H.P. under his thumb, as he put it . . . [12]

After dissolution of the formal Harvard connection, Pickering and Douglass, taking leaves of absence from Harvard, temporarily went on Lowell's payroll. In early March, Douglass went west to test seeing conditions at various sites in Arizona – Tombstone, Tucson, Tempe, Phoenix, Prescott, Ash Fork. In the end, Lowell chose for the site of his observatory an 'opening of the woods' on a hill just west of the town of Flagstaff, where he hoped to get the perfect seeing he was after and which appealed to him not least because of its altitude, 7000 feet. Meanwhile, Lowell and Pickering succeeded in borrowing two refractors, of 18 and 12 inches aperture, for the venture.

What Lowell hoped to accomplish on this expedition is well documented in an address he gave to the Boston Scientific Society on May 22, 1894, which was printed in the Boston *Commonwealth*. His main object, he stated, was to be the study of the Solar System:

> This may be put popularly as an investigation into the condition of life on other worlds, including last but not least their habitability by beings like [or] unlike man. This is not the chimerical search some may suppose. On the contrary, there is strong reason to believe that we are on the eve of pretty definite discovery in the matter.[13]

The strong reason had been provided by Schiaparelli's discovery of the 'canals' of Mars in 1877, with which Lowell was fascinated:

Speculation has been singularly fruitful as to what these markings on our next to
nearest neighbor in space may mean. Each astronomer holds a different pet theory
on the subject, and pooh-poohs those of all the others. Nevertheless, the most
self-evident explanation from the markings themselves is probably the true one;
namely, that in them we are looking upon the result of the work of some sort of
intelligent beings . . . [T]he amazing blue network on Mars hints that one planet
besides our own is actually inhabited now.[14]

Such, then, were Lowell's views before he can be said properly speaking to have put eye
to eyepiece at all – he did not arrive in Flagstaff to begin his series of observations of
Mars until May 28.

3

Lowell's pre-observational remarks to the Boston Scientific Society were not allowed to
pass without criticism from Holden, who in a widely quoted article in the *Publications of
the Astronomical Society of the Pacific* wrote: 'It seems to be the first duty of those who
are writing for [the] public to be extremely cautious not to mislead; and especially to
avoid over-statement. Conjectures should be carefully separated from acquired facts;
and the merely possible should not be confused with the probable, still less with the
absolutely certain.'[15] Perceptively, Holden had put his finger on one of Lowell's most
unfortunate and enduring characteristics – as historian W. G. Hoyt has put it, 'he
expressed [his] conclusions dogmatically, trusting in analogy too much perhaps, and
being too willing to accept the merely plausible for the actual in the absence of any
immediate alternatives.'[16] In his critique, Holden added for good measure a few
trenchant remarks about Pickering's observations of Mars and of the satellites of
Jupiter. Concerning the latter, he commented that the observations of Schaeberle,
Campbell, Barnard and his own at the Lick Observatory contradicted those made in
Peru, and in reference to a note by Pickering in *Astronomische Nachrichten* regarding
the bright belt on Io recalled that Barnard had observed this belt 'during 1893.' He
continued:

> Professor Pickering says that this bright belt is not 'a permanent one, for it certainly
> did not exist at the time of the opposition of 1892.' By this he can only mean that he
> did not see it in Peru with the 13-inch equatorial, and that so far as his observations
> go, it did not exist. If he had turned to the *Publications* of the A[stronomical]
> S[ociety of the] P[acific] for 1891, . . . he would have seen that Professors
> Schaeberle and Campbell regularly observed the belt during September and
> October 1891. Its existence was fully demonstrated at this time, which is the reason
> why there are no further published observations by these gentlemen.'[17]

Though Barnard agreed with Holden about Lowell's views of Mars, the paragraph
just quoted drew his fire. The way in which Holden had referred to his observations of
Io, Barnard complained, had done him 'a great injustice,' as he explained:

In speaking of the white belt you refer to observations of [Schaeberle and Campbell] made in 1891 September and October and refer only to mine of 1893. You have deliberately suppressed the original observations of mine concerning this phenomenon made in 1890 [and published in *Astronomische Nachrichten* at the time]. You thus give the impresssion that two years after their observations I had taken up the matter as something new.'[18]

Barnard insisted on a full correction in which his priority in recognizing Io's white belt was acknowledged. Holden refused. This brought an angry follow-up letter from Barnard: 'This thing must be righted and in a just and honorable manner or I shall be forced to do it myself. You must not quibble any more about this. Do the right thing and the matter shall be dropped, otherwise it will cause trouble.'[19] Nevertheless, despite this warning, as usual the matter ended up before the Lick Observatory Committee of the Board of the Regents, with Holden protesting the harshness of the language Barnard had used in his letters and Barnard complaining that Holden had once more set out intentionally to injure him. The sensitive Tennessean added:

Now Gentlemen, I give you my word, I have tried to live peacefully here as you have requested me to do. I have suffered petty annoyances rather than to make complaint, but when the Director of the Lick Ob[servator]y attacks me in the unfair and insidious manner that he has done . . . my manhood revolts at a quiet submission to such an insult.[20]

The Lick Committee, which at the time consisted of Phelps, Crocker, and San Francisco lawyer Henry Foote, as always tried to take an evenhanded approach to the dispute. On behalf of the Committee, Foote wrote in true lawyerly fashion to Holden of the 'inexpressable sorrow over the differences that have again arisen between you and Prof. Barnard':

Frankly in the opinion of the Committee both Complaints are well founded. We think your failure to mention in your communications the observations of the 1st moon of Jupiter made September 8, [1890], by Prof. Barnard and published in 'Astronomische Nachrichten', was an unfortunate oversight, as we cannot believe it was intentional on your part. [But] considering the strained relations . . . which have existed between you and Prof. Barnard, we think you should have used the most extreme care not to do him an injustice, even unintentionally or by accident; and when your attention was called to the matter we are of opinion you should willingly have given him the most ample amends.

Upon the other hand the communications of Prof. Barnard were couched in highly improper language, and in their whole temper and spirit were not such as we can commend . . . [But] we regard his hasty and improper words as eminating from a highly nervous temperament and a wounded spirit . . . [21]

A copy of the same letter was also sent to Barnard, and only then – after six weeks of mutual recriminations between the director and his staff astronomer – was the matter finally allowed to drop. In this case, at least, one's sympathies lie entirely with Barnard. As Holden himself had passed the original article about the belt on Io on to

Astronomische Nachrichten in 1890 and as he had an unusually retentive memory for such things (when he wished to), it is hard to believe that his omission was not intentional and malicious as Barnard believed. Even if inadvertent, a simple and straightforward correction such as Barnard demanded would have sufficed to set matters right.

4

Though Jupiter's satellites – including especially the fifth, in which he naturally took a paternal interest and measured on every available opportunity – continued to claim much of Barnard's attention during the summer and fall of 1894, the lion's share of his observing time on the 36-inch refractor was devoted to Mars. He began scrutinizing the still tiny disk on May 21, when he recorded a dusky patch in the midst of the south polar cap. The cap had not yet begun its rapid melting with the approach of Martian summer and at the time covered an area which Barnard estimated at about 365 000 square miles. He carefully followed the south polar cap's rapid melting during June and July.

Already by the end of July, Barnard was beginning to have breathtakingly detailed views of the planet's surface. On July 23, for instance, with the Mare Sirenum region on view, he recorded two small dusky spots which appeared 'very feeble and faint when near the middle of [the] disc' but grew almost black as they drew toward the terminator.[22] We now know that these spots were in the positions of the great Martian volcanoes, Olympus Mons and Arsia Mons, though neither Barnard nor anyone else had any idea what they were at the time.

As the planet's apparent diameter increased steadily with its decreasing distance that summer, Barnard attempted to keep pace by sketching it on an ever larger scale – from a scale of $1\frac{1}{2}$ inches to the diameter of the planet in July, to not quite 3 inches in early August, to 5 inches in mid-August and September. Though his customary practice had always been to make sketches of various celestial objects right in the pages of his observing books, he was by early September seeing such a wealth of detail on Mars that he began to make it a point to sketch it on separate pages of higher quality paper, in order to have the best chance of capturing this detail. On September 2 and 3, the seeing became so exceptionally steady at times that it allowed him to use magnifications of $\times 1000$ and more on the great refractor, and he made a dramatic series of 5-inch sketches of the disk. At the same time he wrote in his observing book:

> The past two nights while making drawings I have examined Mars most thoroughly under good conditions. The region of the lake of the Sun [Solis Lacus] has been under review. There is a vast amount of detail . . . I however have failed to see anything of Schiaparelli's canals as straight narrow lines. In the regions of some of the canals near the Lacus Solis there are details – some of a streaky nature but they are broad diffused and irregular and under the best conditions could never be taken for the so called canals.[23]

At his next opportunity to use the 36-inch, on September 9 and 10, he again watched

Mars, drawn by Barnard on the night of September 2–3, 1894 with the 36-inch refractor of the Lick Observatory. The region shown is that of the 'Eye of Mars' (Solis Lacus). Yerkes Observatory

Mars all night long, though because there was no water in the engines, he had to turn the dome and wind the clock by hand – 'dreadfully hard and exhausting work.'[24] The magnificent Hourglass Sea – Schiaparelli's Syrtis Major (the Great Bog) – was in view. What he saw on this night confirmed his earlier impressions, and he confided to Simon Newcomb:

I have been watching and drawing the surface of Mars. It is wonderfully full of detail. There is certainly no question about there being mountains and large greatly elevated plateaus. To save my soul I can't believe in the canals as Schiaparelli draws them. I see details where he has drawn none. I see details where some of his canals are, but they are not straight lines at all. When best seen these details are very irregular and broken up – that is, some of the regions of his canals; I verily believe – for all the verifications – that the canals as depicted by Schiaparelli are a fallacy and that they will so be proved before many favorable oppositions are past . . . It is impossible to adequately draw all that can be seen . . .[25]

5

Barnard's observations of Mars through the 36-inch refractor were of the first importance, and far in advance of anything achieved elsewhere. Some of his observing reports of that fall have 'the ring of discovery,' wrote the distinguished planetary astronomer Gerard P. Kuiper, and he added: 'In my judgment Barnard was the first astronomer who saw Mars as it can be seen with modern equipment.'[26] But though there can be no question that Barnard's views of the planet came to him as revelations, his public pronouncements on the subject at first gave no indication of anything radical. At a benefit lecture in San Francisco in early December 1894 (at which the attendance was 'pruned severely,' as the San Francisco *Examiner* reported, because of a terrific downpour and the blockade of several cable car lines by floods), he still made reference to the light and dark areas of the planet as 'lands' and 'seas,' and alluded to Schiaparelli's views about the canals as they had been presented in a series of articles translated from the Italian by Pickering and running in *Astronomy and Astro-Physics* that fall.[27] The only hint of his true views about the subject came in the comment: 'That these canals exist we must believe or refute the testimony of Schiaparelli and quite a number of other observers who now report seeing them. There are, however, not wanting good observers who cannot see them with any instrument.' In fact, it would be over a year before his views of Mars were published in detail.

But while Barnard, ever conservative, was in no rush to publish, Lowell did not share his scruples. Despite having observed the planet through only a single opposition, he blitzed the press, in a spurt of manic energy, with his sensational results. A series of articles on Mars appeared in W. W. Payne's new journal *Popular Astronomy*, even before the opposition was over.[28] Lowell followed this up with a similar series of articles in the *Atlantic Monthly* and, in February 1895, with well-attended lectures in Boston's Huntington Hall. Finally, in December 1895, he published his first book on the red planet, entitled simply *Mars*, which described in detail his observations and conclusions – that Mars was a desert planet, on which the inhabitants had had to build a vast system of irrigation to transport water from the melting polar caps in order to survive. This, in a nutshell, was Lowell's 'theory' – and needless to say, it created an immediate sensation.

Lowell's theory of intelligent Martian life unleashed a firestorm of controversy. The public was fascinated, while professional astronomers were generally more guarded, and some were openly hostile. While many competent observers recorded the canals – indeed, on the maps of 1890s they had become fashionable – other astronomers, including some using very large telescopes, had never seen them.

Lowell argued that 'Negative evidence is no evidence at all,'[29] and attributed the failure of skeptics to see the canals to the fact that they did not observe the planet in steady enough air. Moreover, he fired a volley at professional astronomers:

> [N]o amateur need despair of getting interesting observations because of the relative smallness of his object-glass . . . In matters of planetary detail size of aperture is not the all-essential thing it is tacitly taken to be . . . A large glass in poor air will not show what a small glass will in good air.[30]

After reading this, the generally amiable Keeler confided to Hale: 'I dislike [Lowell's] style . . . [I]t is dogmatic and amateurish. One would think he was the first man to use a telescope on Mars, and that he was entitled to decide offhand questions relating to the efficiency of instruments; and he draws no line between what he sees and what he infers.'[31] The two men, who were editors of the *Astrophysical Journal* which Hale founded in 1895, later refused to accept any of Lowell's papers for publication.

The most outspoken critic of Lowell's theory was, however, Barnard's colleague W. W. Campbell. Even before the 1894 opposition, Campbell had predicted that, while valuable scientific results would undoubtedly be forthcoming during the opposition, 'It is, unfortunately, perfectly safe to predict that we shall also hear from the sensationalists, astronomers included; and that fact is a source of sincere regret to all healthy minds.'[32] Campbell later published a hostile review of Lowell's *Mars* in which, among other things, he called attention to the fact that Lowell's observations made at Flagstaff supported his 'pre-observational views' as delivered in his lecture to the Boston Scientific Society, and ended by bluntly labeling Lowell an opportunist: 'In my opinion, he has taken the popular side of the most popular scientific question about.'[33]

Campbell himself had studied the planet carefully with the spectroscope during the summer of 1894. While in the 1860s and 1870s pioneer spectroscopists William Huggins, Jules Janssen and Hermann Vogel had announced the spectroscopic detection of water vapor on Mars, Campbell, with a more sensitive instrument and better technique, found that the spectrum of Mars appeared identical with that of the Moon in every respect. Holden, indeed, announced that the opposition of 1894 would be memorable chiefly for having proved the absence of a Martian atmosphere.[34] Here he went too far – the planet does have an atmosphere, though extremely attenuated. Campbell correctly deduced that its thickness was less than that at the summits of the highest terrestrial mountains, a result which was announced in banner headlines in the San Francisco *Chronicle*:

THAT MAN UP ON MARS
Very Thin Air for Him to Breathe.[35]

Lowell, notwithstanding this discouraging news, was undaunted, and observed wittily:

> One deduction from this thin air we must be careful not to make – that because it is
> thin it is incapable of supporting intelligent life. That beings constituted physically
> as we would find it a most uncomfortable habitat is pretty certain. But lungs are not
> wedded to logic, as public speeches show, and there is nothing in the world or
> beyond it to prevent, so far as we know, a being with gills, for example, from being a
> most superior person. A fish doubtless imagines life out of water to be impossible;
> and similarly to argue that life of an order as high as our own, or higher, is
> impossible because of less air to breathe than that to which we are locally
> accustomed, is, as Flammarion happily expresses it, to argue, not as a philosopher,
> but as a fish.[36]

Though privately skeptical of Lowell's pronouncements, Barnard, who did not share Campbell's bulldog temperament, carefully avoided criticizing them publicly. Nevertheless, he was eventually brought into the controversy through the need to respond to Lowell's provocative analysis of the complex questions of atmosphere, aperture, and astronomical 'seeing' which, by implication, presented a broad attack on his work.

6

Among Lowell's points was that in matters of planetary detail small telescopes were superior to large ones. He was not the first to claim this;[37] William F. Denning, of Bristol, England, a gifted amateur astronomer who, in addition to discovering several comets and being a leading expert on meteors, was an avid observer of the planets, had already been arguing this for a decade. In 1885, Denning, whose own telescope was only a 10-inch reflector, called attention to the fact that most of the important planetary discoveries seemed, like Schiaparelli's of the canals of Mars, to have been made in relatively small instruments (Schiaparelli had used an 8.6-inch refractor). At the same time, he pointed out, observers using large instruments had in some instances even yet failed to confirm some of these discoveries completely. He concluded that 'apertures of from 6 to 8 inches seem able to compete with the most powerful instruments ever constructed.'[38]

Lowell, drawing on studies of atmospheric seeing made by Pickering and Douglass,[39] expanded on Denning's basic argument, and protested that except on the rare occasions when the air was exceptionally steady – that is, layered out in undisturbed strata above the observer – cells of turbulence could produce what appeared to be excellent definition in a large glass, and yet the unsuspecting observer would be unable to see any markings at all: 'One observer . . . will mark the seeing perfect while he is unable to detect detail which at another station is visible in air tabulated as mediocre and the world decides which of the two is right by abstract reference to the size of their object glasses, a standard only preferable to that of the diameters of their respective domes.'[40] Lowell himself would regularly use diaphragms to stop down the aperture of the 24-inch refractor which he acquired in 1896, usually to 12 to 18 inches – effectively, so his critics would argue, making his large telescope into a comparatively small one.

Having had to fight his director tooth and nail for time on the large refractor, Barnard found it difficult to be sympathetic to the view that his work with that noble instrument could be regularly annihilated by that carried out by observers using smaller telescopes – in some cases much smaller ones. He knew that there were certain circumstances (for instance, when the object to be viewed was a faint diffuse nebulosity) when a small telescope was always preferable to a large one. Moreover, in the case of the brighter planets, the excessive light gathered by a large telescope was undoubtedly bothersome. Barnard and others realized that at least in part because of the elimination of the excessive brightness, the best planetary views were often to be obtained shortly after sunrise or before sunset rather than against a dark sky. At other times Barnard adopted the expedient of using a cap with a small hole, 0.03 or 0.04 inch, over the eyepiece, which he found a more convenient way to eliminate a planet's excessive glare and to improve the definition than diaphragming the instrument.[41]

He remained steadfast, however, in his conviction of the overall superiority of large telescopes for planetary work. At times, as he knew from first hand experience, the definition in an instrument like the 36-inch refractor could be simply breathtaking. Inevitably atmospheric tremors affected the definition in large telescopes more than small ones, as Barnard explained in an article penned for the San Francisco *Examiner* in December 1894:

> There will be nights on which [the observer] can successfully use a 6-inch glass that will not permit a satisfactory use of a 12-inch, and which would wholly forbid the use of a 36-inch. In this case the tremors present in the air would not be sufficiently magnified by the 6-inch to affect the clearness of the image . . . Such nights have occurred where features could be seen in the 12-inch that were entirely blotted out in the 36-inch. But let the conditions be the best for observing with the air steady, and the 36-inch is far ahead of the 12-inch. It is very seldom, however, that the tremulousness of the air is not more or less apparent in the 36-inch, and under such conditions it is difficult or impossible to use the highest powers of the telescope. One has to wait and watch patiently and snatch a moment here and there of steadiness to do his best work.[42]

In anticipation of the results to be realized with the Yerkes telescope which was then under development, he pointed out that doubtless it would be even more severely affected by atmospheric tremors than the 36-inch telescope of the Lick Observatory. Nevertheless, there would be times when the seeing would allow surpassing results to be obtained with it:

> Now let us increase our aperture to, say, 40 inches. The atmospheric conditions being the same, then this quivering of the air, which has been objectionable in the 36-inch will, through the greater power of the 40-inch, have become far more objectionable. Now let the two instruments remain under the same conditions, but

let the air grow more tremulous. We shall notice the effect soonest on the 40-inch, and after it has become unbearable in that telescope it will still be tolerable in the 36-inch, and, much later, in the 12-inch. Now, let us imagine another telescope still more powerful, say several times as powerful as the 40-inch. The effect of a slight disturbance in the air is multiplied just so many times more, and we should have to look long and often during a year to find a night that would permit only a few hours of good observing with that great telescope. In general, it would be so crippled by the unsteadiness of the air that its effective power would much of the time dwindle down to that of the 40-inch, or even below it. But, when a few hours of the best seeing did come, what marvels that glass would show![43]

Barnard's point was that, though in average or poor seeing, the image in a small telescope might be steadier than in a large one, a moment's true revelation with the full resolving power of a large telescope, on the comparatively rare occasions when the 'seeing' permitted its use, was apt to be infinitely more valuable than years of work with the inferior instrument. Thus, he maintained, the question came down not so much to what small telescopes showed in bad air but to what a large telescope showed in good.

7

The large versus small telescope controversy, simmering for years, finally boiled over in 1895. For two years an English amateur, Arthur Stanley Williams, had been reporting on various faint spots on Saturn which he claimed to have made out with only a $6\frac{1}{2}$ inch reflector.[44] Williams was a skillful observer – his pioneering observations with the same small telescope of spots on Jupiter laid the foundation for his brilliant identification of the nine main currents in its atmosphere, and even today he is regarded as one of the greatest observers of that planet. Nevertheless, his observations of Saturn at once met with greater skepticism – even Denning was dubious, though confirmatory observations were forthcoming from several observers using small telescopes, including Henry McEwen, for many years director of the British Astronomical Association's Mercury and Venus Section, and Fernand Quénisset, one of Flammarion's assistants at Juvisy. Barnard, who had been toiling at the same time on a careful set of micrometrical measures of the ball and ring system with the 36-inch refractor, had seen nothing of these spots, and made no secret of his skepticism about Williams's observations in his account in the *Monthly Notices of the Royal Astronomical Society*: 'The black and white spots lately seen upon Saturn by various little telescopes were,' he wrote, 'totally beyond the reach of the 36-inch – as well as of the 12-inch – under either good or bad conditions of seeing.' And in alluding to his own careful but comparatively bland drawing of the planet, he added: 'It is true the picture appears abnormally devoid of details when compared with drawings with some of the smaller telescopes. I am satisfied, however, to let it remain so.'[45]

There is now little doubt that Barnard was right and that Williams's spots were illusory.[46] However, the discussion about them brought to a head the overriding

Barnard's drawing of Saturn, July 2, 1894, which he published with the comment, 'I have only drawn what I have seen with certainty. It is true that the picture appears abnormally devoid of details when compared with drawings made with some of the smaller telescopes. I am satisfied, however, to let it remain so.' Royal Astronomical Society

question of the relative virtues of large and small telescopes for planetary work, a question that was by no means divided entirely along professional and amateur lines. Among professionals who tended to the small telescope side of the controversy was none other than Simon Newcomb, who had earlier warned Barnard that 'the effect of the secondary spectrum is so enormous in great telescopes that defining power depends mainly upon the focal length, and everything you add to the aperture above a certain limit injures the definition.'[47] Describing as 'almost astounding' the fact that Barnard with his large telescope had seen nothing of Williams's spots, Newcomb urged Barnard to put the issue to a crucial test by diaphragming the 36-inch, to see whether anything would be revealed with the telescope stopped down that was not visible in the full aperture. 'A large objective,' he emphasized, 'is objectionable for such seeing as this, where only definition and not light is required. I should like very much to see the result of careful examination of Mars and Saturn with the great equatorial when this aperture was reduced to 12 inches.'[48] It is clear that he personally expected that the diaphragm would result in a substantial improvement in the definition.

Barnard at once accepted the challenge and did, to the best of his ability, perform the experiments Newcomb called for. Mars, unfortunately, was on the other side of the Sun, but Venus and Saturn were well placed for observation. Despite the extreme inconvenience of getting at and diaphragming the object-glass of the great telescope,

Barnard took the pains necessary to do so, and observed both planets on several occasions with the full 36-inch aperture and with the aperture stopped down to 12 inches. The result, published in the *Monthly Notices* for January 1896, was in his view definitive, and agreed completely with the conclusions he had reached in his earlier San Francisco *Examiner* article: 'I am convinced that everything that can be seen with the telescope diaphragmed down can be better seen with the full aperture when the air is steady.'[49] He then added, almost as an afterthought, a summary of the remarkable revelations that he had had of Mars with the great telescope in 1894.

Following a statement of his skepticism regarding the drawings of observers using small telescopes, many of whom had covered the planet with the canal network – 'If we are to take the testimony of the drawings themselves,' he wrote, 'the smaller the telescope the more peculiar and abundant are the Martian details' – he described at considerable length his very different experiences with the 36-inch refractor:

> On several occasions during that summer [1894], principally when the planet was on the meridian shortly after sunrise – at which time the conditions . . . are often exceptionally fine at Mount Hamilton – its surface with the great telescope has shown a wonderful clearness and amount of detail. This detail, however, was so intricate, small, and abundant, that it baffled all attempts to properly delineate it. Though much detail was shown on the bright 'continental regions,' the greater amount was visible on the so-called 'seas.' Under the best conditions these dark regions, which are always shown with smaller telescopes as of nearly uniform shade, broke up into a vast amount of very fine details.[50]

In his struggle to understand the nature of what was being revealed to him on the surface of this strange and distant planet, Barnard turned to analogy to help him. Indeed, just as Percival Lowell had arguably received much of the inspiration for his Martian deserts, lonely and desolate almost beyond imagining, from the Arizona deserts south of Flagstaff,[51] Barnard also drew his hints from the local terrain.[52] 'I hardly know how to describe the appearance of these "seas,"' he wrote:

> To those, however, who have looked down upon a mountainous country from a considerable elevation, perhaps some conception of the appearance presented by these dark regions may be had. From what I know of the apearance of the country about Mount Hamilton as seen from the observatory, *I can imagine* that, as viewed from a very great elevation, this region, broken by canyon and slope and ridge, would look just like the surface of these Martian 'seas.' During these observations the impression seemed to force itself upon me that I was actually looking down from a great altitude upon just such a surface as that in which our observatory was placed. (italics supplied)[53]

These comments make clear that Barnard did not claim to have made out 'canyon and slope and ridge directly (indeed, under usual conditions true topographical differences are beyond the reach of Earthbased observers), but was inferring their possible existence. In doing so, he was seeing by what William Herschel once called the 'mental eye' – the 'eye of reason and experience.'[54] He concluded, firmly, that 'No straight hard sharp lines were seen on these surfaces, such as have been shown in the average drawings of recent years.'[55]

Though he criticized what he called the 'average drawings' of the planet, he was not satisfied with his own either; thus most of his drawings from the 1894 opposition did not appear during his lifetime – and for that matter have remained unpublished to this day. He later wrote that 'every precaution was taken to represent as faithfully as possible the proper proportions and locations of the details, and I think they fairly represent the appearance of the planet so far as my limited artistic skill would permit. The difficulty lay [however] not in the scarcity or illusiveness of the details but in their great variety and abundance. So much was this so that it was impossible to draw what was easily seen. To do this would have required an exceptionally skillful artist . . . [The details] in the dark portions were specially rich, but the untrained hand could not adequately deal with them from a pictorial point of view.'[56]

As soon as Barnard's article was published in the *Monthly Notices*, Newcomb wrote to him: '[A] few months ago I asked W. H. Pickering how sure he was about [the canals] at the Flagstaff Observatory. He assured me that there could be no doubt whatever; that he had repeatedly seen them with entire distinctness':

> The only theory I can form to explain these discrepancies with your views is that there are certain arrangements or features of the markings, which, with the imperfect means at the command of Schiaparelli and Pickering, are sufficiently near to the appearances of long, dark lines to make the eye judge such to be the effect.[57]

Newcomb would later develop this 'optico-psychological' theory of the canals in greater detail, and so did others – notably England's E. W. Maunder, and later Lowell's assistant Douglass and at least in certain moods Pickering himself. The point is that Barnard's observations of 1894 marked an important sea change – or rather, see change. Henceforth the canals, at least in the highly regular form in which they were depicted on many maps of the planet, were increasingly regarded as an illusion, a sort of shorthand 'summing up' by the eye of the more complex and irregular natural features of the surface, though it would be some years before this view achieved its final triumph.[58] With Barnard's *Monthly Notices* article, the canalists, though not yet openly in retreat, began to assume the defensive.

8

Meanwhile, in 1896–7, Lowell dropped another bombshell on the astronomical community. He announced the discovery of a new set of markings on Venus that were completely unlike anything seen at Mt. Hamilton or, for that matter, anywhere else.

In August 1896, just after he had installed his new 24-inch Clark refractor, Lowell trained it on the inner planets, Mercury and Venus. A few years earlier, Schiaparelli had announced that both planets had a captured rotation with respect to the Sun. He was quite sure of the result in the case of Mercury, less so in that of Venus, as his conclusion regarding the latter had been based mainly on a few 1877 observations of a pair of round bright spots near the cusps which had appeared nearly stationary over a period of several weeks and thus suggested to him a long rotation. He deduced from this highly uncertain evidence that Venus's rotation was probably equal to its period of revolution around the Sun – 225 days. If true, this meant that the planet always held the same face toward the Sun.

Lowell had no difficulty convincing himself that Schiaparelli was correct about Mercury, finding on the planet markings which he regarded as 'so unmistakeable' that within a day or two 'the rotation period of the planet was patent.'[59] Though his map of the planet was idiosyncratic in appearance, being covered with a linear grid, there was nothing particularly startling about his conclusions. However, his work on Venus was a different matter. This planet had hitherto frustrated the most skillful observers – Barnard, for all his pains, had never seen anything more than the usual vague shadings, and he was far from being alone. Under the circumstances, the rotation period was in considerable doubt, and the periods given ranged from just under 24 hours to Schiaparelli's 225 days. Lowell claimed to have discovered a set of markings as apparently distinctive of the planet's physical appearance as the canals were of Mars. Long fingerlike streaks, they, 'in the matter of contrast as accentuated, in good seeing, as the markings on the Moon and owing to their character much easier to draw . . . They are rather lines than spots . . . A large number of them, but by no means all, radiate like spokes from a certain center.'[60]

These unexpected markings were seen not only at Flagstaff but at Tacubaya, near Mexico City, where Lowell, who was always in search of the best possible atmospheric conditions for his planetary work, temporarily moved his observatory at the end of 1896. Nor was Lowell alone in seeing them. His assistants, who then included Douglass, Daniel Drew, Wilbur Cogshall, Thomas Jefferson Jackson See, and also his secretary, Wrexie Louise Leonard, tried their hand at sketching them, although actual inspection of their observing notebooks shows that only in the case of Leonard was the spoke pattern rendered with a boldness approaching Lowell's own. Lowell emphasized that in contradistinction to the usual markings of a more or less indefinite character, those seen at his observatory 'were of a different order entirely. They were absolutely distinct and absolutely definite with contours in some cases wonderfully clearly cut against their surroundings; and in good seeing they were always the same in the same places.' Their appearing in the same places meant that the planet always held the same face toward the Sun, and Lowell concluded that 'they therefore . . . may be said to have put the rotation period [of Schiaparelli] beyond even unreasonable doubt.'[61]

At once Lowell again found himself at the center of controversy. The symmetry of his markings was too radical, too peculiar, and he was accused of canalizing Venus. 'Mr. Lowell's observations,' opined John Ritchie, Jr. in the Boston *Evening Transcript*, 'will not be accepted by astronomers as final,' while Camille Flammarion noted with icy politeness that they were 'entirely at variance with all that has gone before.'[62] Captain William Noble criticized them more sharply at a meeting of the Royal Astronomical Society – 'I do not know whether Mr. Lowell has been looking at Mars until he has got

Mars on the brain,' he declared, 'and by some transference transcribed the markings to Venus.'[63] This broadside drew from Lowell an admirably restrained reply: 'For skepticism, of course, I care nothing; people are naturally prone to disbelieve what they have not seen, and the distinctness of the markings in good air raises them above criticism. It is merely a question of time when they and their consequences shall be generally recognized.'[64]

Needless to say, astronomers at Mt. Hamilton were dumbfounded. Barnard, who whenever possible tried to avoid offending others even when he disagreed with them, publicly expressed his doubts in the most general terms:

> This rotation controversy [regarding Venus] has raged for upwards of two centuries, with fitful periods of quiescence – after some observer more combative than the rest had definitely 'settled the question' – only to break out again with renewed virulence when a new champion for rotational honors entered the field . . . [The] discrepancies are due in the main to the difficulty . . . of seeing the markings which really exist on the surface of Venus.[65]

Holden suggested that the disputed markings might be due to a 'strain on the glass, induced by an overtight condition of the adjusting screws or of the objective in the cell'[66] – a comment that evoked a sharp retort from the maker of the Flagstaff refractor, Alvan G. Clark. Privately, however, Holden was less confident, and queried Douglass:

> I have been rather skeptical, as you know, about the markings on Venus, and I still am (you must not, should not, take criticisms as personal which are not meant so. We are all after the truth and nothing else).
>
> How do you personally explain Barnard's and Schaeberle's and my inability to see any of your *class* of markings on Venus? Barnard and Schaeberle are skilled observers. I began to observe Venus in 1873 and missed no chance for several years.
>
> Not once, as far as I know, have any of us even suspected your *class* of markings . . .[67]

To Lowell, the question came down once more to the all-important matter of seeing. As he lectured Barnard: 'I am not surprised that you should have failed with Venus. The markings are among the very most difficult of planetary detail and require emphatically good air for their first identification. Also high powers will show nothing and a very low power indeed is best . . . The markings are more difficult than the "canals" of Mars and unless the air be good enough to show these *well* not diffusely it were futile to attempt anything on Venus . . . [I]n the best air they are as clear-cut as one could wish.'[68] See also wrote to Barnard in similar vein: 'The fact that the markings . . . cannot be seen by you only proves your misfortune in being in a location where the air is like the curly glass in the doors of office buildings – it is translucent, not transparent, and you can't expect to see anything difficult through it. I am fully convinced that our telescope in this atmosphere is the most powerful in the world, for our results prove it . . . Time will prove that with all its show and talk and waste of money, the Yerkes Observatory will not get half the results we shall reap here.'[69]

9

To Barnard's failure to make out the markings on Venus and the spots on Saturn was soon added his inability to detect a class of extremely narrow linear markings on Jupiter's satellites Ganymede and Callisto. The markings were seen by Douglass with the 24-inch refractor in Mexico in February and March 1897 and on his return to Flagstaff in July 1897.[70]

The publication of Douglass's results led to a prompt reply from Barnard. Noting that Douglass's sketches of the satellites of Jupiter, with their linear markings, bore a striking resemblance to those made at the same observatory on Mars, Venus and Mercury, Barnard placed on record his own observations of the satellites with the 36-inch refractor in 1893–4–5:

> Though I have observed and drawn markings on these satellites, I have not seen any narrow straight lines or similar markings on any of them. The markings I have seen have always appeared to be large and more or less diffused – except in the case of certain white polar spots . . .[71]

He added some scathing comments about contemporary planetary drawings in general:

> It has always appeared to me that when representing any planetary detail it should be drawn as nearly accurate, as to appearance, as possible. If the marking is vague and uncertain it should be made so on the drawing and no definite outlines should be given where none exist. Some observers seem to be in the habit of giving a definite boundary to markings whether such are really seen or not. This is very misleading and no object should be shown with a definite outline unless it really has such.
>
> This diagram method of drawing is specially noticeable in a good many drawings of Mars, and it is perhaps through the study of these and not by any inspection of the planet itself that so many queer and unnatural ideas have been propagated concerning the physical appearance of Mars. Many of the regions shown on drawings and maps of Mars as definitely bounded, are in reality very diffused and uncertain in their outlines.[72]

Despite the sharp tone of these remarks, Barnard and Douglass remained personally on the best of terms, as indeed did Barnard and Lowell. In May 1898, Barnard accepted an invitation from Douglass and See to come to Flagstaff to see for himself. Douglass was then acting director, as Lowell had been sidelined since the return from Mexico because of a nervous condition which the doctors of the day diagnosed as 'neurasthenia.' In accepting the invitation, Barnard told Douglass that the markings on Venus were uppermost in his mind:

> I am intensely interested in this matter. I do not have any ill feelings – why should I have? It is simply to satisfy my own eyesight that I want to come. Dr. See and Mr. Lowell have always been friendly towards me and if for nothing else, I should want to come so that I shall cease to do them an injustice, for I doubt not that my failure at L[ick] O[bservatory] to see these things so easily seen from Flagstaff may have had some sort of influence against the observations.[73]

Come to Flagstaff Barnard did, arriving on May 30, 1898 at a season when See had predicted the best seeing. He remained until June 4, but the observational results were disappointing in the extreme – and in addition he found evidence that there was dissension among the staff on Mars Hill. One of the Lowell assistants, Samuel P. Boothroyd, resigned, citing his deep dislike of See, while Douglass confided in Barnard the results of some experiments he was then performing with artificial planet disks which led him to suspect that many of the Lowellian markings were illusory.[74] The following extracts are from Barnard's notebook:

> May 30. High wind blowing – more or less threatening clouds. Observing Jupiter power 750. Especial attention was paid to the 1st satellite. The disc was seen once in a while, but poorly. Sometimes the image would separate into two alike. They would slowly recede from each other . . . and then join again.
> Uranus. The image was poor . . . I could see nothing on its surface but Dr. See and several others saw a belt on it.
> Saturn. The image so bad that I could only faintly see the division in the ring.
> May 31. 10:30 A.M. No clouds; considerable S.W. wind . . . Examined Mercury . . . The image occasionally was fairly well defined; but it was constantly jumping so that it was not possible to see the planet steadily . . . Spent about one hour on this watch. There was no markings seen . . .
> May 31. Venus . . . The jumping was constant so that nothing was seen.
> June 2nd. Venus. Image jumping excessively. High S.W. wind. The seeing continued very variable, running from a blurred mass to a fairly well defined limb, but the image was jumping badly . . . Dr. See saw a dark line from . . . limb to the center of the planet . . . I saw nothing – not even the old vague markings which I had seen in former years.
> June 3rd. 4:30 P.M. Examined Venus up to 5:30 . . . The image was quite steady; I saw no markings; Mr. [G. A.] Waterbury, however, saw the usual Flagstaff markings.
> June 4th. Fearful high N.E. wind. 6:00 P.M. Venus. Mr. Waterbury . . . saw the usual Flagstaff markings easily – I could not see them.[75]

Because Lowell had not been present, Barnard decided not to publish his account, and confided his impressions of his visit only to the private pages of his notebook: 'My going to Flagstaff was purely a desire to verify Mr. Lowell's observations of Venus and Mercury – Mars was out of reach,' he wrote. 'I had almost no prejudice at this time and had a sincere hope that I could give a favorable report of my visit.' He regretted the fact the conditions were not the best and that even the regular observers admitted that they were unusually poor for that season. He could not, therefore, say what they would be like at other times, but he ventured a guess:

> If I were to judge of the conditions from a mere inspection of the surroundings of the Lowell Observatory, I would not suppose that it was a very suitable site . . . It is on the edge of a Mesa or bluff some hundreds of feet high. On one side (to the East) is a broken country and the hot desert; on the other a vast forest of pines – the observatory being on the edge of the forest. Very near – to the North – are the San Francisco peaks some 12 000 ft. high, which have snow on them most of the year.[76]

To the end of his life Barnard remained skeptical of Lowell's markings on Venus, and indeed they never received general acceptance among the astronomical community. As for Lowell's conclusion that Venus always held one face toward the Sun, Barnard was no less cautious, writing in 1915: 'I have no observations to negative this idea, at the same time I have none that would offer it the slightest support.'[77]

10

So much for Barnard's well-known failure to confirm the Martian canals and other linear markings drawn by some of his contemporaries on the planets and their satellites. What, though, of the Martian craters? There is a famous story that he saw them seventy years before their existence was recognized from the spacecraft photographs. Is it fact or legend?

The story rests entirely on the authority of John E. Mellish. The careers of the two men were intertwined, in that Mellish became in some ways Barnard's protégé. Their friendship began in April 1907, when Mellish, who lived on a farm three miles south of the village of Cottage Grove, Wisconsin (near Madison), discovered a comet with a homebuilt 6-inch reflector. It later proved that the comet had been found independently by another astronomer in New Zealand, and it was given the name Grigg-Mellish. Young Mellish, then twenty-one years old, must have reminded Barnard of himself in younger days – like Barnard, the 'boy astronomer of Cottage Grove' had had little formal education, grew up in disadvantageous circumstances, and won fame as a discoverer of comets.[78] Barnard, at any rate, sent Mellish a note of congratulations and invited him to pay a visit to the Yerkes Observatory. Mellish did so that fall, by which time he had discovered another comet. This was the first of several visits in which he was to be the guest of the Barnards in their home on the shore of Lake Geneva.

Though Mellish confided to a reporter from the *Wisconsin State Journal* that he hoped to become a professional astronomer, his dream went unrealized. For the next several years he worked hard on the farm and busied himself building telescopes, some of which were sold. By 1913, he had built a 16-inch reflector about which he boasted to Barnard that, for lunar and planetary observing, it outperformed the $15\frac{1}{2}$-inch Clark refractor at the Washburn Observatory at Madison.[79] He had also resumed comet-seeking in earnest. His efforts paid off again in February 1915 with the discovery of his third comet, which led to an invitation to come to Yerkes as a Volunteer Research Assistant. It was the turning point of his life. He would never again return to the farm, and within only a few months of his arrival in Williams Bay, he had married, discovered yet another comet, and – so the story goes – made some of the most remarkable observations ever made of Mars.

According to Mellish, he observed Mars with the 40-inch refractor in November 1915, when the planet was still three months from its aphelic opposition of February 9, 1916 and showed as a small gibbous like the Moon three days from full. One morning he caught the planet in an exceptionally steady air just after sunrise and noted with magnifications of × 750 and × 1100, as he recalled in a 1935 letter to amateur astronomer Walter Leight, 'something wonderful about Mars, it is not flat but has many craters and cracks. I saw a lot of the craters and mountains one morning with the 40″ and could hardly believe my eyes.'[80] That morning Mellish went to Barnard and confided to him what he had seen. Barnard, so the story goes, was pleased and fetched his own drawings of Mars from an old trunk, drawings that, according to Mellish's later recollection, had been made with the Lick 36-inch refractor in 1892–3. Mellish referred to these as 'the most wonderful drawings that were ever made of Mars.' They showed

> the mountain ranges and peaks and craters and other things both dark and light that no one knows what they were, I was thunder struck and asked him why he had never published these and he said no one would believe him and [others] would only make fun of it . . . Barnard took whole nights to draw Mars and would study an interesting section from early in the evening when it was coming on the disk until morning when it was leaving and he made the drawings four or five inches [in] diameter and it is a shame that those were not published.[81]

After working as a volunteer at Yerkes for fifteen months, Mellish and his wife, who now had an infant daughter to support, left for Leetonia, Ohio, where Mellish started an optical business, in part with a loan from Barnard. He later moved to Wilmette, Illinois, then to St. Charles, Illinois, Escondido, California, and finally Cape Junction, Oregon. He never published his drawings of Mars, which unfortunately were destroyed in a fire in his optical workshop in 1964, six years before his death, while Barnard's drawings were also thought to have disappeared.

Then, in 1987, Richard Dreiser and I discovered that they were still at Yerkes. Some of them are indeed on a scale of five inches to that of the planet as Mellish remembered, and there are series of them that were made during the course of a single night. Mellish was wrong about the dates, however; most of the good ones date from 1894 rather than 1892. A number show the 'oases' as large dark circular patches, and in a few cases these patches can be identified with actual relief features such as the volcanic calderas Olympus Mons, Arsia Mons, and Ascraeus Mons. Barnard also rendered parts of the Valles Marineris canyon system. But there are no craters or mountains – indeed, no topographical features as such. On the whole the drawings are not as spectacular as Mellish's comments ('the most wonderful drawings that were ever made of Mars') might lead one to expect, and Barnard may have decided against publishing them simply because he realized that they were not all that remarkable.

The legend that Barnard saw the craters on Mars must, therefore, be laid to rest – not that his fame cannot withstand the loss of such a distinction. Mellish's claim, in the absence of his drawings, falls within the provenance of what Barnard said near the end of his life which is completely characteristic: 'There are a great many facts about Mars that have been discovered by the astronomers that will live; but there are some things unattainable. Some of the speculations at present circulating may be pretty near the truth, but you can rest assured there are many of them that are not worth a moment's consideration. If you want to make a statement, make it so that no one can disprove it, then it will stand. To countenance any other kind of statements is bad.'[82]

1   See Michael J. Crowe, *The Extraterrestrial Life Debate 1750–1900: The Idea of a Plurality of Worlds from Kant to Lowell* (Cambridge, Cambridge University Press, 1986), p. 400

2   E. S. Holden, 'The Lowell Observatory in Arizona,' *PASP*, 6 (1894), 160–9 165

3   ibid.

4   W. H. Pickering, 'The Planet Jupiter and Its Satellites,' *AA*, 12 (1893), 193–203:195–6

5   W. H. Pickering, 'The Planet Jupiter and Its Satellites,' *AA*, 12 (1893), 193–203; 'Jupiter's Satellites,' ibid., 390–7; 'The Rotation of Jupiter's Outer Satellites,' ibid., 481–94

6   Pickering, 'The Planet Jupiter and its Satellites,' p. 196

7   E. E. Barnard, 'On the Forms of the Discs of the Satellites of Jupiter as seen with the 36-inch Equatorial of the Lick Observatory,' *AA*, 13 (1894), 1–7. See also E. S. Holden, 'Recent Observations of the Satellites of Jupiter,' *AA*, 13 (1894), 356–7 and E. E. Barnard, 'Recent Observations of the Satellites of Jupiter,' ibid., 14 (1894), 1–4

8   See Joseph Ashbrook, 'W. H. Pickering and the Satellites of Jupiter,' *Sky and Telescope*, 26 (1963), 335–6

9   E. E. Barnard, 'Review of Pickering's *The Moon*,' *Ap J*, 20 (1904), 359–64

10 W. G. Hoyt, *Lowell and Mars* (Tucson, University of Arizona Press, 1976), p. 12

11 SWB to EEB, November 26, 1892; SLO. This letter reads: 'This will introduce Mr. Percival Lowell, who is on his way to Japan with one of Clarke's telescopes. I know you will be glad to meet him, and to show him all you can about Mt. H[amilton]. He will understand and appreciate it.'

12 SCC to ESH, September 4, 1894; SLO

13 Boston *Commonwealth*, May 26, 1894

14 ibid.

15 Holden, 'The Lowell Observatory in Arizona,' p. 160

16 Hoyt, *Lowell and Mars*, p. 311

17 Holden, 'The Lowell Observatory in Arizona,' pp. 167–8

18 EEB to ESH, August 3, 1894; VUA

19 EEB to ESH, August 7, 1894; VUA

20 EEB to Board of Trust of the Regents of the University of California, August 1894; VUA

21 Lick Observatory Committee (T. G. Phelps, C. F. Crocker and H. S. Foote) to ESH, Sept. 1894; SLO

22 EEB, observing notebook; LO

23 ibid.

24 ibid.

25 EEB to SN, September 11, 1894; LC

26 G. P. Kuiper, 'On the Martian Surface Features,' *PASP*, 67 (1955), 271–85:275

27 G. V. Schiaparelli, 'The Planet Mars,' (trans. W. H. Pickering), *AA*, 13 (1894), 635–40, 714–23

28 P. Lowell, 'Mars,' *PA*, 2 (1894), 1; 'The Polar Snows,' ibid., 2 (1894), 52; 'Spring Phenomena,' ibid., 2 (1894), 97; 'Atmosphere,' ibid., 2 (1894), 153; 'The Canals. I,' ibid., 2 (1895), 255; and 'Oases,' ibid., 2 (1895), 343

29 P. Lowell, *Mars* (Boston, Houghton, Mifflin & Co., 1895), p. 7

30 P. Lowell, 'Mars,' *PA*, 2 (1894), 1–2

31 JEK to GEH, December 27, 1894; YOA

32 W. W. Campbell, 'An Explanation of the Bright Projections Observed on the Terminator of Mars,' *PASP*, **35** (1894), 103

33 W. W. Campbell, review of *Mars*, by Percival Lowell, *PASP*, **8** (1896), 217–18

34 A. E. Douglass, 'The Lick Review of "Mars,"' *PA*, **3** (1895), 199–201:201

35 San Francisco *Chronicle*, August 26, 1894

36 Lowell, *Mars*, pp. 74–5

37 For a thoughtful discussion of some of the issues involved, see John Lankford, 'Amateurs versus Professionals: The Controversy over Telescope Size in Late Victorian Science,' *Isis*, **72** (1981), 11–22

38 W. F. Denning, 'Jupiter and the Relative Powers of Telescopes in Defining Planetary Markings,' *Observatory*, **8** (1885), 79–80

39 W. H. Pickering, 'Climate as Related to Astronomical Observations,' *PA*, 3 (1896), 465 and A. E. Douglass, 'Atmosphere, Telescope and Observer,' ibid., 5 (1897), 64

40 P. Lowell, 'Atmosphere: In Its Effect on Astronomical Research,' lecture text, *ca* spring 1897; LOA

41 A similar expedient had first been employed by Burnham in his double star work.

42 E. E. Barnard, 'Nearer to the Stars,' San Francisco *Examiner*, December 16, 1894

43 ibid.

44 A. S. Williams, 'On the Rotation of Saturn,' *MNRAS*, **54** (1893), 297

45 E. E. Barnard, 'Micrometrical Measures of the Ball and Ring System of the Planet Saturn, and Measures of the Diameter of his Satellite Titan. Made with the 36-inch Equatorial of the Lick Observatory,' *MNRAS*, **55** (1895), 367–82

46 A. Lenham to W. Sheehan, personal correspondence, January 8, 1993: 'Concerning Williams's dark spots on Saturn, e.g. in 1893 latitudes 17 to 37° N., rotatation periods about 10h 14m, similar to his results for equatorial zone spots: the Voyager spacecraft results show very large velocity shears across these latitudes and between the two North Equatorial Belt components. Yet Williams talks of pairs of spots (one on each NEB component) moving along together. This does not occur even for the N and S edges of belts on Jupiter let alone different components. I'm afraid Williams was mistaken.'

47 February 27, 1894; VUA

48 SN to EEB, July 6, 1895; VUA.

49 E. E. Barnard, 'Micrometric Measures of the Ball and Ring System of the Planet Saturn, and Measures of the Diameter of his Satellite Titan . . . With some Remarks on Large and Small Telescopes,' *MNRAS*, **56** (1896), 163–72:165

50 ibid., p. 166

51 In *Mars and Its Canals* (New York, Macmillan, 1906), p. 149, Lowell described the analogy between the telescopic appearance of Mars and that of the landscapes near the observatory as seen from the San Francisco Peaks north of Flagstaff: '[T]he resemblance of its lambent saffron to the telescopic tints of the Martian globe is strikingly impressive. Far forest and still farther desert are transmuted by distance into mere washes of color, the one robin's-egg blue, the other roseate ochre, and so bathed, both, in the flood of sunshine from out of a cloudless burnished sky that their tints rival those of a fire-opal. None otherwise do the Martian colors stand out upon the disk at the far end of the journey down the telescope's tube. Even in its mottlings the one

expanse recalls the other.'

52   In addition, W. W. Campbell had argued from the terminator projections observed
      since 1888 that Mars was a mountainous planet. Lowell interpreted the same
      projections as owing to cloud. In this case, Lowell was right and Campbell wrong.

53   Barnard, 'Micrometrical Measures,' p. 166

54   W. Herschel, quoted in *The Scientific Papers of Sir William Herschel*, J. L. E. Dreyer,
      ed. (London, Royal Astronomical Society, 1912), vol. 1, p. xxxviii. Also L. J. Martin to
      W. Sheehan, personal correspondence, July 6, 1994

55   Barnard, 'Micrometrical Measures,' p. 167

56   EEB, untitled MS on Mars, *ca* 1910; YOA

57   SN to EEB, March 9, 1896; VUA

58   For detailed discussions of these issues, see Crowe, *The Extraterrestrial Life Debate*,
      chapter 10, and W. Sheehan, *Planets and Perception: Telescopic Views and Interpreta-
      tions, 1609–1909* (Tucson: University of Arizona Press, 1988)

59   P. Lowell, 'Mercury,' *MNRAS*, **57** (1897), 148

60   P. Lowell, Boston *Evening Transcript*, November 28, 1896

61   P. Lowell, 'Mascari, Cerulli and Schiaparelli on Venus' Rotation Period,' *PA*, **4**
      (1897), 389

62   C. Flammarion, 'Some New Views as to the Rotation of Venus-II,' *Knowledge*, **20**
      (1897), 258–61:260

63   'Proceedings at Meeting of the Royal Astronomical Society,' *MNRAS*, **56** (1896), 420

64   Unsigned, 'The Markings on Venus,' *MNRAS*, **57** (1897), 113

65   E. E. Barnard, 'Physical and Micrometrical Observations of the Planet Venus, made at
      the Lick Observatory with the 12-inch and 36-inch refractors,' *ApJ*, **5** (1897),
      299–304:299

66   E. S. Holden, 'Mr. Lowell's Observations of Mercury and Venus,' *PASP*, **9** (1897),
      92–3

67   ESH to A.E. Douglass, August 19, 1897; LOA

68   P. Lowell to EEB, August 2, 1897; VUA

69   T. J. J. See to EEB, April 7, 1898; VUA

70   A. E. Douglass, 'Drawings of Jupiter's Third Satellite,' *AN*, **143** (1897), 412–14

71   E. E. Barnard, 'On the Third and Fourth Satellites of Jupiter,' *AN*, **144** (1897),
      321–30:321

72   ibid.

73   EEB to A. E. Douglass, May 5, 1898; LOA

74   WSA to EEB, October 29, 1905; VUA

75   EEB, notebook, YOA. Barnard later drew up an account of the trip for W. W.
      Campbell, 'Observations at the Lowell Observatory in May and June 1898'; SLO

76   ibid.

77   EEB to David Wilson, August 24, 1915; VUA. Lowell's spokelike markings have
      generally been regarded as illusory. Nevertheless, they have been seen by a few
      observers since, and it may be that the last word has not been said. A fascinating
      recapitulation of the Venus spoke problem, with at least the suggestion that they may
      be genuine features, is given in J. J. Goldstein, *Absolute Wind Measurements in the
      Lower Thermosphere of Venus using Infrared Heterodyne Spectroscopy*, NASA Con-
      tractor Report 4290 (Washington, DC, National Aeronautics and Space Administra-

tion, 1990), Appendix 4.

78   The most complete biographical account of Mellish is Paul Thompson, 'The Boy
     Astronomer of Cottage Grove,' *Wisconsin Academy Review*, December 1979, 34–40

79   J. E. Mellish to EEB, June 22, 1913; VUA

80   J. E. Mellish to W. Leight, January 18, 1935; a copy of this letter was provided to me by
     Rodger W. Gordon

81   ibid.

82   Quoted in Toronto *Daily Star*, March 24, 1920. A number of others who have used
     large telescopes under a variety of conditions are skeptical of Mellish's claim. See, for
     example, A. P. Lenham, *Journal of the British Astronomical Association*, **97** (1987), 191:
     'I never saw craters on Mars with the 82-inch [of the McDonald Observatory] or the
     40-inch [of Yerkes] even when the planet was gibbous.' This agrees with my own views
     with large telescopes; see, for example, W. Sheehan and S. J. O'Meara, 'Exotic
     Worlds,' *Sky & Telescope*, **85** (1993), 20–4:22.

16

Nature's true artisan

1

In planetary observation in the late nineteenth century, the eye reigned supreme. It was swift where the photographic plate was ploddingly slow, and could follow the moment to moment gyrations of an image disturbed by the ever-present waves in the ocean of air. The plate, by contrast, in the few seconds it needed to build up an image, blurred the finer planetary details. On the other hand, like the proverbial tortoise, the plate was slow but cumulative; as long as the shutter was kept open it continued to register faint light, and an exposure of several hours showed stars and nebulae far beyond what even the most sensitive eye could capture.

While in the 1880s and 1890s photography was still too slow to produce what Barnard called 'convincing pictures' of the planets,[1] in the deep sky it had already achieved a massive superiority over the eye. Photographs taken by Paul and Prosper Henry at the Paris Observatory showed not only the long disputed Merope nebula in the Pleiades (discovered visually in 1859 by Wilhelm Tempel), but further revealed that the entire group of stars was filled, as Barnard wrote, 'with an entangling system of nebulous matter which seemed to bind together the different stars of the group with misty wreaths and streams of filmy light, nearly all of which is entirely beyond the keenest vision and the most powerful telescope.'[2] Equally impressive vistas were revealed in 1880s photographs of the Great Nebula in Orion by Henry Draper and of various clusters and nebulae, but especially of the Great Nebula in Andromeda, by Isaac Roberts, an English amateur who after making a fortune as a builder became an avid deep sky photographer with a 20-inch reflector, which he set up in 1885 at Maghull, near Liverpool, and in 1890 moved to a more favorable site at Crowborough, Sussex.

Among all the objects which properly speaking can be said to belong to the deep sky, however, the Milky Way, that circling zone of light that had teased the imagination of the ancients, was – and is – by far the most spectacular. As an object of naked eye contemplation, it appears as 'a richly textured pattern of bright star clouds, complex mottlings, softly luminous arms and dark lanes, with broad dim outliers.'[3] These forms and structures were, when Barnard began his pioneering work of photographing the Milky Way at the end of the nineteenth century, elusive, evocative, and still largely unexplained.

The late nineteenth century view of the Milky Way is invoked by Thomas Hardy in his novel *Two on a Tower*, published in 1882. The budding astronomer Swithin St. Cleeve is pondering the night sky with his patroness, Lady Constantine. 'You would hardly think, at first,' he tells her,

> 'that horrid monsters lie up there waiting to be discovered by a moderately penetrating mind – monsters to which those of the oceans bear no sort of comparison.'
>
> 'What monsters may they be?'
>
> 'Impersonal monsters, namely, Immensities. Until a person has thought out the stars and their interspaces, he has hardly learnt that there are things much more terrible than monsters of shape, namely, monsters of magnitude without known shape. Such monsters are the voids and waste places of the sky. Look, for instance, at those pieces of darkness in the Milky Way,' he went on, pointing with his finger to where the galaxy stretched across over their heads with the luminousness of a frosted web. 'You see that dark opening in it near the Swan? There is a still more remarkable one south of the equator, called the Coal Sack . . . In these our sight plunges quite beyond any twinkler we have yet visited. Those are deep wells for the human mind to let itself down into . . .'[4]

Though Hardy made a point of keeping himself informed about the latest scientific developments, he was here recalling a view that went all the way back to William Herschel's first attempts to plumb the depths of the sidereal universe a century before. It was Herschel who had first written about these voids and waste spaces in the sky.

By way of background, Herschel in 1783 had begun his pioneering 'star gauges,' by means of which he hoped to map the outline of the Galaxy. Taking soundings in over six hundred directions in space, he counted the number of stars appearing in the field of view of his 20-foot reflector, which had a mirror 18.7 inches in diameter. In some fields he saw scarcely a single star; in other fields so many hundreds of stars were crowded together that he was able to count only half a field, or even a quadrant. When he began this work, he made two assumptions: that his telescope was able to reach to the edge of the Galaxy in all directions, and that the stars were spread more or less uniformly throughout space, in which case the number of stars he counted in a field would be proportional to the actual depth to which the sidereal universe extended in that direction. If these assumptions were true, his star counts would give relative distances, and on this basis he produced, in 1785, a cross-sectional diagram of the Milky Way, which showed the sidereal system in the form of a flattened disc – Herschel's 'grindstone' model of the Galaxy.

However, by 1789 Herschel was forced to abandon his outline map. He was then using a larger telescope – his mammoth 40-foot reflector, with a mirror 48 inches across – and realized that it showed many faint stars that had been invisible with the 20-foot reflector, which disproved his earlier assumption that his earlier star counts had reached to the edge of the Galaxy. Moreover, he knew that the stars were far from being uniformly distributed; in many places they were grouped into clusters. Perhaps the stars had once been distributed evenly; but if so, he suggested, over the aeons they must have been drawn together into these clusters by gravitation, the 'great gathering power,' as he called it. As the stars were brought together in some places, in other places they left, 'as a natural consequence . . . great cavities or vacancies by the retreat of the stars towards the various centers which attract them.' Among these cavities or vacancies, one in particular had seized him. While sweeping across Scorpio, his telescope had suddenly plunged into what had seemed an unfathomable abyss – 'a hole in the heavens!' he had exclaimed to his sister Caroline. It could hardly be a matter of chance, he believed, that the same region was rich in clusters, including the globulars M4 and M80, the latter of which he described as 'one of the richest and most compressed clusters of small stars I remember to have seen.'[5]

William Herschel's son John, who set up a large telescope at the Cape of Good Hope between 1834 and 1838, later extended these pioneering studies of the Milky Way to the southern hemisphere. But in contrast to his father, he almost despaired of determining the structure of the sidereal system. 'How very difficult it must necessarily be to form any just conception,' he wrote, 'of the real, solid form, as it exists in space, of an object so complicated, and which we see from a point of view so unfavorable.'[6] Concerning the apparent vacancies of the Milky Way, he was at first inclined to think that there were 'vast chimney-form or tubular vacancies' through which one seemed to peer into even more unfathomable spaces beyond the stars.[7] Later he changed his mind, suggesting that 'it would seem less probable that a conical or tubular hollow traverses the whole of a starry stratum, continuously extended from the eye outwards, than that a *distant* mass of comparatively moderate thickness should be simply perforated from side to side.'[8]

2

Various astronomers during the nineteenth century produced maps of the intricate structure of the naked-eye Milky Way, including Eduard Heis, a German, and Benjamin Apthorp Gould, an American who was for many years director of the Cordoba Observatory in Argentina and editor of the *Astronomical Journal*. The most suggestive such map was made by Otto Boeddicker, an assistant of the fourth Earl of Rosse at Parsonstown, Ireland. Boeddicker spent five years in the shadow of the great Rosse reflectors observing and sketching the naked-eye Milky Way, and at last finished his careful delineation in 1889. When he displayed his drawings in the rooms of the Royal Astronomical Society at Burlington House, London, late that year, they created a minor sensation. Among those who viewed them was Arthur Cowper Ranyard, who was fascinated by the mysterious labyrinth shown therein, consisting of 'whisps and streams of light with very numerous dark channels, having more or less sharply defined edges.' The dark markings of the Milky Way impressed Ranyard not as true vacancies and holes as the Herschels had thought, but as 'opaque matter, dust clouds or fog-filled space, which cut out the light of the bright streams' of stars and bright nebulosities beyond it. He concluded with the hope that this impression would soon be confirmed 'by the unerring eye of the camera.'[9] He did not then know that Barnard had already obtained the first photographs of the Milky Way in August 1889 with the 6-inch

Willard lens. He first learned of them in March 1890, when, as discussed earlier, two of Barnard's photographs of the Milky Way and one of the Andromeda nebula appeared in the *Monthly Notices of the Royal Astronomical Society*.

Ranyard would later become a vigorous promoter of Barnard's Milky Way photographs and an important stimulus to his thought about the structures they revealed. Born at Swanscombe, Kent, in 1845, he moved to London in his early youth, and later studied mathematics at University College, London, and Pembroke College, Cambridge. After taking his degree, he was called to the Bar at Lincoln's Inn. Though he continued in the practice of law for the rest of his life, his real passion was for astronomy. He was especially interested in the Sun, and at his own expense he traveled to Colorado for the total eclipse of 1878, which he observed (and attempted to photograph) at Cherry Creek, near Denver, with C. A. Young's party, and mounted another eclipse expedition to Egypt in 1882.[10]

Following the death of Richard Proctor in 1888, Ranyard succeeded him as editor of the London scientific magazine *Knowledge*. One of his first innovations was to publish, as a sort of *hors d'oeuvre* to the magazine's popularly written articles on science, large-scale photogravure reproductions of the finest astronomical photographs of the day. Many of Roberts's photographs appeared for the first time in his pages,[11] and as soon as Ranyard saw Barnard's photographs in the *Monthly Notices*, he was eager to run them in *Knowledge* as well. They appeared there in July 1890, together with Ranyard's detailed commentary. The impression of some kind of 'opaque matter' partially screening the star clouds and brighter nebulae of the Milky Way, only hinted at in Boeddicker's drawings, was, at least in Ranyard's view, strongly reinforced in these photographs. 'After looking at this . . . photograph,' he wrote of Barnard's August 1, 1889 exposure on Sagittarius,

> one can have no doubt as to the aggregation of stars in cloud-like swarms or masses . . . [The structure of this part of the Milky Way] seems to consist of many streams of stars overlaid by dark absorbing masses . . . The many small isolated dark patches on a comparatively uniform background of light are much more easily accounted for as produced by the absorption of foggy areas, or dark bodies, than as due to holes or gaps in the stellar clouds.[12]

Barnard's work on the Milky Way was interrupted for three years because of the unavailability of the Willard lens, but others carried on where he had left off. The first was Henry C. Russell, director of the Sydney Observatory of New South Wales, who began photographing the southern Milky Way with a 6-inch portrait lens in 1890. A year later Max Wolf of Heidelburg used a $5\frac{1}{4}$-inch portrait lens to photograph the Milky Way in Cygnus. On his plates he discovered the gaseous nebula near Deneb (NGC 7000), which he named the 'America' nebula, though Barnard later suggested that the name be changed to 'North America' nebula,[13] and also recorded the Loop nebula near ε Cygni, of which the brightest part (NGC 6992) is the well known 'Veil' nebula.[14]

Barnard had seen the Loop visually in his comet sweeps years before, but instead of a nebulosity he had supposed it to be merely 'diffused light from dense masses of distant and faint portions of the Milky Way.' On finding out the reality, he began to think back to other regions where he had encountered 'a singular dulling of the ordinarily black sky, as if a thin veil of dust intervened.'[15] There was the region north of the Pleiades, another surrounding the star 15 Monocerotis, and the 'singularly blank region' northwest of Antares. Above all there was the vast stretch of sky extending for many degrees west of the Milky Way, 'between it and Hercules and in Ophiuchus,' of which he wrote to Ranyard in early 1892: 'I have supposed this dull dead look of the sky in that region to be due to a very distant diffused portion of the Milky Way, but the revelations of the photographic plate elsewhere in the sky would now lead one to believe it to be an immense nebulosity intimately connected with the Milky Way.'[16] In these regions William Herschel had in his earlier sky sweeps suspected the existence of great masses of diffused nebulosity. Thus there were intimations of wonders that had thus far been seen only darkly, obscure wonders from which the photographic plate alone could remove the veil of mystery, and by the early 1890s, Barnard was eager to chase them with his camera.[17]

3

As earlier mentioned, after taking his pioneering Milky Way photographs in the summer of 1889, Barnard noticed a marked deterioration in the images produced by the Willard lens. This meant that the lens needed refiguring, and it had to be sent away to John A. Brashear's optical shop in Pittsburgh. On its return to Mt. Hamilton, the reworked lens was to be mounted on its own equatorial and set up in a small observatory on Mt. Hamilton, using funds provided by Colonel C. F. Crocker. At Holden's request, Barnard worked out plans for this as early as February 1891,[18] but much to the eager young astronomer's dismay – he later accused Holden of intentionally stalling his Milky Way work as a means of persecuting him[19] – the project was not finished until June 1892. Thus Barnard had been unable to carry forward his Milky Way photography for three years.

Barnard's design for the dome called for an unusually wide shutter needed to accomodate the telescope's 10 or 12° field of view and to make unnecessary too frequent shifting of the dome during long exposures. It was set up near one of the observatory's water tanks on the small outcrop on Mt. Hamilton known as 'Huygens Peak,' close by the cottage in which the Barnards were still living at the time (and continued to live until the summer of 1894, when they moved into a two-story brick residence on Ptolemy Ridge). Adjoining the small dome was a dark room, partitioned into two areas, one of which was used for changing and filling the plate holders, the other for actually developing the plates.

Barnard's first plates with the Crocker telescope were trial exposures of 45 min made on June 17, 1892 (one was exposed on what later proved to be one of the most fascinating parts of the sky, the region around ρ Ophiuchi, but it was too short to show anything remarkable). A few nights later he began work in earnest, obtaining a 4 hr 30

Barnard wearing reindeer skin coat at Crocker photographic telescope. Mary Lea Shane Archives of the Lick Observatory

min exposure on the so-called small Sagittarius star cloud, south of the nebula popularly known as the Swan or Omega – M17. The improvement over the results of 1889 was dramatic. Among the objects shown were two striking 'black holes,' from which ran 'diverging semi-vacant lanes.'[20] The more prominent of these black holes (B92 in his later catalog of these objects) was so sharply outlined on its eastern edge that Barnard remarked that 'if drawn with a brush and India ink it could not be any more definite.'[21]

On June 25, Barnard exposed a plate to the incredibly wild and dramatic landscape of the Milky Way near θ Ophiuchi. This plate first showed the very singular small S-shaped 'vacancy' two degrees north of θ (B72, the 'Snake') and also the vast 'rift' or 'chasm' which runs southwest of the star. The latter, known as the 'Pipe,' is one of the greatest vacancies in the sky visible from the northern hemisphere – it is readily seen with the naked eye from a clear dark site if one looks above the tail of the Scorpion and $2\frac{1}{2}°$ south of θ Ophiuchi. The 'bowl' of the pipe is a 2 by 3° rectangle of darkness (B78), and is adjoined to a long thin lane extending westward for 5° which makes up the 'stem' (different parts of which were later cataloged separately by Barnard as B59, B65, B66, and B67).

As these images emerged from the chemicals of the developing tray, Barnard was well aware that he was looking for the first time upon many mysterious and hauntingly beautiful objects in the sky. His feeling of excitement in making important scientific discoveries was mingled with a feeling of religious awe. Recalling his earlier comet sweeps with his small telescope in Nashville, he wrote in the midst of photography of the Milky Way in 1893, 'I have never swept over [the Milky Way,] that "broad and ample road whose dust is stars" without the keenest appreciation of the grandeur of creation and the infinite glory of the Creator.'[22]

The plates, of course, spoke eloquently for themselves – yet Barnard's fanciful, at times even poetic, descriptions of them show the extent to which they caught hold of his imagination. In these tracts of sky Barnard found some of Hardy's 'monsters of immensity.' On a plate exposed just inside the western edge of Sagittarius and about 8° northeast of the Scorpion's tail, he discovered a sharply defined irregular black spot some 12' of arc wide which reminded him of a 'parrot's head,' with a 10th magnitude star for the 'eye' (the object was later cataloged as B97). Of another formation on the western edge of the great Sagittarius star cloud he wrote that

> the great spray of stars takes on, to the fanciful eye, the form of some huge feline monster crouching, ready to spring from an eminence on to some object below . . . The mass of stars north of the head represents well the crouching body of the weird beast, while the two nebulous ropes running up from the eastern side of the head seem to hold the creature in leash.[23]

Perhaps his most poetic description was, however, reserved for the very heart of the great Sagittarius star clouds.

> These magnificent star clouds are the finest in the sky. They are full of splendid details. One necessarily fails in an attempt to describe this wonderful region of star masses. They are like the billowy clouds of a summer afternoon; strong on the side towards the Sun, and melting away into thin atmosphere on the other side. Forming abruptly at their western edge against a thinly star strewn space, these star clouds roll backwards into the general sky . . .[24]

4

After exposing sixteen plates on the Milky Way that first summer, Barnard's work with the Crocker photographic telescope was interrupted during his absence from the observatory during the spring and summer of 1893 (when he was making the grand European tour following the discovery of Jupiter's fifth satellite), but from his return that fall until the end of his tenure on Mt. Hamilton he carried it on almost continuously, except when prevented by clouds or moonlight.

One of the rare interruptions occurred in October 1893. Engaged since the absence of the September Moon in Milky Way photography, he took time out to photograph a small comet (1893 IV) which had been discovered by Brooks very low on the eastern horizon on the morning of October 17. Earlier having examined the comet with the 12-inch refractor, Barnard noted that 'there was nothing . . . to suggest any special results in photographing it, especially as the comet could only be seen for about half an hour and was badly obscured by the dense atmosphere near the horizon.' In a plate exposed on October 21, the main tail was straight and presented 'a rather graceful appearance,' but by the following morning there had been an unexpected change:

> To say the least the resulting picture was astonishing. It presented the comet's tail as no comet's tail was ever seen before. The graceful symmetry was destroyed; the tail was shattered. It was bent, distorted and deflected, while the larger part of it was broken up into knots and masses of nebulosity, the whole appearance giving the idea of a torch flickering and streaming irregularly in the wind.[25]

These photographs recorded for the first time what later became known as a tail disconnection event. Barnard strongly suspected that the comet's tail in its flight through space had suddenly been shattered on encountering 'some kind of resisting medium – a cosmical cloud – a swarm of meteors – certainly a region of resistance of some form.'[26]

'With these thoughts paramount,' Barnard continued, 'it was with considerable anxiety that the developments of the next morning were awaited.' The pre-dawn sky was cloudy but 'the clouds were breaking and flying in the face of almost a hurricane . . . The little observatory rocked in the wind and the dome threatened every moment to fly away in the direction of San Francisco.'[27] Despite the horrific conditions, Barnard obtained a useful exposure in the moments when the flying clouds permitted the image of the comet to fall on the plate. It confirmed the disturbance of the previous morning:

> The tail was broken and seemed to hang in irregular cloud masses, deflected out of line with the stem of the tail near the head. A portion of the end of the tail was completely detached and was drifting off as an independent comet.[28]

Thereafter clouds and moonlight interfered until the morning of November 2, at which point Barnard obtained yet another remarkable plate. Now the tail appeared strongly

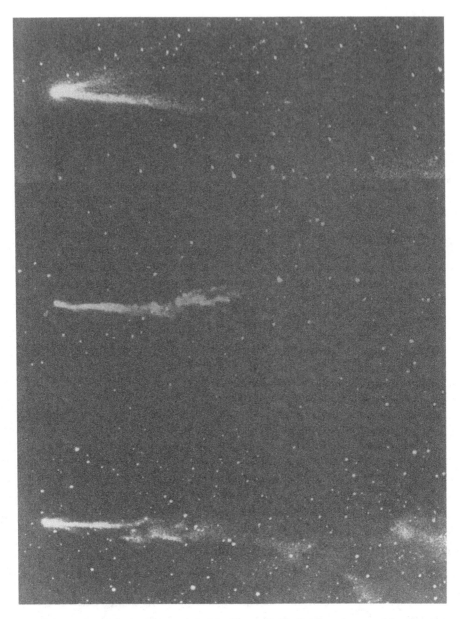

Brooks's Comet (1893 IV), photographed by Barnard with Crocker photographic telescope on October 20, 21, and 22, 1893. This series shows a tail disconnection event which occurred on October 21. UCO–Lick Observatory

concave in the direction of motion, as if it were 'beating against a current of resistance.' The end was suddenly bent backward almost at a right angle.

Barnard kept up his series of photographs of the comet through November 19, and later regarded them as priceless. In his view – and in this he was correct – they initiated a new era in the study of comets. Whereas only a few years before, Brooks's Comet would have passed as a small, entirely unremarkable affair – indeed, it never attained naked eye visibility – the photographic plate showed it to be one of the most interesting objects on record and, Barnard suggested, pointed to the possibility of using comets' tails to probe the interplanetary medium.[29]

5

Ranyard was always eager to publish Barnard's latest photographs in *Knowledge*, and was almost alone in prodding the struggle to understand what they were revealing. He agreed with Barnard that the changes in the tail of Brooks's comet were probably accounted for by its encounter with some kind of resisting medium, though differing in some details in his interpretation. As for the dark markings of the Milky Way, Ranyard struck out in an independent direction, seeing in them streams of opaque nebulous matter, heated but not luminous and ejected like the solar prominences across space. This dark matter seemed in its outer portions to condense into stars or masses of luminous nebulosity.[30]

By remarkable intuition, Ranyard had seen to the heart of the matter, but as yet his view lacked proof. Barnard, for one, was unconvinced. Still looking at his plates through traditional Herschelian categories, he saw in the dark markings actual vacancies and holes in the stars. Concerning a plate exposed to a region in Cepheus on October 13, 1893 – it showed a prominent series of irregular dark lanes ramifying in various directions – Barnard wrote to Ranyard:

> When viewed at a distance of a few feet the effect is enhanced; it is then seen that the sky (or Milky Way) is broken up into numerous black cracks or crevices. Looking at these peculiar features, I cannot well see how one can avoid the conclusion that they are necessarily real vacancies in the Milky Way, through which we look out into the blackness of space. I am aware that you are opposed to this view, and I would like to have your opinion of the real nature of these apparent crevices in the Milky Way, as shown on this particular plate.[31]

While Barnard held cautiously back, W. H. Wesley, Savilian astronomer at Oxford University and assistant secretary of the Royal Astronomical Society, eagerly embraced Ranyard's ideas. A respected authority on astronomical photographs, Wesley pronounced: 'That the dark spaces in the Milky Way are caused by dark . . . matter, lying between us and the stars appears a reasonable supposition, and in many cases it seems difficult to account for the appearance observed in any other manner.'[32]

Wesley's comments inspired Barnard to photograph again, on July 6, 1894, the wild regions of the Milky Way near θ Ophiuchi. He at once sent the plate to Ranyard, and in describing the dramatic structures therein once more invoked imagery of vacancies and chasms. 'It is,' he wrote, 'essentially a region of vacancies. There is a great chasm here in the Milky Way [the 'Pipe' described above] . . . It will be noticed that in many of these vacancies there are "deeper depths" yet, which almost suggest that the appearance of diffused nebulosity over the region is real nebulosity, and that these dark and black places in it are thin places and actual holes.' As strong as were his own convictions at the time, he eagerly awaited Ranyard's views.[33] In *Knowledge* for November 1, 1894, Ranyard responded with his strongest arguments against Barnard's belief that the dark markings were actual vacancies:

> The dark vacant areas or channels running north and south, in the . . . picture referred to by Prof. Barnard, seem to me to be undoubtedly dark structures, or absorbing masses in space, which cut out the light from a nebulous or stellar region behind them . . . It is comparatively easy to conceive of a narrow stream of dark nebulosity or foggy matter cutting out the light of a uniform background; while if the narrow dark regions correspond to thin places or holes in the nebulosity, they must be holes or thin places extending in a direction away from the earth. The probabilities against such a radial arrangement with respect to the earth's place in space seem to my mind to conclusively prove that the narrow dark spaces are due to streams of absorbing matter, rather than to holes or thin regions in bright nebulosity.[34]

When Ranyard wrote this, he was suffering from cancer and had less than two months to live. He died on December 14, 1894, at the age of only forty-nine. For Barnard, the memory was still fresh of the jaunt that the two of them had taken to France a year earlier, and he wrote sadly to Wesley:

> My wife and I were sincerely grieved when finally we received notice of the sad end. We came more in contact with Mr. Ranyard by far than with any other person cross the ocean. In every way as a scientific man and as a perfect gentleman, we learned to admire and to love him. It is not only a great loss to English science, the death of a man like him but it must be felt by the entire scientific world. Such eminently honest and able men are so few that the world is in every way a loser when they die.[35]

Nevertheless, for all his heartfelt expressions of regard for the man, Barnard omitted almost all mention of him and his ideas from his own writings about the dark markings of the Milky Way for the next quarter of a century.

At the moment, and for some time to come, Barnard clung to the orthodox view and remained convinced that he was seeing in his plates reflections of the established order of things – the grand Herschelian cosmology. The Milky Way, which at the time was still regarded as the entire universe, remained for him as it had been for William Herschel in 1802, not 'the mere visual effect of an enormously extended stratum of stars, but . . . an actual aggregation, highly irregular in structure, made up of stellar clouds and groups and nodosities,'[36] together with corresponding voids and vacancies.

Aside from the obvious influence of ingrained Herschelian concepts of the universe, the most compelling reason that Barnard tended to see the dark markings on his plates as holes and not as obscuring masses had to do with his assumptions, at the time, about what a nebula was. Barnard still tended to think of nebulae as by nature luminous rather than dark; they were like comets, tenuous and transparent. If so, then as he had himself witnessed in the case of several comets which had passed in front of stars, the stars should continue to flicker through them and not be obscured as Ranyard supposed.

From Ranyard's argument that it was improbable that vacancies, if they existed, should be distributed radially away from the Earth, Barnard found refuge in John Herschel's concept of the star clouds of the Milky Way being arranged in relatively thin, perforated sheets, which consisted of stars that were actually, and not only apparently, small in size and crowded together. Thus he wrote: 'Some of [the pictures] show strong evidence that the general body of the Milky Way may be made up of small stars . . . Many parts of the Milky Way appear to be comparatively thin sheetings of stars with relatively no very great depth, *for it is not possible otherwise to explain the black holes and rifts shown in them.*'[37]

In 1894, when Ranyard died, Agnes Clerke could still write of the Herschelian view of the Milky Way that 'all the facts [thus far] ascertained fit in with this conception.'[38] It did not yet face a crisis. As a result, the time was not yet ripe for the discovery of dark obscuring masses (and Ranyard's intuition, however bold, did not yet a discovery make). Rather, as Thomas Kuhn has argued in *The Structure of Scientific Revolutions*, the discovery of a new phenomenon is not 'a single simple act assimilable to our usual (and also questionable) concept of seeing,' but involves 'both observation and conceptualization, fact and assimilation to theory . . . [It is thus] a process and must take time.'[39] Ranyard's intuition was only a point of departure, a first step in this process of discovery. The hard going – and most of the credit for the discovery – would, as we shall see, ultimately belong to Barnard himself.

6

If as yet Barnard was conservative in his interpretation of his photographs, one of the reasons may have been that at the moment he was more concerned about taking the photographs than in worrying about their meaning. Many interesting regions had still to be explored, and among the regions of sky to which his earlier comet-seeking had suggested 'pointings' for the camera was that north of the Pleiades. Unfortunately, the mounting of the Crocker telescope did not permit an exposure to be carried through the meridian, and as a result very long exposures had to be made over successive nights. With such a marathon exposure on two nights in December 1893 – the combined time was 10 hr 15 min, the longest Barnard ever attempted – he first recorded the 'exterior nebulosities of the Pleiades.'[40]

Another suspicious area was that around the star 15 Monocerotis, where Barnard had suspected nebulosity visually with the Lick Observatory 12-inch refractor in 1888. A plate exposed for three hours in February 1894 confirmed the existence of this nebulosity. It spread out for 3°, 'in a weak, diffuse light with rifts in it and irregularly terminated along the edges of a vast vacancy in the Milky Way.'[41] The most condensed part of this nebulosity was only a few minutes of arc in diameter – much too small to be shown to advantage on the small scale of the Willard plates – yet even so it appeared peculiar in shape, and Barnard suggested that it might well repay study with a larger photographic telescope. He was right. When Roberts captured it with his 20-inch reflector a year later, he found on his plate 'a conical dark space bounded by a rim of nebulosity'[42] – one of the most astounding and beautiful objects in the deep sky, known today as the Cone Nebula.

Of all the areas, however, which had attracted his attention in earlier years, none intrigued Barnard more than that just north of Antares in Scorpio. 'For many years this part of the sky troubled me every time I swept over it in my comet seeking,' he wrote; 'though there seemed to be scarcely any stars here, there yet appeared a dullness of the field as if the sky were covered with a thin veiling of dust, that took away the rich blackness peculiar to many vacant regions of the heavens.'[43] Barnard was not the first to wonder about this particular region, since it had been here that William Herschel had encountered his 'hole in the heavens.' Much later, Richard Tucker, a meridian circle observer at the Cordoba Observatory in Argentina who afterwards spent many years at Lick, was making observations in this part of the sky for the star catalog *Cordoba Durchmusterung*. He set his telescope just north of Antares and prepared to record the transits of stars as they passed through the field, but no stars came, and he concluded that the sky had clouded over. However, on looking up he found it perfectly clear – his telescope had simply been pointed to one of the blank lanes in this part of the sky![44]

Strangely, given his longstanding interest in this part of the sky, Barnard did not get around to photographing this region (with the exception of the unsatisfactory June 1892 trial plate) until March 23, 1895. The resulting exposure of 2 hr 20 min was one of the most important he ever took. When he removed the negative from the developing tray, he found 'a vast and magnificent nebula, intricate in form and apparently connected with many of the bright stars of that region including Antares and Sigma Scorpii . . . [T]his nebula from its dimensions, its individual peculiarities, and its occupying a region almost devoid of stars – which is the center of great radiating lanes or vacancies among the stars – has scarcely an equal for interest in the entire heavens.' The greatest portion of the nebulosity was condensed about the 4th magnitude star ρ Ophiuchi. 'Perhaps as remarkable as the nebulosity itself,' Barnard wrote,

> is the region is which it is placed. The entire nebula occupies a vast vacancy among the stars here, from which great sharply defined vacant lanes run eastward for from ten to twenty degrees. With the exception of the great vacancies, this part of the sky is covered with a uniform sheeting of very small stars, and the lanes which pass eastward from the nebula through these are singularly sharply defined at their borders, especially so in the case of the great south lane . . . If these lanes are examined carefully it will be seen that they are not uniformly dark – there are still darker regions in them as if they were filled with feeble nebulosity, in which are holes and rifts, permitting the blacker sky beyond to be seen.[45]

7

Barnard often emphasized that his success in photographing the Milky Way was a direct result of the Willard lens's wide field and effective light gathering power. In 1894, he began experimenting with another short-focus lens. Taken from a cheap oil projecting lantern, Barnard purchased it for only seven dollars. This 'magic lantern lens,' as he called it, was $1\frac{1}{2}$ inches in diameter and $3\frac{1}{2}$ inches focus, and attached to the Crocker telescope mounting it allowed the photography of even wider fields of sky than the Willard – fields on the order of 20 to 25° across.

With this lens Barnard was able to take in nearly the entire constellation of Orion, and on images made in October 1894 he captured 'an enormous curved nebulosity encircling the belt and the great nebula, and covering a large portion of the body of the giant.'[46] Like the Loop nebula in Cygnus, this partial wreath of nebulosity is now believed to be a supernova remnant. Today it is referred to as 'Barnard's Loop,' since it was Barnard's work which made it well known, although in fact it had been discovered by W. H. Pickering in photographs taken from Mt. Wilson in 1889.

Barnard also rephotographed parts of the Milky Way with the magic lantern lens, noting with satisfaction that the prominent cloud forms were already beginning to register with exposures of only ten minutes. And on March 30, 1895, only a week after his first Willard lens photograph of the region, he used the magic lantern lens to extend his explorations of the remarkable celestial landscape north of Antares in the Scorpion. An exposure of two hours showed the ρ Ophiuchi nebula and the dark lanes extending to the east, and in the wider view, he also captured a new large irregular nebula involving the star ν Scorpii.[47]

Two years later, A. A. Common, in bestowing upon Barnard the Gold Medal of the Royal Astronomical Society, remarked:

> Professor Barnard has shown us with what modest means good results can be obtained – with not only a small but a cheap lens, such as may be found in any magic lantern. No doubt his early experiences . . . were valuable in giving him this knowledge of what can be done with humble apparatus; but we must certainly admire, not merely the skill, but the courage of a man who could, under the very shadow of the great 36-inch refractor, demonstrate the merits of a lens which could be bought for a few shillings.[48]

His was, indeed, the most remarkable versatility. At the very moment that he was showing the crushing superiority of large telescopes to small ones for planetary observations, he was devising investigations suitable for this modest lens – surely among the most humble instruments ever used successfully for astronomical research. Of it he could say without exaggeration: 'It is so inexpensive, and its results are so wonderful, that every amateur, and every professional observatory, should possess one.'[49] All the more remarkable that this small lens should prove perfectly suited to capturing the features of the excessively large – the vast tracts of nebulosity and the massive star clouds of the Milky Way, in whose forms might be deciphered nothing less than the grand plan of the Galaxy itself.

ρ Ophiuchi region, 4 hr 35 min exposure taken by Barnard with the Crocker photographic telescope, April 19, 1895. Barnard's writing on the back of the original plate, is reversed on this print. UCO–Lick Observatory

1   San Francisco *Chronicle*, December 8, 1894

2   E. E. Barnard, 'The Development of Photography in Astronomy – I,' *Science*, New Series, **8** (1898), 341–53: 351

3   Joseph Ashbrook, 'The visual Milky Way,' in *The Astronomical Scrapbook: Sky-watchers, Pioneers, and Seekers in Astronomy* (Cambridge, Mass., Sky Publishing Corp., 1984), p. 373

4   Thomas Hardy, *Two on a Tower* (London, Macmillan & Co., 1960), pp. 33–4

5   William Herschel, 'On the Construction of the Heavens,' in J. L. E. Dreyer, ed., *The Scientific Papers of Sir William Herschel*, (London, Royal Astronomical Society, 1912), vol. 1, 223–59: 253. The 'small, miniature cluster' described here is yet another globular – NGC 6144.

6   John Herschel, *Outlines of Astronomy* (Philadelphia, Blanchard and Lea, 4th ed., 1856), p. 449

7   John Herschel, *Results of observations made during the years 1834/5/6/7/8 at the Cape of Good Hope; being a completion of a telescopic survey of the whole surface of the visible heavens, commenced in 1825* (London, 1847), art. 335, quoted in Michael Hoskin, 'John Herschel's Cosmology,' *Journal for the History of Astronomy*, 18 (1987), 1–34: 22

8   John Herschel, *Outlines of Astronomy*, p. 449.

9   A. C. Ranyard, 'Drawings of the Milky Way', *Knowledge*, 12 (1889–90), 6–7: 7

10 See G. E. Hale, 'Arthur Cowper Ranyard,' *Ap J*, 1 (1895), 168–9; W. H. Wesley, 'Arthur Cowper Ranyard and his Work,' *Knowledge*, 18 (1895), 25 7

11 On one of Roberts's photographs of the Andromeda nebula, published in February 1889, Ranyard noticed that the outer arms appeared to resolve partially into stars. Since the star images were slightly out of focus owing to the long exposure necessary to register them, he concluded, wrongly, that these star images were themselves 'nebulous.' Obviously he was still blinded by the apparent implications of the nova of 1885 and by the prejudice of the time that the Andromeda nebula was a nearby and relatively small swirl of gas and dust rather than, as we now know, a vast system of stars like our own Milky Way galaxy.

12 A. C. Ranyard, 'On the Distribution of Stars in the Milky Way,' *Knowledge*, 13 (1890), 174–5:175

13 E. E. Barnard, 'Diffused Nebulosities in the Heavens,' *Ap J*, 17 (1903), 77–80:77

14 Both had originally been seen, though only as diffused nebulosities in the Milky Way, by William Herschel.

15 E. E. Barnard, 'The Great Nebulous Areas of the Sky,' *Knowledge*, 15 (1892), 14–16:14

16 ibid., p. 15

17 ibid.

18 EEB to ESH, February 18, 1891; SLO

19 EEB, 'notes for Young's letter,' June 5, 1892; SLO

20 E. E. Barnard, 'Photographs of the Milky Way and Comets,' *LOP*, 11 (1913), notes to plate 55

21 E. E. Barnard, *Atlas of Selected Regions of the Milky Way*, E. B. Frost and M.R. Calvert, eds. (Washington, DC: Carnegie Institution, 1927), vol. 1, notes to plate 31. This plate had been published by A. M. Clerke in her *History of Astronomy During the Nineteenth Century* (London: Adam & Charles Black, 3rd ed., 1893), p. 508, with the comments: 'It will be noticed that the bright mass [of stars] near the centre of the plate is tunneled with dark holes and dusky lanes.'

22 E. E. Barnard, 'How to Find Comets,' San Francisco *Examiner*, February 5, 1893

23 Barnard, 'Photographs of the Milky Way and Comets,' notes to plate 45

24 ibid., notes to plate 49

25 E. E. Barnard, 'Photographs of Brooks' Comet,' *PA*, 1 (1893), 145–7:146–7

26 E. E. Barnard, 'Photographs of a Remarkable Comet,' *AA*, 13 (1894), 789–91:790

27 ibid., p. 146

28   ibid., pp. 146–7

29   E. E. Barnard, 'Photographs of Comets, and of the Milky Way,' *MNRAS*, **59** (1899), 355

30   A. C. Ranyard, 'What is a Nebula?' *Knowledge*, **16** (1893), 10–12

31   EEB to A. C. Ranyard, November 14, 1893; published in *Knowledge*, **17** (1894), 17. The plate referred to is no. 82 in Barnard, 'Photographs of the Milky Way and Comets.'

32   W. H. Wesley, 'On the Distribution of Stars in the Milky Way,' *Knowledge*, **17** (1894), 179–82:182

33   E. E. Barnard and A. C. Ranyard, 'Structure of the Milky Way,' *Knowledge*, **17** (1894), 253

34   ibid.

35   EEB to W. H. Wesley, February 2, 1895; RAS

36   Clerke, *History*, p. 506

37   E. E. Barnard, 'The Development of Photography in Astronomy-II,' *Science*, New Series, **8** (1898), 386–95:387

38   Clerke, *History*, p. 506

39   Thomas Kuhn, *The Structure of Scientific Revolutions* (Chicago, University of Chicago Press, 2nd ed., 1970), p. 55

40   E. E. Barnard, 'On the Exterior Nebulosities of the Pleiades,' *AA*, **13** (1894), 769–70

41   E. E. Barnard, 'Photographs of the Milky Way near 15 Monoceros [sic] and near ε Cygni, *Ap J*, **2** (1895), 58–9.

42   Isaac Roberts, 'Photograph of the Nebula near 15 Monocerotis,' *MNRAS*, **55** (1895), 398–9

43   E. E. Barnard, 'The Great Nebula of Rho Ophiuchi and the Smallness of the Stars Forming the Ground Work of the Milky Way,' *PA*, **5** (1897), 227–32:227

44   As recounted by E. E. Barnard, 'On a Great Nebulous region and on the Question of Absorbing Matter in Space and the Transparency of the Nebulae,' *Ap J*, **31** (1910), 8–14: 14

45   ibid., p. 228

46   E. E. Barnard, 'The Great Photographic Nebula of Orion, Encircling the Belt and Theta Nebula,' *AA*, **13** (1894), 811–14:813

47   E. E. Barnard, 'On a Great Photographic Nebula in Scorpio, near Antares,' *MNRAS*, **55** (1895), 453–5

48   A. A. Common, 'Address Delivered by the President . . . on presenting the Gold Medal to Professor E. E. Barnard,' *MNRAS*, **57** (1897), 321–8: 326

49   Barnard, 'On a Great Photographic Nebula,' p. 455

17

A tide in his affairs

1

After Burnham and Crew left Mt. Hamilton in 1892, there were frequent rumors that Barnard would be next. Just after he set out on the grand tour of the Eastern United States and Europe in February 1893, the San Francisco *Examiner* announced: 'Now that Professor Barnard has been well started on his six months' ramble . . . the report is being circulated that he intends leaving the Sick Observatory [sic] at the expiration of his vacation and accepting a more congenial, if not more lucrative, post at some Eastern university.'[1] Burnham lent credence to the report, being quoted in the newspapers when Barnard arrived as far as Chicago: 'If the people of California don't do something they will lose Barnard sure as fate. We will keep him here unless it is decidedly for his advantage to remain in California, and that certainly will not be unless there is a radical change at Mount Hamilton.'[2]

In San Jose, where Barnard was an immensely popular figure with the general public and the press, every rumor of his imminent departure was treated with alarm. The San Jose *Report* suggested that no remedy should be overlooked in the effort to keep Barnard on Mt. Hamilton, including Holden's prompt dismissal:

> Professor Barnard is a fine character, above littleness, above money motives. He refused compensation for his lectures before the University Extension Club. His sense of honor is nice. He is not self assertive, but he is a man for all that, and . . . we lose a gentleman and an astronomer and a man famous while fame shall last, if we lose Edward Emerson Barnard.[3]

Barnard, of course, did return to the Lick Observatory that August, though as we have seen, his relations with Holden were if anything even more troubled than they had been before. By March 1894, rumors of his departure were once again flying. Though he was deeply touched to receive a petition from students of the University of California at Berkeley that he remain in the state, he refused to close any doors:

> Be assured that no other petition could have more influence on my actions than yours. [However,] it is now the time of my life when I must definitely decide under what conditions the remaining portion of it must be passed . . . It is scarcely necessary for me to remind you, for you yourselves will understand, that there is a tide in the affairs of men which taken at flood leads on – not necessarily to fortune – but certainly to contentment and happiness. This tide is at flood with me now and it is for me to decide, before too late, for my future happiness.[4]

Burnham was eager to see him in Chicago, and knowing him better than almost anyone else, was confident of the outcome. 'So far as E. E. B. is concerned,' he assured the observatory's future director, George Ellery Hale, 'I think that between us we can control his movements when the time comes.'[5] Already, within only a few months of Charles Tyson Yerkes's announcement of his gift of the world's largest refracting telescope to the University of Chicago, Hale had, through Rev. Arthur Edwards, a prominent Chicago divine and editor of the *Northwestern Christian Advocate*, formally recommended Barnard to president Harper for the Yerkes Observatory staff.[6] Despite the fact that the University of Chicago was operating on a tight budget at the time – after the panic of 1893, Chicago remained under the cloud of an economic depression, and Yerkes had suffered such devastating losses that he was becoming increasingly resistant to requests for more funds for his observatory – Hale wanted another $15 000 for astronomers' salaries, and hoped to add both Barnard and Keeler immediately to the staff. Harper reminded him that 'astronomy is *one* department of the university,' and proposed, in December 1894, that Barnard be the only addition to the staff.[7]

As Barnard wavered, Burnham and Hale began barraging him with letters in the attempt to help him make up his mind. 'My hopes for the reputation of the Yerkes Observatory . . . make it almost absolutely necessary that you should come,' Hale cajoled. 'A refusal from you would put us in a difficult position, for there is no one else in America or anywhere else who would fully satisfy our requirements.'[8] Burnham, knowing Barnard's hysterical dislike of Holden first hand, was even more effective in applying the pressure, taking every opportunity to inflame Barnard's perpetual dissatisfaction with conditions on Mt. Hamilton. Noting Barnard's chronic complaints about his health – largely psychosomatic, as Barnard himself realized, and a result of accumulated fatigue, worry and stress – Burnham, in February 1895, urged: 'My judgement is that the best thing you can do for yourself is to throw up your hand at once and leave the L[ick] O[bservatory] to the dogs, come East and take the Summer to build up into a first class Condition . . . By all means do it if you are still under the weather . . . You know as well as I do that much the best, if not the only way, is to leave the grinding and wearing associations of L[ick] O[bservatory].'[9] When Barnard replied that he intended to persevere at Lick at least through the summer in order to take advantage of the prime observing season there, Burnham persisted: 'If you are wearing down, getting worse, or no better, it may result in a physical complete collapse, or something worse, and then your work will be ended. You will then see how shortsighted it was to sacrifice health and years of astronomical work for the sake of continuing there a few months longer.'[10]

Aside from Barnard's understandable desire to stay on Mt. Hamilton in order to complete work in progress, one of the other issues that had to be overcome was his not entirely irrational fear of Midwestern winters, which astronomically promised to be dismal. But Burnham, who had done the site testing on Mt. Hamilton and was also influential in the decision to locate the Yerkes Observatory at Williams Bay, Wisconsin, on the shore of Lake Geneva, again tried to be optimistic: 'While some of the time it is colder [in winter here], I am not sure it is not a better place to live at that season [than

Mt. Hamilton] . . . In an astronomical way it is infinitely better . . . One could do here 10 times as much with a telescope in the 3 mos. of winter as at Mt. H.'[11] Time, alas, would show that that prediction was completely fallacious!

Barnard continued to hesitate, despite the occasional flurry of press reports that he had made up his mind. In February 1895 the San Francisco *Call* announced, 'Barnard Will Go,' and indicated that Barnard had accepted not only a staff position but the directorship at Yerkes.[12] The report was, of course, completely unfounded, and Barnard issued prompt denials. He also reassured Hale: 'There is one thing I wish you would not pay any attention to – for I am not responsible for it. That is the statement that keeps getting into the papers that I am to have charge of the Yerkes Observatory. To a person not as well balanced as you are this would become very offensive, but I know you will not attribute it to me.'[13]

By early March 1895, Barnard was definitely leaning toward accepting Harper's offer. He wrote to Hale,

> I have been very much distressed in trying to make a decision in the matter. I am confident that I can do better work in a purer social atmosphere. There is no harmony here. When Mr. Burnham was here every hour of work was a happy one, now a great deal of it is unhappy work . . . Just now I am strongly inclined to throw up my hands here for good – the sooner the better . . . I think you can surely count on me, but I must have a little longer time to decide. My health is better now. The mountain is a trying place – especially on ones nerves, and I have been sick a good deal during the winter[,] not laid up, but unwell.[14]

By now, alas, there were additional complications. Barnard was making every effort to collect the comet and Milky Way photographs he had taken with the Willard lens for a volume which he hoped would be published by the Lick Observatory. In approaching Holden in the matter, he urged that they were 'of the highest scientific value and would, I am sure . . . be a great credit to the Observatory.'[15] However, if they were to receive their due, the photographs would be very expensive to publish – Barnard himself estimated that the cost of reproducing sixty to a hundred of his finest plates would range between $1500 and $2500, not a trivial sum at the time considering that Barnard's annual salary at the time was only $2400.[16] Holden conveniently recalled that Barnard had just a few months earlier submitted many of the same photographs to the Royal Astronomical Society for possible publication in the *Monthly Notices* and smoothly suggested: 'I suppose nothing should be done here until we know whether the R.A.S. will accept or refuse your offer. If the Society declines to print your negatives I should like to know on what grounds they do so, before I make any recommendation to the Regents.'[17] With the Lick Observatory strapped for funds, this suggestion was not unreasonable, but Barnard was infuriated by what he regarded as yet another of Holden's attempts at petty persecution. Under mounting strain, he penned a turbulent letter to Timothy Guy Phelps, the chairman of the Lick Observatory committee, urging the Regents to intervene, as they had so often done, in his behalf. His sixteen photographs of Brooks's Comet of 1893 were, he told Phelps, 'the most wonderful and important set of photographs of a comet in existence.' His star photographs were equally important – and infinitely more so than Holden's photographs of the Moon, which, he reminded Phelps, had already been published by the Lick Observatory.[18] Holden defended the publication of his Moon photographs, pointing out that this had been paid for entirely with outside funds 'so that no one should be able to say that I had used observatory money to print my personal work,'[19] and continued to argue to the Regents that there could be no possible harm in delaying publication of Barnard's photographs until the Royal Astronomical Society had reached a decision. 'There is no doubt that Prof. Barnard's photographs should be published somewhere, preferably by the L[ick] O[bservatory],' he advised Phelps. 'It does not, however, appear to me to be desirable to print them at an expense of $1500–$2500 if they are also to be published by the R.A.S.'[20]

Barnard was only too keenly aware that if at this moment anything were to leak out about the Chicago offer, it would ruin his chances with the Regents of getting his photographs published. He had personally said nothing about the Chicago negotiations to anyone in California, with the exception of his wife. He now implored Hale to do likewise and keep the matter completely under wraps. 'I wish to state here,' he told Hale,

> so that no mistake may be made, in reference to my desire for delay in closing with Dr. Harper. It could be made (and would be) very disagreeable for me here if I should take immediate steps in the matter. I have petitioned the Regents to publish my star pictures. The Director is fighting it tooth and toe nail. The matter is now in the hands of the Regents for decision. It is important that I publish these pictures for if left behind they will go to the dogs.[21]

2

Holden had every reason to believe that Barnard was sitting on an offer from the University of Chicago and was likely to accept it, so that, as Osterbrock suggests, 'he was trying to delay the decision on the publication for the very same reason that Barnard was trying to press it to a successful conclusion.'[22] Barnard realized this, and wrote to Hale, 'I have good reason . . . to know that the Director has been informed of the matter.'[23] Given the extent to which he himself was guilty of ambiguity and evasiveness at the time, he was, however, hardly in a position to complain too much of the director's motives. Indeed, he was probably projecting some of his own equivocations onto Holden in the paranoid belief that the director knew more than he was letting on and was using this knowledge behind his back to undermine his position with the Regents. Thus on May 11 he challenged Holden: 'If any communications have been sent by you to anyone concerning the matter now before the Regents about my celestial photographs . . . that copies of all such be supplied to me . . . A simple negative statement from you is all that is necesary.'[24] As there were in reality no such communications, Holden, through his secretary Charles Dillon Perrine, could in good conscience pass along to Barnard just the statement requested: 'No.' Barnard now found this statement ambiguous, and asked for a further clarification. When Holden ignored his letter, Barnard asked again.[25] This time Holden replied tersely:

> You asked for an answer and said 'a simple negative statement is all that is necessary.' On May 11 the Secretary of the L.O. told you that the answer to your letter was 'no.' On May 11 you again wrote saying: 'If you have sent any such please notify me of the fact, for your answer is ambiguous.' . . . On May 23 you write again saying that the answer you received was evasive and ambiguous. I reply that the answer to the questions in your three letters is no, I have sent no such letters.[26]

By now Harper and Hale were also beginning to lose patience with Barnard. Barnard sensed this, and did everything in his power to try to smooth things over. 'You cannot understand how much I have the subject of printing these pictures at heart,' he appealed to Harper. 'If I do not get this done before I go they will be a loss both to astronomy and to me for they will be permitted to sink into oblivion. They consist of comet nebular and Milky Way pictures. They are the most wonderful in existence . . . For the salvation of this work it must be printed – it will be a great credit to me and a lasting benefit to Astronomy.'[27] He wrote in similar vein to Hale: 'I am expecting to hear, everyday, from the regents about my pictures. If I could see them on the way I should be happy.'[28]

On May 22 Barnard still hesitated over the Chicago offer, and that day wrote to Harper: 'There is one important point that you should clearly see. When I leave here it is for good so far as the Lick Observatory is concerned. There is no return to this Observatory during the present administration. I therefore must be sure before I take any step towards a change.'[29] He wrote in much the same vein to Burnham, who repeated the same arguments he had urged three months earlier on his indecisive friend: 'You will live a great deal longer, enjoy life better, and accomplish more by clearing out. So I don't see any reason for hesitation, and you may as well make the break.'[30]

Hale had warned Barnard that the Chicago offer could not be kept secret indefinitely, and finally, on May 31, things began – or more likely were allowed – to unravel. The San Francisco *Chronicle* announced that Barnard had turned in his resignation that very day, followed, as usual, by Barnard's denials. 'Any such resignation I shall send through you,' he assured Phelps. 'Therefore you know that the statement that I had sent in my resignation is a mistake.'[31] Phelps, though not without suspicions, was willing to accept Barnard's denial at face value. He told Barnard that he had first learned of the *Chronicle* story when questioned about it by a reporter in San Francisco:

> I told him I had seen you only a few days before, that you had not then resigned and that I had no knowledge of your intention to do so . . . We were deeply pained to loose Profs Burnham and Keeler; and shall be even more pained to loose you, if we do loose you, but we shall continue to hope we shall not do so to the last . . . Personally I want to see you do what is best for yourself, but unless to your certain advantage to go I hope you will stay with us.[32]

Barnard was still vigorously lobbying Phelps and the Regents to publish his Milky Way and comet photographs when, on June 6, events overtook him. The University of Chicago issued on that day its catalog of summer courses; Barnard was listed on the first page as a professor of astronomy in its department of mathematics and astronomy. When questioned about the entry, Harper explained that he had been 'satisfied in his own mind that Prof. Barnard was coming, although his official acceptance had not been received.' He added that he had held the catalog back from the press for several days in expectation of receiving this acceptance, and when it was not forthcoming, he had ordered the reference struck out, but it had been passed anyway through a proofreader's blunder.[33] That Harper's actions were as innocent as this account suggests seems unlikely, but the catalog, in any case, was soon all over Chicago, and news of Barnard's inclusion in it was promptly telegraphed to the West Coast. Holden, who had been expecting such news for fully two years, promptly telegraphed a friend: 'The printed programme of the Department of Astronomy of the University of Chicago just issued contains the name of Mr. Barnard. This is public news therefore.'[34] One can only guess how disappointed Phelps must have felt after Barnard's repeated assurances that he would be first to know of such developments – he first heard of the announcement from a newspaper reporter.

Barnard reacted to the embarrassing leak with predictable emotion. 'I vividly recall what happened [that] day,' Schaeberle recalled long afterwards. 'A most pathetic figure, poor broken-hearted Barnard came to me, rested his hands and head on my shoulder, broke down and cried convulsively like a child.'[35] To the end of his life, Schaeberle remained convinced that Barnard had never really intended to leave California. Certainly Barnard's vacillations in the matter may have given this impression to a colleague who never enjoyed his full confidence. Schaeberle, in any case, quickly penned a letter to Phelps suggesting that under the right circumstances Barnard might still be persuaded to stay on. But too late. By this time Phelps was tired. He told Schaeberle that as much as the Regents desired to have Barnard remain, the matter had progressed so far that it would be best to let him go.

There was nothing left for Barnard to do but to submit his formal resignation to the Regents and his acceptance to Harper.[36] The same day he announced to Hale: 'I have thrown in my lot with you.'[37] To the Regents and especially Phelps, who had some reason to feel betrayed by him, he commented for the record: 'As it has several times been stated . . . that I had accepted this position and it had so by mistake been announced from Chicago, I wish to state that though the offer has been under consideration by me for some time I have not up to this moment accepted the position . . . I state this so that no misunderstanding may be had in the matter.' And he added: 'It has been a long and bitter struggle on my part, extending over many months, to make this final decision, so attached have I become to California and its people.'[38] The cat and mouse game was finally over.

3

Phelps informed Barnard of the vote of the Regents to accept his resignation on June 11, and their decision to table, for the moment, the critical vote on the publication of his star pictures.[39] Barnard's mood was far from euphoric. 'I expect to freeze out this coming winter at Chicago!' he joked grimly to Davidson. 'Its all over now and I feel much better for it – as [one might say in taking] the final drop . . . Soberly, I feel very sad at leaving this Gods own country, but I shall not let it worry me too much because when I get old I'm coming back here to die – not to work, but just simply to die, for I can't think of dying in the cold and gloomy east. I want to die in the sunshine of California.'[40]

Barnard's resignation from the Lick Observatory was due to take place effective as of October 1, 1895. A compulsive worrier, he no sooner made the decision to leave Mt. Hamilton than he began to worry about the conditions to be encountered in the Midwest. Originally it had been hoped that the Yerkes Observatory would be ready by October, but that timeline proved to be overly optimistic – Burnham, who had carefully suppressed the discouraging news previously, now informed Barnard frankly: 'Hale . . . says he is very doubtful of the [40-inch] instrument being ready for use as soon as they are talking of, October. Of course the architect and builders do not talk this way, but . . . it is a big affair, and some allowances must be made [for delays].'[41] Pending the completion of the large telescope, Hale offered to put the 12-inch refractor of his Kenwood Observatory at Barnard's disposal. Burnham, however, thought – somewhat unrealistically, given Barnard's observaholic tendencies – that until the large glass was ready a suitable alternative might be for him to 'knock around the country . . . and loaf or lecture.'[42]

With the memory still green of his struggle with Holden to gain access to the 36-inch refractor, Barnard also worried about having enough observing time on the Yerkes refractor. Thus, when Hale floated a plan to invite visiting astronomers to use the telescope, Barnard was strictly opposed. 'I think it would be very unwise to make any rules for this purpose,' he told the future director, 'as such would imply a kind of standing offer of the instrument to outside observers and would rather invite applicants for its use':

> These applications would come during the good observing weather and if the requests were granted would be an injustice to the regular observers as there will be little enough time for them to do their best work in this climate.[43]

Barnard's title at the University of Chicago was to be Professor of Practical Astronomy, and his salary, though generous, was hardly munificent – $3000 a year, $600 more than his salary at the Lick Observatory. Always insecure about financial matters, he worried about having enough for his old age, and in negotiating the final terms with Harper he confided: 'I wish to be assured that my position shall be permanent and that my salary shall not at any time be reduced below $3000. I would hope that it might some time be raised above this, to enable me to put away a living for that time when age shall prohibit ones earning his own living further.'[44]

A final concern was the availability of suitable accommodations at Yerkes Observatory. He and Rhoda had settled quite comfortably into the two-story brick house on Ptolemy Ridge which had been theirs since the summer of 1894, and they were disappointed to learn that the University of Chicago had no immediate plans to construct anything similar. 'We are distressed about the house situation,' he confided to Hale. 'To board is dreadful – we have always lived alone.'[45]

4

Meanwhile, Barnard faced an even more pressing concern – the daunting prospect of trying to tie up all of his work in progress in only four months. That summer he worked himself at a frantic pace. He spent almost every clear and moonless night in the small Crocker dome extending his photographic survey of the Milky Way, and as usual carried on simultaneously several other lines of research.

His most noteworthy investigation was a search with the Crocker telescope for faint satellites of the Moon, which he thought might be detected during a total lunar eclipse. He had made a first attempt at the eclipse of March 10, 1895 but it had been spoiled by haze. He tried again at the eclipse of September 3, 1895, and this time conditions were excellent. Barnard guided on the small round lunar 'sea,' Mare Crisium, which he kept carefully and precisely bisected by the wires of the finder telescope. He found that the guiding 'required constant attention as the motion of the Moon was considerable.'[46] In the resulting plates the image of the Moon was perfectly sharp, and Barnard declared, 'I have certainly never seen such exquisite pictures of the Moon . . . The details of the surface are clearly and beautifully shown and the Moon stands out from the sky like a beautiful globe.'[47] Nevertheless, Barnard concluded that none of his photographs showed anything that could be taken for a lunar satellite, and he gave as his opinion that 'a further search for it . . . appears quite unnecessary.'[48]

In addition to his photographic work, Barnard hurried to complete a long series of measures he had been making of the dimensions of the bodies in the Solar System with the filar micrometer on the 36-inch refractor. As an astronomer trained in the methods of the old school, much of Barnard's work had always involved making measures with the micrometer – during his years on Mt. Hamilton he had made thousands of such measures. As earlier noted, one of the chief frustrations for Barnard had always been the lack of a wind break for the observing slit of the 36-inch – one was finally installed only in August 1895. Until then, work with the micrometer was especially frustrating, and Barnard complained:

> The free access of wind to the tube of the 36-inch kept it often in constant vibration, so that accurate measures are obtained only by waiting patiently for moments of steadiness of the image. This was extremely trying on the time and nerves of an observer.[49]

Despite the obstacles, Barnard obtained a benchmark set of measures of the planets, the rings of Saturn, the four Galilean satellites of Jupiter, and Saturn's largest satellite, Titan. Among his notable results, he was the first to show conclusively that, contrary to previous belief, Uranus is larger than Neptune. (And incidentally, though he never lost an opportunity to scrutinize the discs of these remote planets for detail, at no time was he able to make out anything definite.) The four largest asteroids – Ceres, Pallas, Juno, and Vesta – also came within the scope of his investigations. With a magnifying power of × 1000 on the 36-inch refractor, he found that he was able to make out their tiny discs and to place the wires of the micrometer across them. These measures revealed that Ceres, which he put at 485 miles (780 kilometers) across, is the largest asteroid, not Vesta, which had hitherto been so regarded. His figures, for want of anything better, remained the standard until 1970.[50]

Apart from measuring the dimensions of the principal bodies of the Solar System, Barnard also continued his series of measures of the positions of the fifth satellite of Jupiter – aside from himself, only Hermann Struve, using the 30-inch refractor of the Pulkowa Observatory, had up to this time succeeded in measuring its position, so that knowledge of the satellite's orbit was almost entirely based on Barnard's own observations. He also obtained numerous positions for the other faint satellites known at the time – the two moons of Mars, the satellites of Uranus, and Neptune's moon Triton. While measuring Uranus's moons Titania and Oberon, he found, to his surprise, that whenever Titania was in the uppermost position, Oberon appeared brighter, while when Oberon was uppermost, the reverse was true.[51] This 'horizontal-vertical illusion of brightness' – 'two stars that are equal when the line between them is parallel to my eyes, will differ by half a magnitude when they are in a vertical line; the lower one being always the brighter,' as Barnard later elaborated on it[52] – obviously had important consequences for any work that depended on visual brightness estimates, and was later shown to be an interesting subjective illusion. If the observer can persuade himself to think of the two stars as horizontal the inequality of brightness disappears.[53]

Another object which captured Barnard's attention during his last months on Mt. Hamilton was Hind's variable nebula (NGC 1555) in Taurus. Located southwest of the variable star T Tauri, this nebula had been discovered in 1852 by the English astronomer John Russell Hind with a 7½-inch refractor. It was recorded in the same position by other observers over the next several years, but in 1861 Heinrich d'Arrest found that it had disappeared. In 1868 Otto von Struve was unable to find it even with the 15-inch refractor at Pulkowa, but recorded another small nebula nearby, in a position where previously nothing had existed (it was later cataloged by Dreyer as NGC 1554). Burnham scrutinized the region with the 36-inch refractor in 1890, and found T Tauri itself surrounded by a fairly bright elliptical nebula, but otherwise the field was unremarkable except for an excessively faint breath of nebulosity southwest of the star and in the same position where Hind had recorded his nebula in 1852 – Burnham felt that this faint nebulosity would not have been visible in a telescope much smaller than the Lick refractor. Barnard, too, had a look, and noted that 'the small star [T Tauri] was

placed in an elongated nebula just as [Burnham] describes.'⁵⁴ On returning to the
region in February 1895, Barnard found that the faint nebulosity southwest of the star
was still visible but that T Tauri itself had undergone a remarkable change – it had
become perfectly stellar, and in place of the bright nebula in which it had been invested
in 1890 was surrounded only by a very faint diffused nebulous glow. Finally, in
September 1895, he noted yet another change:

> To my surprise no trace of Hind's nebula now exists – it seems to have entirely
> vanished! I have examined the place of this object several mornings . . . under the
> very best conditions, but the variable nebula could not be seen . . . Every means was
> tried to see [it]: T Tauri was occulted, the place was examined with averted vision,
> but no strain of eyesight could detect any trace of the nebula. If it had been as bright
> as at the observations in February last it would have been easily visible, as the
> conditions for seeing it were better. This proves . . . that this nebula still fluctuates
> in its light. It is certainly now invisible in the 36-inch. The place of this object
> should therefore receive careful attention with powerful telescopes to see when the
> nebula reappears.⁵⁵

Though a number of nebulae had at various times been suspected of light
fluctuations, Hind's nebula was the first case in which variability was convincingly
demonstrated. But above and beyond the fascination inherent in the object itself, the
nebula's disappearance in 1895 had further significance, in that it nudged Barnard a
step closer to recognizing the true explanation of the dark markings of the Milky Way.
Up until then he had assumed, along with most astronomers, 'that a nebula remains
luminous, and finally develops into a star or system of stars – that is, its ultimate destiny
is a stellar condition. This opposes any supposition that a nebula may become dark by
the loss of its light.'⁵⁶ Hind's nebula showed that this was not so. A nebula could indeed
lose its light; there could be such a thing as a dark nebula.

5

Barnard always worked alone on Mt. Hamilton. He preferred it that way, and never
took a student. Largely self-taught himself, he saw no reason why others could not do as
he had done. Moreover, as E. B. Frost later wrote, 'Barnard could not bring himself to
lose time at the telescope in having a pupil take part in measurements, which he could
himself make so much better, and he begrudged the possible loss in quality of a
photograph if some one less skilled than himself took some part in the guiding.'⁵⁷
    Despite working alone and the fact that there were occasional dangers such as the
mountain lion he met with in the gray dusk one morning as he returned to his cottage
after a night's work, he usually regarded it as unnecessary to use a lantern to find his way
home at night. Even if unable to see his way, he could feel the trail or 'chicken walk'
beneath his feet. Thus it was his custom, when through with work, to place the lantern
which was no longer needed at the door of the elderly janitor, Mr. Curtis, who slept in a

small room on the north side of the large dome. Barnard admitted that he took some comfort during his long night vigils in the dome in knowing that the janitor slept there:

> [A]t times it was pretty lonesome. For when one stopped to think, the dead body of James Lick lay under the pier of the great telescope only a few feet away. This was specially trying after the servants had reported seeing Mr. Lick looking through their window at them one dark night. The proximity of a live human being, though sound asleep, did much to offset the equally close proximity of a dead human being who according to the above reports might not be so sound asleep as is customary under such conditions. [58]

One night Barnard made an exception to his usual rule of making his way without a lantern. A storm had come up suddenly on the mountain, making it 'one of the blackest nights I have ever seen . . . a thick black wet mist driven by a terrific southeaster that threatened to hurl you into the canyon. After falling over various things in an endeavor to find my way, I finally returned to the observatory and lighted my lantern. At last I reached my home in safety.'[59] Since it was unusual for him to take a lantern home with him, Barnard was afraid that he would forget to return it the following morning. His mnemonic device in such cases was to lay across the threshold of his door anything he did not want to forget. The method would have worked on this occasion had it not been for the fact that early the next morning Rhoda, who was awake ahead of him, removed it from the threshold and placed it outside the door. In the morning Barnard went to his office, 'blissfully ignorant of the lantern's whereabouts.' On later being asked by the janitor about it, he acknowledged that he had taken it home the night before and offered to go down to his house and fetch it immediately. However, the janitor replied that this was unnecessary and that it would do if Barnard brought it back with him after lunch. However, Barnard again forgot it, and this time the janitor upbraided him for his carelessness. Barnard apologized, and at once went home to fetch the missing lantern. That night, the storm having passed, Barnard was again observing in the large dome from dark until daylight. At about 1 a.m. he became aware of a light shining through the door of the janitor's room above. He was struck by the fact that this was several hours before the janitor's usual rising time, and a little later became aware of the aroma of freshly made coffee permeating the atmosphere:

> On these nights from dark until daylight, I did not stop to eat any lunch and so carried none with me. The rich smell of that coffee was tantalizing. Presently the door opened and the old gentleman raising his head above the elevating floor said in a gentle and pleading voice, 'Oh Professor, wouldn't you like a nice cup of hot coffee?' Poor old man! His conscience had worried him and he could not sleep, and he wanted to make peace. I gladly accepted his offer and we sat together and ate a few crackers and drank a cup or two of excellent coffee. Neither of us mentioned the disagreeable affair of the lantern. [60]

6

The strain of his last summer on Mt. Hamilton told on Barnard. With so much on his mind, he was perhaps more careless than usual, and an accident with graver consequences than the incident of the lantern occurred at the beginning of August, when he fell from a high rock on the side of the mountain. 'I just missed being killed,' he reported to Hale, 'and am a mass of bruises . . . I am too sick to write more. I can scarcely move my body is so terribly bruised.'[61] On learning of the accident, Burnham advised him: 'Until you get all right you had better stop observing. I don't know any harder work on the back than using an Equatorial, and especially yanking the 36-in. around for several hours a night.'[62] However, Barnard had too much to do to take to a sick bed for long. The day after the accident, though 'in a shattered condition,'[63] he went to San Francisco to start a subscription for funds to publish his comet and Milky Way photographs. The Regents had voted against publishing them because of the expense, but had agreed to pay for binding and issuing them as a volume of the Lick Observatory provided that Barnard was able to raise the rest of the funds from outside sources, and he was anxious to get started.

Schaeberle helped get the fund-raising effort started by making a first contribution, and was instrumental in securing several others. Most of the money was contributed by a few wealthy individuals, including Colonel Crocker and Miss Catherine Wolfe Bruce of New York, but everyone on the Lick staff, with the exception of Holden, who presumably was not asked, subscribed at least nominally to the effort. The staff at the time included spectroscopist W. W. Campbell, Allen L. Colton, who had been hired by Holden in 1892 to assist with the lunar photography project, Richard H. Tucker, the meridian circle observer who worked closely with Schaeberle, and Charles Dillon Perrine, then the observatory secretary who later became a skilled astronomer in his own right and discovered ten comets between 1895 and 1902. Within a few weeks Barnard was able to raise what he considered to be a sum sufficient for the purpose, $2225.00 – enough, he told Phelps, to reproduce his photographs in a 'first class manner.'[64] He regarded this as a personal triumph over 'the contemptable opposition of the Director of the L.O.'[65]

Naturally Barnard wanted to supervise the reproduction of the plates himself, and for this purpose hoped to make copies of his plates to take along with him. However, Holden refused to pay the costs for making such copies out of observatory funds: 'I do not think the L.O. should provide the plates to make copies for your personal use,' he told Barnard.[66] He did offer to allow Barnard to take the original plates with him for study as long as they were returned when he was finished with them, and in addition he offered Barnard 'a selection of Moon negatives and positives – good ones – and if you want more, you are welcome to them.'[67]

Though willing to take the original plates, Barnard naturally wanted copies of them for permanent use. He protested to Holden:

> I cannot see why the L.O. should not supply me with plates to make these copies for
> my personal use . . . As for the . . . expense you mention you will perhaps

remember that great quantities of these plates have been annually wasted in the photographic work with the big telescope. I can remember very well, when I used to develop such plates, that as many as 25 or 30 10 × 8 plates were exposed on the moon for many nights together . . . and scarcely one of these was good . . . What I have asked for, for purely scientific purposes, is less than these few nights waste plates. In the name of justice, who is there to whose personal use these copies of my star pictures are more justly due? You have permitted them to be made – are they to lie here on the shelves and rot for the want of scientific use? . . . Who will get more out of them scientifically than the one who through long tedious nights patiently made them? and who is more justly entitled to them? The Observatory has failed to publish these pictures so that astronomers could study them. You can explain why this has not been done . . . I am asking no favor in this matter. I can at least stand on the same footing of many an outsider who has asked for these things and got them.[68]

After continuing for several more pages in the self-righteous vein which he generally adopted in addressing Holden, he added a footnote: 'You say that I am welcome to more moon pictures. Why am I not welcome to my star pictures?'[69]

Meanwhile Barnard had again appealed to Phelps, asking that the Regents decide whether he was entitled to plates for copying his star pictures.[70] The Regents sided with him, but Holden took his time in notifying Barnard of their decision – incurring Barnard's wrath yet again. Reflecting that 'My stay here is passing away rapidly and the opportunity to do the work is passing with it,' Barnard chided Holden on September 16: 'Your promptness in thus notifying me of the official action of the Regents in this extremely important and urgent case, five or six days after the event is certainly to be commended!'[71]

By this time Barnard's nerves were totally shot. Anxiety about his pictures, urgency about finishing other work in progress and the bustle of preparing for the upcoming move were wearing him down. In addition, he was agitated over Holden's latest project – the refurbishing of the 36-inch Crossley reflector. Originally built by the Newcastle engineer and amateur astronomer A. A. Common and used by him in 1883 to obtain what were then the most marvelous photographs of the Orion Nebula, this telescope was later sold to Edward Crossley, a wealthy English amateur, who finding that he had hardly used it in ten years in his possession and was no longer interested in it, informed Holden in April 1895 that he was willing to let the Lick Observatory have it for a mere $1000, which was no more than the cost of dissassembling the instrument and its dome and having them shipped from England to Mt. Hamilton. In hindsight, the procuring of this instrument would be regarded as the shining hour of Holden's directorship, but when the telescope arrived in July 1895 its potential value was far from clear. Most astronomers of the day, including Barnard, were skeptical about reflecting telescopes in general, and the Crossley hardly inspired confidence; its mounting was a mechanic's nightmare, and it was only too obvious that it would require enormous effort to get the instrument into working order. The man to whom Holden assigned that thankless task was William J. Hussey, who had once studied under Schaeberle and Campbell at Ann Arbor and who later, as assistant professor of astronomy at Stanford University, had spent summers observing on Mt. Hamilton – among other things, he had obtained some fine photographs of Comet Rordame-Quénisset with the Crocker telescope in the summer of 1893, when Barnard was away in Europe. After Barnard's resignation from the Lick staff in June, Hussey had been hired as his replacement, his appointment being confirmed hardly more than a week before the mirror and other optics of the Crossley arrived on Mt. Hamilton.

Hoping to have the telescope operational as soon as possible, Holden began unpacking the parts on Ptolemy Ridge at the beginning of September. For Barnard, some of his deep-seated resentment against Holden was displaced to the Crossley – he later told Davidson that it was 'no good old junk' and that he would not pay $5 for the telescope.[72] Exhausted from overwork and not yet recovered from his fall, he complained bitterly to the director that work on it was interfering with his sleep:

> My sleep is constantly being broken in the mornings by the observatory wagon going back and forth by our house to the Crossley telescope. To one working all night this loss of sleep is distressing. This occurred again this morning, after I had got but little sleep, by the heavy team coming over to get a lot of packing cases that have lain there for a month or so. I respectfully request that this unnecessary nuisance be stopped . . .[73]

The schedule of work on the Crossley was changed, but Barnard still continued to find it impossible to rest. On September 24, he told Hale: 'I am rushed to death and tired out.'[74] Nevertheless, he never considered giving up any of his precious observing time – his observing notebooks show that he worked himself as hard as ever on the telescopes to the very last. His very last night of observing was September 30, 1895. One guesses that it must have been an emotional moment for him, for he would never again spend another night in the great dome where he had had so many memorable experiences. But if so, he gave no hint of his personal feelings in his observing book; he simply recorded in this, the eighty-eighth and final volume of the series that he begun on his arrival on Mt. Hamilton in 1888, a set of routine measures of double stars – κ Pegasi, 24 Aquarii, η Pegasi, 85 Pegasi, γ² Andromedae, and 20 Persei. The same night he also exposed his last plate with the Crocker telescope – a twenty minute exposure on the Pleiades. Thus, and without fanfare, ended Barnard's great era on Mt. Hamilton – so turbulent personally, so creative and productive scientifically.

Since Holden justifiably insisted that the notebooks containing the observations of the staff astronomers were the property of the Lick Observatory, Barnard went to the great expense of hiring a man to copy the notebooks for him, and had intended to check these copies by comparison with the originals before leaving Mt. Hamilton. However, in the rush to the finish, he found that he did not have time to do so, and notified Holden just before his departure that he was taking the original notebooks as well as the copies with him 'and will return them without expense to L.O. when through with them.'[75] Thus, with observing books and star pictures in tow, he and Rhoda departed by stage for San Jose on October 1, 1895. It must have been a sad journey along the winding Mt. Hamilton road which they had taken so many times before, and he and Rhoda no doubt paused frequently to cast a longing lingering look back, for they both hoped to return one day and live out their final days in this 'delightful land.'[76]

Barnard's house on Mt. Hamilton road – his residence from summer 1894 until he left Mt. Hamilton in 1895. Mary Lea Shane Archives of the Lick Observatory

In San Jose, Barnard received as a token of appreciation an inscribed silver vase from the faculty and students of the San Jose State Normal School where he had often lectured. Then, on October 7, 1895, after bidding farewell to friends in San Jose and San Francisco, he and Rhoda left California via Vancouver for Chicago.

By the time they reached Chicago they were 'badly worn out,' Barnard wrote to Hale, who was away in Cambridgeport, Massachusetts, at the Clarks' workshop, where the finishing touches had just been put to the great 40-inch lens.[77] Nevertheless, Rhoda especially was anxious to see Lake Geneva, and escorted by Burnham, they made a hasty trip there to see how the observatory was progressing.[78] Hale invited Barnard to come out to Cambridgeport and assist in further tests of the great lens. Barnard must have been tempted, but he really was exhausted and instead decided on Nashville for a few weeks' rest. Interviewed by a newspaper reporter there, he declared that the new Yerkes Observatory was 'progressing favorably,' with the walls of the building almost complete, so that the roofing in and finishing of the interior was all that was required to fit it for the great telescope. He added that 'the building, architecturally, is one of the most beautiful ever constructed for any purpose, and so far as an observatory is concerned there is nothing in the world to approach it in magnificence.'[79] He expected that the new observatory might be in working order in another four or five months, but this was wishful thinking – in reality it would not be finished for another two years.

7

Holden must have been relieved to see his old nemesis finally gone, and probably thought that his troubles were finally over. Lewis Boss, wrote to him apropos of the fact:

> I rather think you are to be congratulated on the departure of Mr. Barnard. B[arnard] is a good man gone (temporarily) wrong, and it is probably better for him to take a fresh start, and certainly it must be better for the L.O. to get rid of such a [malcontent], though I doubt if you are likely to find soon a man of such indefatigable energy and ocular acuteness. Mind you, what I say against B[arnard] relates purely to his cantankerousness at L.O. in certain ways (more than one) and does not prevent me from entertaining considerable esteem for the man for his immense earnestness and industry. I have heard that his course, and his literary peculiarities, are most probably due to the former influence of another.[80]

However, Hussey proved to be even more insubordinate than Barnard had been, and became the leader of a revolt against the director in which Colton, Perrine, and even the director's longtime ally Campbell, who was now steadfastly turned against him, were joined. Much of the trouble, though by no means all of it, had to do with conflict over the Crossley. While Holden maintained that its addition, when it was in working order, would help the Lick maintain its superiority even over Yerkes, the rest of the staff and Hussey in particular regarded the instrument, as the San Francisco *Chronicle* reported, as 'antiquated as Noah's ark,'[81] and Hussey did everything he could to drag his feet. Barnard was by no means surprised at the news of continuing strife on Mt. Hamilton, and told Davidson: 'The time has come when the deplorable condition of things . . . should cease – It has been permitted long enough.'[82] Once started on the subject, he could never be brief, and proceeded with his usual litany of complaints against the director:

> There is one thing that cannot be remedied and it is the most to be regretted . . . [H]is whole work at the L.O. has been to calmly and deliberately write a false history of the workings of the Observatory and put them on record for future historians to work from. He has written me letters and copied them which were written only for the purpose of putting in that book of records. These letters in many cases were deliberate lies . . .[83]

Holden hung on as long as he could, but the conflict with his staff continued. The Regents were now thoroughly weary of it; Holden could no longer count on their support, and eventually Hussey and the rest of the staff were able to bring the director down. On September 18, 1897, as the Regents were preparing to meet to decide his fate,

Holden slipped away, clearing out his office and apartment and leaving farewell cards in everyone's mailbox. Hussey noted in his diary: 'Shortly after the Observatory conveyance is seen driving down the Mountain Road. It is Edward S. Holden leaving Mt. Hamilton forever.'[84]

Thus laconically was noted the literal descent from the heights of the embattled director, who ten years earlier had stood at the top of the astronomical world and who now, at just past fifty, faced the bitter prospect of rebuilding his career on the ruins. After his departure Holden sent in his official resignation to the Regents, to take effect as of January 1, 1898, and Schaeberle became interim director.

Holden, in fact, had long since made up his mind to leave, and had for sometime been casting around for other positions. Still having the backing of such powerful figures as Simon Newcomb and E. C. Pickering, he aimed for such prestigious posts as the presidency of the Massachusetts Institute of Technology and the head of the US Coast and Geodetic Survey. In connection with one of these applications, Newcomb solicited Barnard's opinion. Barnard was unable to forgive or forget, and responded sharply:

> I have avoided the subject since leaving the Lick Observatory because it is a painful one to me. You have asked me my opinion of him and I shall plainly, and with as little prejudice as possible, give you that opinion. He is the most contemptable man I have ever known.[85]

This time Holden's powerful allies went for nought, for nothing came of any of these applications. For a while Holden supported himself in New York City as a freelance writer – among other things, he published a primer on heraldry and a book on the poetry of Omar Khayyam, and he was also a prolific contributor to the *Cosmopolitan* magazine. Finally, in 1901, he obtained a position as librarian at West Point. In his true element at last, he enjoyed as a librarian the success that had eluded him as a research scientist and adminstrator. Brashear, the optician, saw him there, and wrote to Crew: 'He is much mellowed and does not seem like the same Holden to me.'[86] He died in 1914. After his departure from Mt. Hamilton, he and Barnard never communicated again.

As for the Crossley reflector, it went from its dismal beginning under Holden to become a remarkably productive research instrument. It was finally brought into working order by Holden's successor as director, James E. Keeler, who had been lured back to Mt. Hamilton from Pittsburgh's Allegheny Observatory. Keeler returned to Mt. Hamilton on June 1, 1898, and by the following November was using the Crossley to photograph the latest Comet Brooks. In spring and summer 1899 he photographed M51 (the Whirlpool Nebula), M81, and M101, all well-known spirals from Lord Rosse's work. In addition, he found that almost all of the faint nebulae that had been added in recent years by Swift, Barnard and others were actually small spirals, and on nearly every plate he recorded many more. Keeler, along with most astronomers of the day, thought that these nebulae were rotating clouds of gas and dust, though this was only a guess and had not yet been proved. Barnard, commenting in 1898 on an enlargement of one of his Willard photographs of the Andromeda nebula which showed that 'it consists of a central condensation surrounded with nebulous rings, crudely like the rings of Saturn,' suggested a cautious approach to these objects:

> Some astronomers have professed to see in this mighty nebula, and its rings, a verification of the nebular theory – the condensing nebula having so far developed as to have thrown off the world-producing rings from the central body, which, itself, is destined to become a mighty sun, like our own. The eye, however, easily pleases and satisfies the minds of those who have firmly rooted and preconceived ideas of things.[87]

In reality, of course, the Andromeda nebula and the other spirals were extragalactic star systems, though no one knew this at the time. But though no one fully grasped their implications, Keeler's photographs opened a new era in the study of the spiral nebulae.[88] When in June 1899 Keeler sent Hale some of his nebular photographs, which included diffuse and planetary nebulae as well as spirals, Hale reported that Barnard 'was simply delighted . . . and stood for hours in front of them admiring their details.'[89] To Hale, Keeler's photographs 'created a genuine sensation and showed to many who had been skeptical regarding the advantages of reflectors what the instrument is capable of doing in the right hands.'[90] With the great Yerkes refractor hardly finished, he was already dreaming of the great American reflectors to come.

1   'Why Barnard May Leave,' San Francisco *Examiner*, February 24, 1893
2   ibid.
3   San Jose *Report*, March 1, 1893
4   EEB to George H. Boke, Frederick Derrick, Hellen Ballard and others, students of the University of California at Berkeley, March 27, 1894; VUA
5   SWB to GEH, July 9, 1894; YOA
6   Rev. A. Edwards to WRH, January 24, 1893; RL
7   WRH to GEH, December 6, 1894; RL
8   GEH to EEB, January 24, 1895; VUA
9   SWB to EEB, February 1, 1895; VUA
10 SWB to EEB, February 6, 1895; VUA
11 SWB to EEB, February 1, 1895; VUA
12 San Francisco *Call*, February 11, 1895
13 EEB to GEH, June 23, 1895; YOA
14 EEB to GEH, March 5, 1895; YOA
15 EEB to ESH, April 10, 1895; SLO
16 ibid.
17 ESH to EEB, April 11, 1895; SLO
18 EEB to TGP, April 22, 1895; SLO
19 ESH to TGP, April 22, 1895; SLO
20 ibid.
21 EEB to GEH, April 23, 1895; YOA
22 D. E. Osterbrock, 'The Rise and Fall of Edward S. Holden, Part 1,' *JHA*, 15 (1984), 81–127:119

23 EEB to GEH, April 23, 1895; YOA

24 EEB to ESH, May 11, 1895; SLO

25 EEB to ESH, May 11, 1895 [2nd letter]; also May 23, 1895; SLO

26 ESH to EEB, May 24, 1895; SLO

27 EEB to WRH, May 22, 1895; RL

28 EEB to GEH, May 23, 1895; YOA

29 EEB to WRH, May 22, 1895; RL

30 SWB to EEB, May 30, 1895; VUA

31 EEB to TGP, May 31, 1895; SLO

32 TGP to EEB, June 3, 1895; VUA

33 'Barnard Is In It – Appears in a Catalogue of Chicago,' San Francisco *Chronicle*, June 7, 1895. Incidentally, the University of Wisconsin had used a similar ruse in 1878 to force James Watson's decision to leave Ann Arbor for Madison.

34 ESH to A. H. Babcock, June 6, 1895, telegram; SLO

35 JMS to EBF, March 14, 1923; YOA

36 EEB to the Board of Regents of the University of California, June 7, 1895; VUA. EEB to WRH, June 7, 1895; RL

37 EEB to GEH, June 7, 1895; YOA

38 EEB to the Board of Regents of the University of California, June 7, 1895; VUA

39 TGP to EEB, June 11, 1895; VUA

40 EEB to GD, June 13, 1895; BL

41 SWB to EEB, June 3, 1895; VUA

42 ibid.

43 EEB to GEH, January 29, 1895; YOA

44 EEB to WRH, May 22, 1895; SLO

45 EEB to GEH, May 23, 1895; YOA

46 E. E. Barnard, 'On a Photographic Search for a Satellite to the Moon,' *Ap J*, 2 (1895), 347–9:348

47 ibid.

48 ibid. Further searches were, however, mounted, most notably by Clyde Tombaugh, then at New Mexico State University, Las Cruces, though with no success. See Richard Baum, 'The Search for a Satellite of the Moon,' in *The Planets: Some Myths and Realities* (Newton Abbot: David & Charles, 1973), pp. 19–47.

49 E. E. Barnard, 'A Micrometrical Determination of the Dimensions of the Planets and Satellites of the Solar System Made with the 36-inch Refractor of the Lick Observatory,' *PA*, 5 (1897), 285–302:301

50 Barnard's values were: Ceres 485 miles, Pallas 305 miles, Juno 118 miles and Vesta 245 miles. By comparison, the currently accepted values (1993) are: Ceres 625 miles, Pallas and Vesta 310 miles, and Juno 153 miles.

51 E. E. Barnard, 'Micrometrical Observations of the Four Satellites of the Planet Uranus, and Measures of the Diameters of Uranus,' *AJ*, 16 (1896), 73–8:74–5

52 E. E. Barnard, 'Micrometrical and Visual Observations of Nova Cygni (1876) made with the 40-inch Refractor of the Yerkes Observatory,' *MNRAS*, **62** (1902), 405–19:417

53 A. S. Eddington, 'A Horizontal-Vertical Illusion of Brightness,' *Observatory*, **39** (1916), 185. For Eddington's authorship of this note, see A. S. Eddington to EEB,

April 1, 1916; VUA

54   E. E. Barnard, 'On the Variable Nebulae of Hind (N.G.C. 1555) and Struve (N.G.C. 1554) in Taurus, and on the Nebulous Condition of the Variable Star T Tauri,' *MNRAS*, **55** (1895), 442–51:446

55   E. E. Barnard, 'Invisibility of Hind's Variable Nebula (N.G.C. 1555),' *MNRAS*, **56** (1896), 66–7:66

56   E. E. Barnard, 'Some of the Dark Markings on the Sky and What They Suggest,' *Ap J*, **43** (1916), 1–8:2

57   E. B. Frost, 'Edward Emerson Barnard,' *BMNAS*, **21** (1926), 16

58   EEB, MS draft, undated; VUA

59   ibid.

60   ibid.

61   EEB to GEH, August 5, 1895; YOA

62   SWB to EEB, August 21, 1895; VUA

63   EEB to GEH, August 5, 1895; YOA

64   EEB to TGP, September 20, 1895; VUA

65   EEB to GD, September 20, 1895; BL

66   EEB to ESH, August 30, 1895; ESH to EEB, August 30, 1895; SLO

67   ESH to EEB, August 30, 1895; SLO

68   EEB to ESH, August 31, 1895; SLO

69   ibid.

70   EEB to TGP, August 31, 1895; SLO

71   EEB to ESH, September 14, 1895, September 16, 1895; SLO

72   EEB to GD, May 26, 1897; BL

73   EEB to ESH, September 2, 1895; SLO

74   EEB to GEH, September 24, 1895; YOA

75   EEB to ESH, October 1, 1895; SLO

76   EEB to WWC, June 21, 1921; SLO

77   EEB to GEH, October 25, 1895; YOA

78   ibid.

79   Nashville *Banner*, October 19, 1895

80   Lewis Boss to ESH, August 21, 1895; SLO. The 'other' was presumably Burnham.

81   Quoted in Remington P. S. Stone, 'The Crossley Reflector: A Centennial Review-I,' *Sky and Telescope*, **58** (1979), 307–11:309

82   EEB to GD, September 27, 1897; BL

83   ibid.

84   Quoted in D. E. Osterbrock, 'The Rise and Fall of Edward S. Holden, Part 2,' *JHA*, **15** (1984), 151–76:169

85   EEB to SN, undated draft from 1897; VUA

86   JAB to HC, July 4, 1908; NUA

87   E. E. Barnard, 'The Milky Way and the Great Nebula of Andromeda,' *American Annual of Photography* (1898), 1–8:8

88   See D. E. Osterbrock, 'The Observational Approach to Cosmology: U.S. Observatories pre-World War II,' in R. Bertotti, R. Balbinot, S. Bergia and A. Messina, eds., *Modern Cosmology in Retrospect* (Cambridge: Cambridge University Press, 1990), p. 250

89   GEH to JEK, October 6, 1899; YOA. Barnard later expanded on his remarks, and contrasted the forms of these nebulae as shown in the photographs with the earlier impressions of visual observers. Some of the nebulae, he wrote, 'show structural details and . . . forms that are very remarkable. Perhaps some of these do not look as extraordinary as the early drawings show them to be, but what the photographs show is sometimes more extraordinary than what the observer imperfectly saw and more imperfectly drew. For instance the photograph's view of the "owl nebula" [M97 in Ursa Major] is less bizarre than the drawings of it made with Lord Rosse's telescope, but there are others that out rival anything that Lord Rosse's observers ever saw, or thought they saw, with his great telescope. Some of these photographs . . . show very extraordinary features that lead one to wonder at their strangeness.' EEB, unpublished MS draft; YOA.

90   GEH to JEK, October 6, 1899; YOA

18

Yerkes Observatory

1

Though Hale favored Lake Forest, north of Chicago, for the Yerkes Observatory, Harper in December 1893 decided on the north shore of Lake Geneva, near its western end, about a mile from the center of the small town of Williams Bay – not least because many well-off Chicagoans had summer homes on the lake, which Harper hoped would help his fund-raising activities. However, the deed to the land was not acquired by the University of Chicago until the fall of the following year, and construction of the buildings did not begin until April 1895. The architectural plans were freely adapted by Chicago architect Henry Ives Cobb from those that had been used for the Lick Observatory and the Astrophysical Observatory in Potsdam, Germany. The main building was to be in the form of a Latin cross, with three domes and a meridian room; the long axis of the building was to measure 326 feet, and the large tower at the western end, to contain the 40-inch telescope, was to be 92 feet in diameter.

Apart from the telescopes, Hale planned that the observatory should include its own physical laboratory, with equipment for all kinds of spectroscopic, bolometric, photographic and optical work. Cobb worked out the design in the Romanesque style; the building was constructed of brown Roman brick, with gray terra-cotta ornaments – architecturally it is superb.

The small staff that Hale assembled was led by senior astronomers Barnard and Burnham, the latter continuing in his job as a court reporter in Chicago and coming out by train to use the 40-inch refractor every weekend. Others were Frank L. O. Wadsworth, who had worked under physicist A. A. Michelson but proved to be a scientific mediocrity at Yerkes and left soon afterward for another post at the Allegheny Observatory, and two young men, Ferdinand Ellerman and George Willis Ritchey, who had been Hale's personal assistants at the Kenwood Observatory. Ellerman was a handyman with skill in machine-tool work and photography who had been with Hale at Kenwood since 1892. Ritchey, a cabinet-maker and self-taught optician who was teaching woodworking at the Manual Training School in Chicago when he met Hale in 1891, was hired by Hale as Yerkes Observatory optician. He had already completed a 24-inch mirror in the workshop in his South side home by the fall of 1896, and on coming to Williams Bay was eager to start work on an even larger mirror, using a 60-inch glass disk that Hale's father had purchased for him for $2000.[1]

Yerkes 40-inch dome under construction. Yerkes Observatory

2

As had happened at Lick, the completion of the Yerkes Observatory was delayed, and Barnard spent several months living in Chicago, near Hale's Kenwood Observatory. This period occupied an unwelcome hiatus in his observational activity – Hale's 12-inch refractor, which Barnard had counted on using until the 40-inch was ready, was no longer available, having already been dismantled for shipment to Williams Bay.

Barnard filled some of his time writing reports on his observations from Mt. Hamilton. His first priority, however, continued to be the publication of his Milky Way and comet photographs, about which he made inquiries in Chicago and also visited firms in Philadelphia, Boston, and New York. Various methods of obtaining the best reproductions were tried. By spring 1896, Barnard decided to hand the work over to the Chicago Photogravure Company, which specialized in the collotype process.

While in New York, he paid a visit to the fashionable Fifth Avenue address of Catherine Wolfe Bruce, an octagenarian heiress whose father, George Bruce, had made a fortune as a type-founder. A cultured woman who spoke several languages and had once studied painting, she was a generous benefactor of astronomy, and altogether made grants of over $200 000 to support various astronomical projects. One of the largest sums, $10 000, had been given in 1894 to equip Barnard's rival in Milky Way photography, Professor Max Wolf of Heidelberg, with a triple photographic refractor with two 16-inch Brashear lenses and a 10-inch objective, perhaps as an expression of gratitude for the fact that Wolf, on making the first photographic discovery of an asteroid in December 1891, had named the object after her – 323 Brucia. Miss Bruce had already demonstrated her generosity to Barnard, having contributed $500 toward the publication of his Milky Way and comet photographs in 1895. Ostensibly his reason for visiting her on this occasion was to express his appreciation for her earlier support. By this time Miss Bruce was an invalid, and access to her – generally granted only to close relatives – was jealously guarded by her younger sister, Matilda W. Bruce. Barnard, however, was so well known that an exception was made and he was allowed into the inner sanctum.

Barnard's real motive in visiting her was that he wanted a new telescope with which to continue his Milky Way and comet photography, and he hoped that she would be willing to fund it. However, on meeting her, he found her in such frail condition that he 'did not have the heart to mention it,' and only intimated very indirectly his need for such an instrument. Apparently she did not take the hint, for as we shall see, he had to frame his plea more directly later.

In the summer of 1896, Barnard and his wife moved from Chicago to a cottage on the shore of Lake Geneva. From there they could personally supervise the construction of a house on land which they had purchased adjacent to the observatory grounds and on the hill overlooking the lake. By October, though the observatory was still surrounded by a sea of mud, Barnard and the rest of the staff were able to move into their offices, and Barnard reported optimistically: 'The observatory is progressing slowly but – I suppose surely. The [large] dome is nearly all covered in. The offices are receiving their finishing touches. Our residences are very nearly finished.'[2] By December, the Hales had moved into their much larger house on the observatory grounds, and the 12-inch refractor and dome which had been shipped from the Kenwood Observatory were installed on the northeast tower of the observatory, so that Barnard was able to do some observing at last.

Yerkes Observatory, May 1897. Hale's house at extreme right. Yerkes Observatory

3

At about this time, Barnard received word from Sir William Huggins in England that he had been awarded the Gold Medal of the Royal Astronomical Society. 'I value very highly any expressions that may come from the Society,' he replied to Huggins. 'I have tried to do good work and to find that it is appreciated by English Astronomers is one of the keenest pleasures that I could possibly experience.'[3] Huggins and the Council of the Royal Astronomical Society hoped that Barnard would travel to England to receive the medal personally at its February 12 meeting. Right up to the last minute, however, Barnard was undecided about whether to go. 'My wife and I talked the matter of my going to England over last night,' he wrote to Hale early on the morning of January 30, 1897, 'and at a late hour decided that it was probably best for me to go. I have decided to leave it with Mr. Burnham as to whether I ought to go or not, and I start this morning, not knowing whether I shall take my trip or stay at home, but prepared for the trip if Mr. Burnham favors it. I should have spoken to you yesterday about it but I had then made up my mind not to go.'[4] In Chicago, Burnham succeeded in making up Barnard's

Barnard's house on the shore of Lake Geneva. Yerkes Observatory

mind for him. 'I started Barnard off for London [last] evening,' Burnham informed Hale the next day, 'and he leaves New York tomorrow – which was a good thing to do, as I am sure you will agree.'[5] Hale did agree: 'I am very glad . . . It was, of course, the proper thing for him to do, and the vacation will be exactly what he needs.'[6]

Thus, on February 2, Barnard departed New York harbor on the *Trave* of the North German Lloyd line. The manners of his fellow-passengers were not pleasing to him – one of them got hold of the only nutcracker, and 'cracked his nuts *deliberately* one by one'.[7] Because of a 'hurricane' in the North Atlantic, his passage was delayed; he did not reach Southampton until after midnight February 12, by which time the meeting of the RAS was already over.[8] He had missed a fine tribute to him given by the president, A. A. Common, and the formal presentation of the Gold Medal. However, he lectured at Oxford and Cambridge, and attended an afternoon party given in his honor by Agnes Clerke in London. Among the others there were Alexander Herschel, Sir John's son, who specialized in meteors; Sir George Baden-Powell, brother of the future Lord Baden-Powell, hero of the Boer campaign and founder of the Boy Scouts, and Baden Baden-Powell, a retired army officer who took over the magazine *Knowledge* in 1902 and lost a great deal of money on it; and E. Walter Maunder of the Greenwich Observatory. Clerke later jotted her impressions of Barnard to Sir David Gill: 'He is a simple-hearted straightforward man with something of nobleness under his Nashville exterior . . . His photos excited raptures.'[9] Finally, before his departure, a special meeting of the RAS was arranged for March 2, at which Barnard exhibited some of his Milky Way and comet photographs, followed by a reception for him at the Criterion Restaurant. Next day he set sail again for New York.

4

He was now thinking constantly of his need for a photographic telescope with which to continue the work begun with the Willard lens, and as soon as he returned to Williams Bay he wrote to Catherine Bruce's sister, Matilda:

> I have been away from the Lick Observatory now going on two years. During all that time I have had no photographic instrument to work with, and the work I loved so much has necessarily been neglected for the want of a portrait lens. I have been tempted to apply to your sister to see if she could not help me, but it has been a rather delicate thing for me to do. It might look rather hard to apply to her when Mr. Yerkes has founded this great Observatory. The impression has got out that it is abundantly supplied with instruments. Such is not the case. Mr. Yerkes so far has given only one instrument, the great 40 inch. We have besides this, Professor Hale's 12 in. Equatorial from his Observatory in Chicago, and a small 6 inch equatorial borrowed from Warner & Swasey. But we have no lens to continue my work on the Milky Way, the nebulae and the comets.[10]

Barnard, who was obviously experiencing intense feelings of deprivation, had become 'tired and desperate' waiting for such a lens. Moreover, he was jealous of Max Wolf, who had received such an instrument when he, the pioneer, had had to go without:

> I was the first person – and with the crudest of appliances – that succeeded in photographing the Milky Way, and its wonderful structure; this work has been followed up by Dr. Max Wolf and a few others. You will see from the President's address in awarding me the gold medal of the R.A.S. that this fact is recognized. It seems pretty hard that after having originated the work I should not have instruments to carry it on with, and yet others who followed me in the work are more than bountifully supplied with the best of instruments. If I only had *one* of the *two* 16 in lenses your sister gave to Dr. Max Wolf, I should be delighted beyond measure: but I have absolutely nothing . . . I can not live for ever in anticipation.[11]

Barnard estimated that an instrument with a lens of from ten to twenty inches in diameter, 'by the best makers,' and a dome to house it, could be built 'for $15 to 20 000 up to $50 000.'[12] His entreaties did not go unheeded. In the summer of 1897, Miss Bruce offered $7000, with the understanding, as Barnard informed Harper, that 'all plans, specifications and contracts for the instruments and building are to be approved by me in writing and that all work to be done under my supervision and to my entire satisfaction.'[13] Unfortunately, Barnard's 'entire satisfaction' was not to be easily achieved.

5

As these negotiations were taking place, the 40-inch telescope was nearing completion. On the evening of May 19, 1897, the finished lens was brought by Alvan G. Clark, his two daughters, and Carl Lundin, the foreman of the Clark optical workshop (who probably did most of the actual work on figuring it) by special train to Williams Bay station. Hale and Barnard met them on the platform.[14] The next day the lens was unpacked and fitted to the sixty-two foot tube, and Hale wrote to Harper: 'There is every reason to suppose . . . that we can begin observing tonight.'[15]

The night, however, was cloudy, but the next night being clear, Harper and a group of trustees and officials from the University came out by special train for the first night of observing. They looked at Jupiter. Now the astronomers could begin serious testing, and a few nights later Hale, Barnard and Wadsworth examined a number of objects with the telescope. Barnard called the seeing conditions 'indifferent,' but noted that the 40-inch 'worked admirably.'[16] Hale, the most effective astronomical promoter the world has ever seen, waxed more enthusiastic in a note to Yerkes:

> Such objects as the ring nebula in Lyra, the great cluster in Hercules and the dumb-bell nebula are shown in a surprising way. The brightness of these objects in the telescope is remarkable, and it is already clear that no other instrument in the world will show nebulae in so perfect a manner.[17]

Barnard succeeded in finding close by the brilliant star Vega a tiny star that had never been seen before, though the region had been combed time and time again by Burnham with the Lick telescope under the most favorable conditions. The discovery was of no scientific importance – the star's proximity to Vega was a mere chance alignment, and there was no physical connection between the two. Nevertheless, Hale made the most of it, pointing out to Yerkes that 'although of no strictly astronomical significance, I cannot but regard the discovery as of the very greatest importance as demonstrating conclusively the pre-eminence of the Yerkes telescope.'[18]

Satisfied that his work was done, Clark left for Williams Bay station early on the morning of May 27 – unbreakfasted and on foot, since that morning the Hales had overslept. However, mechanical difficulties still plagued the great instrument. Later that same day Hale complained to Warner and Swasey that 'we are all greatly disappointed . . . in the extreme difficulty of handling the instrument, which is so serious that it greatly handicaps all observational work . . . Mr. Burnham was so much disappointed that I very much doubt whether he would try to do any regular work of observation with the telescope in its present condition.'[19]

Hale had hoped that the great telescope might be open to the public at least one night

40-inch refractor, May 21, 1897, with S. W. Burnham on platform. Yerkes Observatory

a week. However, Yerkes adamantly opposed the idea, telling Harper that 'no one be allowed, except those in charge, to approach anywhere near the glass. It should never be lowered so that visitors could look at it, or be on the same plane with it. In fact, every precaution should be taken to keep visitors away from it . . . [I]t is there for scientific research and for that only!'[20] Yerkes's sensitivity about the instrument, which to build and suitably house had cost him personally close to half a million dollars, was understandable, and its vulnerability to injury was already known – in 1893, the tube and mounting had narrowly escaped damage when a fire spread to the Manufacturers' Building of the Chicago Exposition where they were then on display.

Harper forwarded Yerkes's letter to Hale, and it reached him on May 28. That evening Hale worked on the 40-inch until 12:45 a.m., then yielded the instrument to Barnard and Ellerman. At a little before 3 a.m., in order to observe the 'Swan' nebula (M17) in Sagittarius, they raised the floor to within six inches of its highest point. When day began to break they gave up observing and went home, leaving the floor in this position, as Barnard later recounted, 'for the convenience of some workmen who were to be at work on the tube in the morning.'[21] An hour or so later, at 6:43 a.m., one of the workmen coming to the observatory heard a loud crash. On investigating, he found that the cables suspending the rising floor on the south side had given way, and the floor on that side had fallen forty-five feet, leaving it cantered at a 45° angle. Barnard had narrowly escaped serious injury, or even death. 'A couple of hours either way,' he brooded, 'and death in all probability would have come to one or the other of us. Only a few nights before this accident the President of the University of Chicago and some thirty or more trustees and prominent men of the university had seen through the telescope, and the floor had been up and down with them on it. If it had fallen then, a heavy loss of life would have been almost certain . . . It was providential, then, that the floor fell when it did, for the fault in the attachment of the cables made it certain that it must soon have fallen.'[22] There was perhaps one fatality, however; the strain of the trip out to Williams Bay and the collapse of the floor proved too much for old Clark, who died of a stroke a week later.

Amazingly, despite the catastrophe, the telescope appeared to have survived relatively unscathed. Though some of the iron stairs of the stairway on the south side of the pier had been sheared off, there appeared to have been no other damage. 'It is fortunate,' Hale wrote, 'that the telescope itself was not carried down with the floor, for had not the iron column been exceptionally rigid and securely fastened to the pier, the instrument itself would undoubtedly have been carried away.'[23] However, there was still concern that the object glass had been damaged by the shock when the floor had lurched against the great iron pier. At first no one could tell, since the object glass was still poised a hundred feet in the air. Someone climbed up on the dome and had a look down – to everyone's relief, the object glass was apparently uninjured, though there would be no way of knowing for certain until the telescope could be tested by examining stars through it. Before that could happen, the wrecked floor would have to be ripped out and a new floor put in, and that would take months.

Barnard was heartbroken to lose the instrument just as it had been put in working order and as the favorable summer observing season was approaching, and he could hardly rest worrying about the condition of the lens. But at the moment there was nothing more to be done in Williams Bay, and he left for Nashville for the meeting of the Vanderbilt Alumni Association and to witness the unveiling of a bust of his late mentor, Chancellor Garland, who had died in 1895.[24] By the middle of August the new floor was in place. Though the floor was not yet movable, the telescope could at last be turned toward the stars. The moment of truth had arrived. 'To our consternation,'

Crashed floor of 40-inch dome, May 29, 1897. Yerkes Observatory

Barnard wrote, 'there was a great, long flare of light running through every bright star we examined. This was so strong and conspicuous that it would make the instrument utterly useless. It looked as if the lens had been injured by the shock of the floor against the pier. We examined it in all positions of the instrument, but we could not get rid of this glaring defect. As I had used the glass more than anyone else before the accident, my statement that the defect did not then exist made the matter all the more serious. It was with heavy hearts that we waited for day, to again critically examine the lens.'[25]

When day broke, there proved to be a simple explanation for the telescope's poor performance. Just back of the glass in the tube Hale noticed a thick mass of spider webs stretched across the tube, all running in the same direction. 'Upon comparing notes,' Barnard continued, 'we found that the direction of the spider webs coincided with that of the flare of light seen the night before':

> It seemed that a spider had evidently got in the tube before the object glass was put on by Clark, and had been unable to get out, for there was no opening in the tube. During the time the tube remained at rest, while the new floor was being put in, he

had climbed up to the great glass in the direction of the light, when he found his egress barred by the great window, and as the days went by and he slowly starved to death, he spun his web, perhaps as a signal of distress or maybe in the hope that some unlucky fly might get in through the glass that he could not get out of – anyway, with the result that he caused several astronomers the most uneasy time of their lives. After these webs were swept out by one of the astronomers climbing up in the tube with a feather duster, it was found, when the stars were examined that night, that the flare had vanished and the mighty glass was uninjured.[26]

At last, on October 21, 1897, everything was in readiness for the official dedication of the observatory. The trustees, faculty and guests of the University – seven hundred in all – arrived from Chicago at noon on two special trains. Charles Tyson Yerkes was already on hand; he had arrived the night before, and had stayed over with the Hales. Many of the leading scientific figures of the day were also there, having participated in an astronomical and astrophysical conference during the three preceding days – sterling exemplars of the old astronomy such as Newcomb (who was the guest of the Barnards), Burnham, and Barnard himself, and of the new astrophysics such as Hale, E. C. Pickering of Harvard, and Keeler.

The dedicatory exercises were held in the great dome. Yerkes made a few remarks:

> I . . . with the fullest feeling of satisfaction and pleasure turn over to you this structure with all of its contents, feeling satisfied that it is now in the best of hands, and that the labors here will be serious, conscientious, and thoroughly done. I feel that in your attempts to pierce the mysteries of the universe which are spread before you by our great Creator, the enthusiasm of your natures will carry you to success.[27]

The main address was to have been given by Sir William Huggins, but he was unable to attend, and instead the mantle fell to Keeler. He spoke prophetically on 'The Importance of Astrophysical Research and the Relation of Astrophysics to Other Physical Sciences.' 'There may be some,' he noted,

> who view with disfavor the array of chemical, physical, and electrical appliances crowded around the modern telescope, and who look back to the observatory of the past as to a classic temple whose severe beauty had not yet been marred by modern trappings. So mankind, dissatisfied with present social conditions, looks back with tender regret to the good old times of earlier generations, yet rushes forward with the utmost speed.[28]

Keeler could point to the spectroscopic measurement of the radial velocities of stars and the identification of some hitherto unsuspected binary stars as striking advances already made by the new science. But it promised much more. Indeed, as he spoke, a decided shift was taking place; astrophysical concerns were increasingly coming to the fore – the attempt 'to ascertain the nature of the heavenly bodies, rather than their positions or motions in space – *what* they are, rather than *where* they are,' as Keeler put it.[29] With the new astronomy's increasingly technical investigations into the structure of matter and the nature of radiation (and the corresponding requirements for specialized equipment and expertise), the concerns of professionals and amateurs, for so long overlapping, were beginning to diverge sharply. One result was that the self-made man would find it increasingly difficult to find a foothold in a field where several years of graduate school were increasingly prerequisite. A few years later, such a young man aspiring to a career in astronomy asked E. B. Frost, Hale's successor as director of the Yerkes Observatory, what his chances were for obtaining a position as an assistant and just what education was necessary to begin 'at the bottom.' Frost replied:

> [T]he chances for obtaining a position as assistant, I am sorry to say . . . are not good. Almost all young workers in Astronomy have had at least a college education, and usually two or more years of graduate work. 'Beginning at the bottom' therefore is usually done at school. On the other hand, I can recall to mind four or five men who have begun without much special education: by dint of hard work and great enthusiasm, they have compelled success. The road they have travelled is a hard one, full of much discouragement at first. Such cases are almost doomed to failure unless thay have an unusual gift for making discoveries.[30]

The hard road which Burnham and Barnard had travelled to success in the 1870s and 1880s was still open, but it was even harder now. There were fewer and fewer like them who could still succeed in achieving 'important research results on the basis of sheer drive, technical skill and interest with only a minimal scientific education.'[31]

6

Now that the 40-inch telescope was finally in working order, Barnard threw himself into his usual relentless routine – observing with it two, sometimes three and even four nights a week, from dusk to dawn, never wasting a clear hour. Much of this work was not, however, of the sensational discovery nature of his earlier work. It was, of course, entirely in the classical vein, and consisted for the most part of careful, but on the whole rather routine, measures with the filar micrometer.

Many measures were made of the nebulae. The riddle of the nebulae remained, in the late 1890s, as impenetrable as ever – largely because no one had yet been able to reliably determine the distance to any of them. Drawings were of no help, while earlier measures had been rendered problematic by the fact that the outlines of nebulae were by nature diffuse and ill-defined, thus extremely difficult to fix with precision on the threads of the micrometer. The general sense in the 1890s – and Barnard shared it – was that many of the nebulae might be relatively nearby, and thus that their parallaxes or proper motions might be detectable by classical methods.

The controversy about nebulae came briefly to center stage on September 21, 1898, when a telegram reached the Yerkes Observatory announcing that Seraphimoff at Pulkowa had discovered a star-like condensation at the center of the great nebula in Andromeda. At once suspecting that the Nova of 1885 had reappeared, Barnard eagerly examined the nebula with the 40-inch as soon as darkness fell. However, as he recorded in his observing book, 'The nucleus did not noticeably appear any different to what I

Staff of Yerkes Observatory, 1898. In the rear, left to right, are Ernest F. Nichols, Harry M.
Goodwin, Barnard, Edwin B. Frost, and George E. Hale. In front of them, on steps, are
Ferdinand Ellerman and Frank Schlesinger, and on the wall, second from right, is George W.
Ritchey. Yerkes Observatory

have always seen. I thought that possibly it might be a little brighter, but if so one would
not notice it casually.'[32] Later he concluded that any supposed change must have been
an illusion:

> The nucleus of this great nebula is so affected by the conditions of seeing,
> moonlight, etc. that one can readily be deceived by it, if he is not perfectly familiar
> with its appearance under all conditions. When the air is steady the nucleus is
> almost stellar, especially with low powers.
>
> If the seeing is not good the nucleus is wooly and faint, and often lost sight of
> entirely, so that one familiar with it under these conditions alone would suppose a
> great change had occurred if he saw the nucleus under the best conditions. Its
> distinctness is more easily affected than that of a star of the same magnitude for two
> reasons – first, it is not a stellar point; second, its light would be confused with that
> of the nebula under conditions of seeing that would scarcely affect a star not situated
> in a bright nebula.[33]

With its condensed nucleus, which in the 40-inch refractor appeared 'about 2 or 3" in diameter, but so strongly condensed that under good conditions it can be bisected with almost the same accuracy as the comparison stars,'[34] Barnard decided that the Andromeda nebula was ideally suited for measurement to see whether it had any proper motion or parallax. Beginning in 1898, he began measuring the position of the nucleus with respect to three neighboring faint stars. He kept up the series until 1916, when he concluded that the object was too excessively remote to be within reach of his measures.

Following up on work begun by Burnham with the Lick refractor in 1891, Barnard also made numerous careful measures of planetary nebulae, most of which contained a sharp stellar nucleus whose position could be measured with respect to the neighboring stars as exactly as if it were devoid of nebulous surroundings.[35] The planetaries that were the subjects of his special study included M57 in Lyra, M27 in Vulpecula, and NGC 7662 in Andromeda (the central star of which he suspected of variability). Perhaps the most tantalizing of these objects was M97, the Owl nebula in Ursa Major. In 1848, Lord Rosse had made a classic drawing of it with his great reflector, representing, as Barnard described it, 'a weird bewhiskered grinning face, with two dark spots where the eyes should be. In each of these dark spots or eyes a considerable star, which formed the pupil of the eye, was shown. The whole representation is strikingly like the face of an owl, or some uncanny cross-eyed globlin which only needs a pair of legs to execute some fiendish dance in space.'[36] In the 40-inch, Barnard noted several faint stars in the nebulosity, but none was located in either of the dark spots as Rosse had shown – the brightest, the central star, was now located on the bridge of the owl's nose. If Rosse's drawing could be trusted, it seemed that in the interval since it was made the nebula had moved, but Barnard's micrometric measures between 1899 and 1907 failed to show any motion.

Barnard expended even more time and effort in measuring the globular clusters. 'There is no class of objects in the heavens,' he wrote, 'unless it is the planetary nebulae, that shows the power of a great telescope better than the great globular clusters . . . No photograph, though it may show more stars, can begin to give one an idea of their beauty. The photograph is at best a dead copy of the original. The fire and life given by the myriads of stars are lacking in any picture except that seen in the telescope itself.'[37] Again no one knew the distance to any of them. Barnard felt that the stars in these clusters, especially toward their centers, were too crowded to be measured successfully by photographic means.[38] Hence he undertook the herculean labor of visually measuring hundreds of representative stars in M3, M5, and M13, and later in several more clusters, 'in the hopes of getting motion.'[39]

In the course of this work, he became interested in the variable stars which had recently been found in these clusters by Solon I. Bailey of Harvard on his photographs taken at Arequipa. When Bailey had first announced his findings, some astronomers, especially in Europe, had been skeptical, but Barnard's observations with the 40-inch completely confirmed his results. Over the years, Barnard continued to follow the variations of Bailey's stars, and discovered several himself. His favorite object was Bailey's variable no. 33 in M5, which he found varied between magnitudes 13.4 and

14.6 in a period of just over twelve hours.[40] But his chief object eluded him – he was never able to find any evidence of parallax or proper motion.

7

If Barnard's increasing attention to planetary nebulae and globular clusters was in part indicative of an increased interest to the objects of the larger universe, it also showed a falling off, relatively, of his earlier interest in the Solar System – especially the planets. Whereas in former years he had published more papers about the planets and their satellites than anything else, the proportion henceforth is relatively small. In part this may have been due to the influence of men such as Hale and Keeler, who in leading the field in the new direction of astrophysics were skeptical – and highly critical – of the work of planetary men such as W. H. Pickering and Percival Lowell. In his address at the dedication of the Yerkes Observatory, Keeler had regretted that 'the habitability of the planets, a subject of which astronomers profess to know little, has been chosen as a theme for exploitation by the romancer, to whom the step from habitability to inhabitants is a very short one . . . Science is not responsible for these erroneous ideas, which, having no solid basis, gradually die out and are forgotten. Thus it cannot long suffer from outside misapprehension, while the sustained effort necessary to real progress is, in the end, a sufficient safeguard against the intrusion of triflers into its workshops.'[41] Barnard made it a point, except when he felt personally attacked, to avoid controversy, but it is clear from what he had written about Mars ever since the early 1880s that he had always felt much the same way.

Aside, however, from the disenchantment with planetary observation that began to affect professional astronomers in the 1890s and 1900s, Barnard's tendency to turn to other fields of research owed at least somewhat to the disappointing night time seeing conditions at Williams Bay. Barnard's worst fears about the winter conditions, in particular, were fully realized, as amply documented in his observing notes from the first winter of 1897–8, where he records again and again his frustration. 'The seeing is simply horrible – there is no definition . . . There is a cold heavy wind from the north west – can scarce move the dome against it.' 'The seeing is fearful even with the lowest power – Have tried Jupiter but can make nothing of it.' 'The seeing fearfully bad.' 'The seeing is simply awful – everything one mass of blurry light.'[42]

A few nights before Christmas 1897, Barnard found the shutter of the dome frozen shut. He and his night assistant hiked through the snow, as a brilliant aurora played overhead, to the nearby YMCA camp to bring back a ladder so that they could climb up on the outside of the dome and attempt to clear away the ice. Finally, at just before 1 a.m., as the aurora was dying out, they managed to get the shutter to move – but the effort went for nought. Barnard's further notes that night record: 'Very cold – with cutting North Wind. The seeing is fearful bad.' And a few hours later: 'The thermometer outside is at 5° [Fahrenheit] It is covered with great flakes of frost. The aurora is still fluttering – the seeing is simply abominable.'

Barnard struggled to observe the winter opposition of Mars of 1898-99 with scant success. The following excerpts from his observing book are representative:

> Dec. 13, 1898. Can make out no details on the planet. Have lowered the curtain to the base of the slit – this has improved the seeing some – but it is still bad . . . Bitterly cold wind from the west. Temp. outside −4° [Fahrenheit].
>
> Feb. 6, 1899. The seeing is excessively bad . . . Pretty high cold north wind. Mars – can making nothing of it – the seeing is so bad.
>
> Feb. 11. Temp. in observing chair = −14°; outside −19°. The seeing is too bad to do anything with it. – [Mars] is a great blur of light.
>
> Mar. 29. The image of Mars is excessively bad – fearful blurring – dome [e]ffect.'[43]

The terrible winter, and Barnard's relentless observing routine, inevitably took their toll. On Christmas Day 1898 he wrote to Huggins: 'I am writing this propped up in my sick bed. I have been sick in bed for some time and I do not know when I shall be able to get up again.'[44] In similar vein, he wrote to Davidson:

> I have been unwell much of the past winter. It has been a dreadful winter – something like 30° below zero [Fahrenheit] several times, and below zero a great many times . . . It was the worst winter they have had here for many years – and I hope the worst they will ever have![45]

Barnard was discouraged by the fact that much of the time the bad seeing seemed to be generated within the great dome itself – what he referred to as 'dome bad seeing,' since he frequently found the seeing at the 40-inch telescope 'excessively bad' or 'abominable' at the same time it was good at the 12-inch. In frustration, Barnard had frequent recourse to using a diaphragm with the large telescope, the very practice that he had decried at Mt. Hamilton, often stopping it down to 24 or 30 inches.

The cold Wisconsin nights were often lit up with auroras – beautiful displays which had been uncommon in the more southerly latitudes of Nashville and Mt. Hamilton. For a while Barnard regarded the brighter displays as a nuisance, since they interfered with his other observations, but over time they began to fascinate him. For the rest of his life he kept detailed records of them in his observing notebooks.[46]

Ironically, as mediocre as night time conditions generally were at Williams Bay, during the day the seeing could be fairly good – Hale, who was mainly interested in the Sun, later thought it on average better than at Mount Wilson.[47] On one of these occasions, the morning of August 31, 1900, Barnard obtained his best views ever of Mercury, noting on a disk only 5″ of arc in diameter: 'The markings could be seen for a moment at a time beautifully and I think would have been seen steadily with a diaphragm. The appearance of these markings – was that of large dark areas like those on the moon with the naked eye – Indeed it looked very decidedly like our moon w[oul]d have looked if reduced to 4″ or 5″ diameter.'[48] Unfortunately, such opportunities were a small consolation for Barnard, who unlike Hale had no wish to do most of his observing in the daytime!

8

There were other disappointments for Barnard besides the Wisconsin weather. One was that he was getting nowhere with the publication of his comet and Milky Way photographs. As earlier noted, the work, under the supervision of the A. B. Brunk of the Chicago Photogravure Company, had gotten underway in the spring of 1896. At first Barnard was satisfied with the quality of the reproductions, but soon began to worry that carelessness was creeping in. In October 1896, he wrote to H. H. Turner at Oxford: 'I am having a good deal of trouble with my reproductions. I may have to throw up the work as far as the Chicago firm is concerned . . . I am dreadfully worried, and disappointed about the reproductions of my star pictures.'[49] Two months later he again wrote to Turner: 'I am still waiting about my pictures – I don't know what to do. The Chicago company is making me an offer to continue the work but I feel pretty uncertain about having them continue it. When they were at work on it it was a continual fight to keep them from going wrong.'[50] In the end, the stress proved to be too much for his nerves, and he decided to break the contract. In order to avoid legal complications, he had no choice but to accept the thirty-six editions already completed, which he paid for personally at a cost of over $1000. He took the finished reproductions with him to the Yerkes Observatory and placed them in storage in his office.

A few months later he was close to arranging to have the work finished by a firm in Philadelphia, but then the firm burned down, and that 'threw everything behind again.'[51] Another year passed with nothing further having been accomplished, and he mentioned to Davidson, wearily: 'I am in hopes soon to have [the project] on the way again. Most of the reproductions I have are good. Unless I can get them done properly, I don't want to print them. I have been almost crazy about them. It is the most difficult task I ever encountered.'[52] But the project was not soon on the way again – to the contrary, the frustrating business of publishing his Milky Way and comet photographs would drag on for years.

There were also heartbreaking delays with the Bruce photographic telescope, and for much the same reason – Barnard's impossibly high standards. He wanted a telescope mounted so as to allow long exposures to be carried through the meridian, which had been impossible with the Willard lens on its ordinary equatorial. He personally suggested the idea of making a mount with a bent pier, which would allow the telescope to swing freely under it in all positions of the instrument. But the mount, fashioned on Barnard's plan by the Cleveland firm of Warner and Swasey, proved to be the easy part. The long delay came in making the lens.

Barnard had a good deal of practical knowledge of optics, but theoretically none at all. He therefore had unrealistic expectations for a large, aberration-free telescope. What he wanted was a lens ten inches in diameter, with 'a wide angle and flat field with as short a focus as was consistent with these two qualities.'[53] Barnard's friend Brashear made several good lenses of four inches diameter and upward, but none was ever good enough to satisfy Barnard. Undaunted, Brashear, 'with characteristic faith in his skill,' went ahead and in March 1899, on his own responsibility, obtained glass disks and began working them into a 10-inch Petzval doublet. Barnard, meanwhile, had learned that the English firm of Thomas Cooke and Sons had a triplet lens which he believed would fulfill his requirements, and wrote to Brashear: 'It is impossible for us to go ahead with the lens without looking into the matter . . . I need hardly tell you that it has been my earnest desire all along that you should make the lens of the Bruce telescope . . . mainly . . . on account of the friendship and esteem in which I have always held you. You, yourself, however, must see that my duty not only to Miss Bruce and to the Yerkes Observatory, but to the Science of Astronomy as well, demands that if this lens is what is claimed for it, it is necessary to get it for the Bruce telescope.'[54] Thus Barnard made a brief foray to Europe between December 1899 and March 1900. In the end, the Cooke lens did not satisfy him either, however, and from London, Agnes Clerke informed Campbell at Lick: 'We saw Prof. Barnard last week. He is inspecting the optical resources on this side of the water, and I think he finds those on your side preferable.'[55] After seeing what England had to offer, Barnard crossed the Channel and visited the Palais de l'Optique at the Paris Exhibition, walking with E. M. Antoniadi, Flammarion's assistant at Juvisy, through the 197-foot long tube of the 49.2-inch refractor which was then on display. The tube in this unusual telescope was fixed, and star-light was directed through it by a 79-inch 'siderostat' mirror which tracked the stars as they moved across the sky. Unfortunately, it proved to be a great disappointment – the only scientific results ever obtained with it were Antoniadi's drawings of the Ring nebula in Lyra, NGC 7009 in Aquarius, and a few other planetaries – and after the Exhibition closed the tube was broken up and the optical components placed in storage at the Paris Observatory. Carrying the fresh impression of this unsuccessful telescope, which owed so much to overreaching, Barnard was at last ready to acknowledge that his own ideal for the Bruce telescope had been 'too high and one not attainable with optical skill,' and finally decided on Brashear's doublet after all. It had a 9° field, and though the images were fairly bad at the edges, the definition was sharp for about 7°.[56] With Barnard's go ahead, Brashear was able to finish work on the lens by September 1900, but it was to be almost four more years before Barnard was able to begin his work with the Bruce telescope.

1   For a biography that beautifully intertwines the lives of Ritchey and Hale, see Donald E. Osterbrock, *Pauper and Prince: Ritchey, Hale, and the Great American Telescopes* (Tucson, University of Arizona Press, 1993)
2   EEB to HHT, October 20, 1896; RAS
3   EEB to Sir William Huggins, January 26, 1897; RAS
4   EEB to GEH, January 30, 1897; YOA
5   SWB to GEH, February 1, 1897; YOA
6   GEH to SWB, February 2, 1897; YOA
7   A. M. Clerke to David Gill, March 4, 1897; RGO
8   EEB to HHT, February 13, 1897; RAS
9   A. M. Clerke to David Gill, March 4, 1897; RGO
10 EEB to M. W. Bruce, draft of letter dated May 27, 1897; VUA

11   ibid.

12   ibid.

13   EEB to WRH, July 26, 1897; VUA

14   Helen Wright, *Explorer of the Universe: A Biography of George Ellery Hale* (New York, E. P. Dutton & Co., 1966), p. 127

15   GEH to WRH, May 20, 1897; YOA

16   EEB, observing notebook; YOA

17   GEH to C. T. Yerkes, May 31, 1897; YOA

18   ibid.

19   GEH to W. Warner and A. Swasey, May 27, 1897; YOA

20   C. T. Yerkes to WRH, May 24, 1897; RL

21   E. E. Barnard, 'A Few Unscientific Experiences of an Astronomer,' *Vanderbilt University Quarterly*, 8 (1908), 273–88:286

22   ibid.

23   GEH to CAY, June 1, 1897; YOA

24   'Prof. Barnard in Nashville,' Nashville *American*, June 14, 1897

25   Barnard, 'Experiences,' p. 288

26   ibid.

27   G. E. Hale, 'The Dedication of the Yerkes Observatory,' *Ap J*, 6 (1897), 353–62:359

28   J. E. Keeler, 'The Importance of Astrophysical Research and the Relation of Astrophysics to Other Physical Sciences,' *Ap J*, 6 (1897), 271–88: 277

29   ibid.

30   EBF to Douglas Manning, December 27, 1912; YOA

31   Donald E. Osterbrock, John R. Gustafson, and W. J. Shiloh Unruh, *Eye on the Sky: Lick Observatory's First Century* (Berkeley, University of California Press, 1988), p. 142

32   EEB, observing notebook; YOA

33   E. E. Barnard, 'The Great Nebula of Andromeda,' *Ap J*, 8 (1898), 226–8: 227–8

34   E. E. Barnard, 'The Proper-Motion of the Great Nebula of *Andromeda* (M31),' *AJ*, 30 (1917), 175–6:175

35   His publications on planetary nebulae include: 'On the Probable Motion of the Annular Nebula in Lyra (M57) and the peculiarities in the Focus for the Planetary Nebulae and their Nuclei,' *MNRAS*, 60 (1900), 245–57; 'On the probable Motion of some of the Small Stars in the "Dumb-bell" Nebula (M27, N.G.C. 6853), *MNRAS*, 62 (1902), 466–8; 'The Annular Nebula in Lyra (M57),' *MNRAS*, 66 (1906), 104–13; 'On the "Owl" Nebula, Messier 97 = N.G.C. 3587,' *MNRAS*, 67 (1907), 543–50; 'Variability of the Nucleus of the Planetary Nebula N.G.C. 7662,' *MNRAS*, 68 (1908), 465–81

36   Barnard, 'On the Owl Nebula,' p. 544

37   E. E. Barnard, 'Micrometric Measures of Star Clusters,' E. B. Frost, G. Van Biesbroeck, and M. R. Calvert, eds., *PYO*, 6 (1931), 1

38   E. E. Barnard, 'The Development of Photography in Astronomy–II,' *Science*, New Series, 8 (1898), 386–95:392

39   EEB to GD, May 14, 1899; BL

40   On this star alone, he published the following papers between 1898 and 1922: *AN*, **147** (1898), 243–8; **184** (1910), 273–84; **196** (1913), 11–14; *Publications of the Astronomical*

and Astrophysical Society, **1** (1902), 193–4; **4** (1922), 351; *PA*, **27** (1919), 548–9

41   Keeler, 'The Importance of Astrophysical Research,' p. 276

42   EEB, observing notebook; YOA.

43   ibid. Despite the horrific conditions, Barnard remained as much an observaholic as ever. On Christmas Day 1897, he invited the other staff members and their families to dinner at his house, but when Burnham, who was assigned to the great telescope that night, did not make it out from Chicago, he announced as evening was falling that he 'had to leave the company to observe.' EEB to GEH, December 28, 1897; YOA

44   EEB to Sir William Huggins, December 25, 1898; RAS

45   EEB to GD, May 14, 1899; BL

46   E. E. Barnard, 'Observations of the Aurora Made at the Yerkes Observatory 1897–1902,' *Ap J*, **16** (1902), 135–44; 'Observations of the Aurora Made at the Yerkes Observatory 1902–1909,' *Ap J*, **31** (1910), 208–33. Barnard's observations were later used as the basis for detailed studies of low-altitude aurorae. See A. B. Mcinel, B. J. Negaard and J. W. Chamberlain, 'A Statistical Analysis of Low-Altitude Auroras,' *Journal of Geophysical Research*, **59** (1954), 407–13 and J. W. Chamberlain and H. M. Thorson, 'The Nightly Variation of Auroras at a Subauroral Station,' *Journal of Geophysical Research*, **65** (1960), 133–6

47   O. Struve, 'The Story of an Observatory,' *PA*, **55** (1947), 227–44:243

48   EEB, observing notebook; YOA

49   EEB to HHT, October 20, 1896; RAS

50   EEB to HHT, December 27, 1896; RAS

51   EEB to M. W. Bruce, May 27, 1897; VUA

52   EEB to GD, April 13, 1898; BL

53   EEB to JAB, undated letter probably from 1898

54   EEB to JAB, [1899]; VUA

55   A. M. Clerke to WWC, January 25, 1900; SLO

56   E. E. Barnard, 'The Bruce Photographic Telescope of the Yerkes Observatory,' *Ap J*, **21** (1905), 35–48:37

19

Disappointments and triumphs

1

Almost immediately after his return from Europe in March 1900, Barnard began preparing for another trip – to Wadesboro, North Carolina, for the total eclipse of the Sun of May 28. The group that went included himself, Hale, Frost, Ritchey and Ellerman. Hale and Ellerman planned to measure the heat radiation of the corona with a bolometer, while Frost hoped to obtain spectrograms of the chromosphere and corona. Barnard and Ritchey were charged with taking direct photographs of the corona. The observers were favored with a clear sky, and though in the excitement of totality the mirror in Hale's bolometer was kicked or knocked out of alignment, the rest of the team experienced success. For their photographic work, Barnard and Ritchey used a special telescope called a coelostat, the principle of which had been proposed by H. H. Turner. It consisted, as Barnard explained the principle,

> apart from the lens, of a perfectly flat mirror placed parallel to the Earth's axis and revolving westward by clock-work at one-half the speed of the Earth's rotation. [It] throws a horizontal beam of light through the lens to form the image of the eclipsed Sun on a sensitive plate exposed in a dark room over sixty feet away. The revolving mirror keeps the image of the moving Sun perfectly stationary.[1]

As at the eclipse of 1889, Barnard's involvement with photography meant that he had to deny himself a direct view of the corona; he and Ritchey, in order to assure the accuracy of the exposures, remained in a long, shaded horizontal tunnel that they had set up at the site to minimize thermal air currents, and saw the corona only as it was projected on the photographic film.

Among the other expeditions that had travelled to North Carolina for the eclipse was a group from the British Astronomical Association, who afterwards returned with Barnard to Williams Bay and stayed for two days as his guests. One of the members of this expedition, Rev. J. M. Bacon, wrote:

> All this will long dwell in our recollection, but our most abiding memory will be of the home where at the hands of Professor Barnard and his charming English wife we learned all that true Southern hospitality means. Here is an ideal retreat, for one whose whole life is given up to his work; – a luxuriant garden of Nature's own planting, where the sumac is the undergrowth, and flowers, prized in English gardens, grow as weeds. On the one side the wide reach of Big Foot Prairie, on the

Yerkes Observatory solar eclipse expedition, Wadesboro, North Carolina, 1900. Ritchey is third from left; standing immediately to the right of him are Ellerman and Hale. Frost and Barnard are on the far right. Yerkes Observatory

other, far below, lie the blue waters of Lake Geneva. I cannot recall the scene without picturing a calm, clear evening with the light of the after-glow already fading in the west, and in the distance the retreating figure of a man, nearing middle life, yet hurrying with all the activity of vigorous youth across the grass to his long night's labour.[2]

2

Totality at the May 1900 solar eclipse lasted only a little more than a minute. By contrast, the eclipse of May 18, 1901 would be the event of a lifetime, with totality lasting six and a half minutes on the central line – almost the maximum possible. The shadow of the Moon would first touch the Earth at sunrise near the south end of the island of Madagascar. It would then move eastward over the island of Mauritius, across the long stretch of the Indian Ocean, then touch land again at noon on the west coast of Sumatra. Continuing over the China Sea, it would cross the islands of Borneo and the

Celebes, and reach the coast of New Guinea where it would finally leave the Earth at sunset.

To take advantage of this exceptional opportunity for studying the solar corona, an English team under E. Walter Maunder planned to go to Mauritius; David Todd of Amherst College would go to the island of Singkep, south of Singapore, and English, French, Russian, Dutch, Japanese, and several American parties would head for Sumatra, in the Dutch East Indies.

Barnard went along with the US Naval Observatory expedition led by Professor A. N. Skinner. Among the other members of the party, which included eleven astronomers in all, were Heber Doust Curtis, then a fellow at the University of Virginia, and S. A. Mitchell, who had spent the year 1898–99 as a research assistant at Yerkes and was now at the Naval Observatory. Also traveling to Sumatra was a Smithsonian Institution expedition under Charles G. Abbot, who brought a bolometer with which to measure the heat of the solar corona and a telescope with which to search for intra-Mercurial planets.

Before sailing out of San Francisco Bay, Barnard made a brief visit to Mt. Hamilton. It was the first time he had been back to California since he had left in 1895, and his former colleague Richard Tucker described him as 'nervous and unwell, with no strength, and generally a disappointed man I think.'[3] Tucker was probably right. Always restless and dissatisfied, Barnard at forty-three was no longer young but middle-aged. Compared with the productive but turbulent years at Lick, he had very little to show for six years of work at Yerkes. His attempts to publish his comet and Milky Way photographs were stalled, the Bruce photographic telescope was not yet finished, and his measures of nebulae and globulars had so far shown no discernible parallaxes or proper motions. Moreover, as he confided to Davidson, Williams Bay was 'a mirey climate for a great telescope and discoveries are few and far between.'[4]

The eclipse expedition left San Francisco Bay for Manila on the US Army Transport *Sheridan* on February 16, 1901. As the *Sheridan* passed out through the Golden Gate, a dense fog came up. Later it lifted and the passengers could see the shoulder of Mt. Tamalpais looming up among the clouds in the distance. Within a day they were out of sight from land. As they neared the tropics, they began to see great numbers of flying fish, which Barnard at first took for birds. 'These beautiful little creatures rose from the water with rapidly vibrating wings until under headway; they then seemed to sail without any effort at a few inches distance from the water. They did not alight gracefully as a bird might, but fell like a stone into the sea.'[5]

Three days behind schedule because of a constant head wind and a heavy sea, the *Sheridan* reached Honolulu early on the morning of February 25. 'The magnificent beauty of Honolulu and its surroundings,' Barnard noted, 'opened up before us as the coming dawn brought into mystic relief the harbor and its background of serrated mountains. Such pictures one has frequently seen in a dream, but the reality has seldom been encountered.'[6]

The *Sheridan* continued on to Luzon, the largest island of the Philippines, hugging its west coast by Subic Bay where Admiral Dewey had chased the Spanish fleet during

Barnard on board ship and bound for Sumatra, 1901 eclipse expedition. Vanderbilt Photo Archives

the Spanish–American War three years before, and entered Manila Bay through Corrigidor. Barnard thought the coastal scenery 'very beautiful, the mountains rising densely wooded to their summits, with here and there a wreath of smoke rising slowly from among the trees; and the imagination could see savage beasts of prey and mighty serpents, and no less savage man lurking in these dense forests.'[7]

After a week lay over in Manila, the party boarded the US Gunboat *General Alava* for the rest of the journey to Sumatra. They passed Corrigidor through the south passage as night was settling, and steamed along the west coast of Borneo. After crossing the equator, Barnard noted that 'at noon our shadows fell to the south, for the sun was north of us.'[8]

At last they reached the Strait of Sunda, between Java and Sumatra. On April 2, from early morning until darkness cut off their view, they were in sight of the island of Krakatoa. 'Nothing in the entire journey was perhaps so interesting,' Barnard reflected, 'as this remnant of a mighty catastrophe that blew away three-fourths of the island and destroyed 40 000 to 50 000 lives in 1883, and which sent a tidal wave several times around the Earth and filled the Earth's entire atmosphere with volcanic dust for several

Krakatoa, photographed by Barnard in 1901. Yerkes Observatory

years.'[9] Barnard himself had, as we have noticed, seen the red skies which had marked the aftermath of that tremendous explosion in 1883–4.

The *General Alava* reached Emmahaven (now Tulukbajur), the port of Padang, on April 4, followed closely by a Dutch vessel which carried C. D. Perrine of the Lick Observatory. The astronomers then made the short trip by rail to Padang, the capital of Sumatra, where Perrine remained for the eclipse. Barnard and the rest of the Naval Observatory party traveled further inland, to Solok, a small village beautifully situated in a wide valley amidst groves of cocoanut palms, about four miles distant from the 8000-foot volcano Talang.

The site chosen was an old unoccupied fort at the northwest edge of the town, surrounded with a rampart and a moat and numerous lines of barbed wire fencing. There was a long barracks building that consisted of one very large room, which served for storage of the freight, and several smaller ones which after being equipped with cots and mosquito bars served as sleeping accommodations. Barnard, assisted by Curtis, W. W. Dinwiddie, and N. E. Gilbert at once began setting up his own rather extensive apparatus, which consisted of two portrait lenses of $3\frac{1}{4}$ and 6 inches aperture, mounted together on a small equatorial, which he hoped to use at eclipse time to photograph the solar corona and also any intra-Mercurial planets that might exist within one and a half degrees of the Sun. At night he hoped to put them to work in photographing the southern part of the Milky Way. In addition, he brought along the $61\frac{1}{2}$-foot coelostat telescope which he and Ritchey had used at Wadesboro. Because of the coelostat's large image scale – seven inches to the diameter of the Sun or Moon – he had special plates made, 40 inches by 40 inches, with which he hoped to record out to the greatest extent possible the fainter portions of the corona.

Nothing – except the weather – was left to chance. Barnard practiced his technique constantly until he was certain that the manipulation of these large plates would cause no unexpected problem. Every precaution was taken so that no hitch would occur during totality. Unfortunately, though the meteorological reports had made Solok appear the most favorable of all the stations, in the weeks leading up to the eclipse it was cloudy almost all the time, which forced Barnard to give up his hopes of photographing the southern part of the Milky Way with his portrait lenses. 'It was,' he grumbled, 'absolutely impossible to do this work because of the constant cloudiness of the sky . . . Though the stars were seen on a number of nights, it was always between breaks in the clouds. The same cloudy condition held also in the day time, so that the sky was essentially never free of clouds.'[10]

With the conditions at Solok proving so unfavorable, and in order to lessen the chance of complete failure owing to clouds, Skinner sent Mitchell to Sawahlunto, a place on the railway east of Solok, and dispatched another party consisting of W. S. Eichelberger, W. J. Humphreys, and G. H. Peters to Fort de Kock, near the volcano Merapi. The last site was close to the northern edge of the path of shadow, where meteorological reports had given very little chance of a clear sky.

The cloudy weather at Solok was interrupted for a few brilliant days in early May, allowing the observers a glimpse of a bright comet which was then appearing in the sky close to the horizon just after sunset. This comet had been seen a month earlier farther south (by Viscara at Paysandu, Uruguay), but was first seen at Solok by Dinwiddie on May 3, who at once called Barnard's attention to it. In his diary entry that night, Barnard recorded that the comet sported 'a tail some 5° long and inclined from the horizon perhaps 30 to 40°. It was quite bright in the twilight – it was close to the hills to the west and must have been near to the equator.'[11] The comet was rapidly emerging from the Sun and was seen and photographed by Barnard on subsequent evenings.

3

The total phase of the eclipse of May 18, 1901 was due to occur at about noon at Solok, with totality lasting 5 min 52 sec. Every day before the eclipse, Barnard paid careful attention to the sky conditions at the crucial time of noon. On May 13 he jotted down: 'Today would have gotten something of the corona.' May 14: 'Sun in good clear space but clouds 10° E. Some haze – not very serious.' May 15: 'Splendid blue sky.' May 16: 'Sun hidden in big broken clouds.' May 17: 'Raining heavily.'[12]

At last the nervously awaited day arrived. At dawn, Barnard was desolate to find the sky more or less full of clouds. As the day advanced, there were tantalizing hints of clearing, but then things changed for the worse – by noon the sky was hopelessly covered with thick mackerel clouds. 'Nevertheless,' Barnard wrote,

> it was deemed necessary to go through the program in hopes of getting some traces of the corona through the clouds. At the signal for totality each observer carried out his part as if the sky were perfectly clear. During the longer exposures the plates were carefully examined but no trace of an image could be seen. It was therefore evident that no impression could be obtained. I also went outside the dark room into the open air to examine the progress of the eclipse. The Moon could be faintly seen part of the time, through the clouds in relief against a feeble coronal ring. Venus and Mercury were plainly visible some distance to the east through a thin place in the clouds. The corona was examined with a small telescope . . . but nothing could be made out of it on account of the clouds.[13]

The not quite six minutes of totality passed quickly, and then the eclipse for which Barnard had travelled half way around the world was over. He observed bitterly: 'The best preparations ever made for the observations of a total eclipse of the Sun had come to naught. It was with no light heart that the plates were developed. The small, quick-acting lenses which were under the charge of Mr. H. D. Curtis . . . had secured at most the image of the Moon with a thin ring of light about it and the clouds themselves in the region of the eclipsed Sun. One of the long exposures, indeed, though it showed the clouds well, did not show any trace of the eclipse.'[14] His own large plates, exposed for up to $2\frac{1}{2}$ minutes with the coelostat, showed at most feeble fragments of the chromospheric ring with a few prominences.

Mitchell, at Sawahlunto, was more fortunate, managing to photograph the eclipse through thin haze, while the observers at Fort de Kock enjoyed a perfectly clear sky. 'The most exasperating part of it all,' Barnard commented, 'was, perhaps, the fact that if instead of going away off into the interior to search for good weather, we had simply gone ashore from the gunboat that carried us to Sumatra and put up our instruments on the sea shore or near it, we should have had successful observations, for the ships people had a fine view of the eclipse with nearly twenty seconds longer duration of totality.'[15] Indeed, at Padang, Perrine succeeded in obtaining good photographs of the corona through thin cirrus, as well as images of stars down to the 9th magnitude with his so-called Vulcan camera – he failed to detect any intra-Mercurial planets. The night of May 18, Barnard complained that the sky, 'as if in mockery,' was perfectly clear – the only moonless night that this was so during the entire period the expedition was in Sumatra!

Mitchell, who had been Barnard's companion ever since leaving Chicago and had shared a bed with him in the hotel Orange in Padang, returned to Solok to find Barnard in the depths of depression. 'Never will I forget,' he wrote afterwards, 'the despair and dejection of an astronomer almost heart broken.'[16] On the trip over, Barnard had been in great spirits – the 'life of the whole party,' Mitchell recalled, 'always telling jokes, always interested in the landscape and scenery and strange peoples we saw, and always thrilled with the prospect of doing interesting and valuable astronomical work.'[17] In this euphoric mood, he had taken hundreds of photographs of the scenery and people of the Far East. By contrast, 'the trip back to America was one of deep gloom for Barnard; he was not interested in anybody or anything.'[18]

The eclipse party steamed back into San Francisco Bay on July 16, and Barnard was back at Yerkes on July 26. A few days later Barnard glumly summed up the whole experience to W. H. Wesley, the Secretary of the Royal Astronomical Society:

> I have very little to say about the eclipse in Sumatra, at least in regards to the Solok part of it. It might just as well have not occurred so far as I was concerned, for it was all spoiled by clouds and what I got is perhaps worse than nothing, simply fragments of the chromospheric ring with a few insignificant prominences . . . I believe the preparations with the $61\frac{1}{2}$ foot coelostat were the best ever made for an eclipse and that if it had only been clear the results would have been of the highest importance.[19]

Thus concluded one of the most disappointing chapters in Barnard's life. But though foiled by the astronomer's most feared enemy, clouds, Barnard gradually got over his keen disappointment and was able to admit more philosophically that

> though the scientific results were meager, the opportunity given the astronomers to see one of the most interesting and beautiful countries in the world was of some consolation – not to Science itself however but to the astronomer who had made the long journey and who had faithfully performed his duty though it did result only in failure.[20]

4

While Barnard had been steaming to Sumatra, the first bright nova of the century had flared up – Nova Persei, which the alert Scottish clergyman T. D. Anderson, the discoverer of Nova Aurigae in 1892, first noticed as a 3rd magnitude star between Algol and α Persei on February 21, 1901. Within a few days, the star brightened to rival Capella or Vega, then it quickly began to fade. By the summer of 1901, when Barnard had his first opportunity to observe it, it had fallen to 7th magnitude.

Like its predecessor Nova Aurigae, Nova Persei showed a nebular spectrum as it faded, and Barnard was eager to examine the star carefully with the 40-inch refractor to see whether it, too, would become nebulous. He had earlier discovered that for planetary nebulae the focus of the 40-inch refractor is almost 0.25 inch farther away from the object-glass than for a fixed star – a difference which Hale explained to him as being due to the marked difference in the spectra of the stars and planetary nebulae. However, Barnard was unable to detect any definite signs of nebulosity around Nova Persei nor difference in focus from that of a nearby star used for comparison.[21]

Barnard's first examination of the star with the 40-inch was made on August 12. A few days later, Camille Flammarion and his assistant E. M. Antoniadi at the Juvisy Observatory near Paris announced the photographic discovery of an 'aureola' or luminous shell surrounding the nova.[22] This discovery was shown to be spurious by Max Wolf at Heidelberg, who photographed the nova with his 16-inch twin telescope – the supposed aureola was an optical aberration produced by Flammarion's telescope. However, on his set of plates, Wolf found to his great surprise a series of bright knots and cloudy arcs at a distance of 5″ of arc from the star. These features were confirmed in Perrine's photographs with the 36-inch Crossley and in Ritchey's with the 24-inch reflector at Yerkes. Moreover, the bright knots appeared to be changing in position at the rate of 2″ per day, which meant, at the assumed distance of the star, that they were hurtling outward from it at the speed of light! This rapid motion was at first a complete mystery; only later did the Dutch astronomer Jacobus Cornelis Kapteyn realize that what was being recorded was actually the 'light echo' of the nova – the reflection of the nova's flash off dark clouds in the surrounding space.[23] The actual debris shell thrown off from the nova – a portion of the outer layers of the star, blown off into space during the explosion at a velocity of 750 miles per second – was not yet large enough to be seen from Earth. It was detected only in December 1916, when Barnard discovered it visually with the 40-inch refractor.

5

When not observing with the 40-inch, Barnard was working hard on getting the Bruce photographic telescope ready. As earlier noted, the 10-inch doublet lens had been finished by Brashear in September 1900. Barnard planned to mount two other telescopes, a $6\frac{1}{4}$ inch doublet and a 5-inch guide scope, together with the 10-inch on the same bent pier mounting.

One of Barnard's first priorities with the Bruce telescope was to be the photographic investigation of what he referred to as 'the great nebulous regions of the sky.' Many of these regions had been singled out as 'affected by nebulosity' by William Herschel during his sky sweeps. Herschel's results were, however, called into question by the pioneer English photographer of the night sky, Isaac Roberts. Based on photographs taken with his 20-inch reflector and a 5-inch Cooke lens, Roberts concluded in December 1902 that 'of the fifty-two nebulous regions described by Herschel, the photographs show diffused nebulosity on four of them only.'[24] Among the regions which Roberts agreed was nebulous was that of the North America Nebula discovered by Max Wolf, also the stream of nebulosity extending southward from ξ Orionis. Roberts's plate of part of the latter region, made with his 20-inch reflector on January 25, 1900, is of interest in that it is the first image to show clearly the silhouette of the well-known 'Horsehead nebula' – Roberts described it simply as 'an embayment free from nebulosity.'[25]

Roberts's paper was given to Barnard for comment by the editors of the *Astrophysical Journal*, and his remarks were published together with Roberts's paper early in 1903.

Perhaps partly because Roberts had completely ignored his own photographic work, he bristled at Roberts's presumption in sweeping away Herschel's results: 'It is a little unreasonable to suppose that Herschel, who made so few blunders compared with the wonderful and varied work that he accomplished, should be so palpably mistaken in forty-eight out of fifty-two observations of this kind.'[26] The regions where Roberts had found nebulosity were, he pointed out, already well known. Moreover, Roberts had described one of Herschel's regions as free from nebulosity which was 'really the brightest portion of one of the most extraordinary nebulae in the sky' – the great curved nebula in Orion which is now known as 'Barnard's Loop.' As this omission cast doubt on the rest of Roberts's negative results, Barnard hoped that 'more of these regions given by Herschel may yet be shown to be nebulous with photographic plates,' and cited the fact that he himself had earlier, in the course of his comet sweeping, 'independently come across some of these very regions of Herschel, besides others not noted by him.' Finally, he added: 'It has been a long-cherished desire of mine to investigate them further photographically, and I now hope to be able to put this desire into practical reality within the next twelve months.'[27]

It turned out that the very same investigation was being planned by Curtis, who after completing his fellowship in Virginia had gotten on as an assistant at Lick. This raised a delicate diplomatic point between the two observatories. Hale wrote to Curtis's director, W. W. Campbell, that Barnard 'has been counting on attacking this question at once on the completion of the telescope . . . I hardly feel as though I ought to ask him to give up his long cherished plans. Please tell me frankly whether you think that I am wrong in this feeling.'[28] Barnard also wrote to Campell in the interest of protecting his priority. Apologizing for any inconvenience if his work meant scuttling Curtis's plans, he pointed out that the Bruce telescope

> has been on the way for some six years . . . My aim all along has been to use this outfit for two purposes – the photography of the diffused nebulosities of the sky and for the Milky Way, with the hopes of going south for part of the work. It will of course be necessary for me to carry on this work, for the instrument is not only intended for it but is specially constructed for the purpose . . . I trust this will make the matter plain to you. I did not know any one at the L[ick] O[bservatory] was taking up the work.[29]

In the end, Campbell did politely cancel Curtis's plans to photograph the extended nebulosities with the Willard, and assigned him instead to the radial velocity program. In 1906, Curtis went to South America to take charge of the observatory's Chile station, and three years later he returned to Mt. Hamilton to take over work with the Crossley.

6

Though Barnard optimistically hoped that the transparency of the skies at Yerkes would be as good as at Mt. Hamilton, he admitted that the seeing was seldom as good, and there were other disadvantages. 'Of course interruptions from clouds are against the work here, as also the 5° of greater latitude, and the fearful winters.'[30] Even as Barnard was deploring to Campbell the 'fearful winters' at Yerkes, the restless Hale had gone west. He had managed to get himself appointed to a committee to advise Andrew Carnegie, the steel magnate who in 1902 had endowed the Carnegie Institution with a $10 000 000 gift, on how to best use the money for the promotion of astronomy. Hale used his influence to lobby for a high altitude solar observatory and a large reflecting telescope which, he urged, 'would give extraordinary results, and greatly advance our knowledge of the universe.'[31] Soon afterwards, Carnegie made a $10 000 grant to Hale, Campbell and Lewis Boss of the Dudley Observatory in Albany 'to investigate the proposal for a southern and solar observatory.' W. J. Hussey, the man who had finally brought Holden down at Lick and had since moved on to the University of Michigan, was chosen to search for suitable sites, not only in the United States but also in Australia and New Zealand. Hussey visited Mt. Lowe, Wilson's Peak above Pasadena, and several more isolated sites in California, including Mt. Palomar in the far south, which he described as 'a hanging garden above the arid lands.'[32] Hussey finally decided that the most promising was Wilson's Peak, at elevation 5886 feet. It was one of many peaks forming the southern boundary of the Sierra Madres, and had been of interest to astronomers ever since W. H. Pickering's visit there in January 1889 in quest of a site for the 40-inch telescope which at that time had still been a southern California affair.

In June 1903, Hale himself set out for Pasadena, then a town of only about fifteen thousand inhabitants, to explore the astronomical possibilities of Mt. Wilson. He and Campbell rode burros up the mountain along an old Indian trail leading from Sierra Madre up the canyon of Little Anita stream. Once on the peak, they met Hussey, who had set up a 9-inch telescope. Though a thick fog had settled in the lower elevations, the sky on the peak was clear, and Hale was able to observe the sunspots and solar granulation with a clarity rarely experienced at Yerkes. Campbell and Hussey were concerned with the amount of dust stirred up around the mountain, but Hale was in one of his expansive moods and brushed aside their objections. By the time he returned to Chicago, he was resolved to build an observatory on the site. Comparing the night seeing with that on Mt. Hamilton, he effused that 'no observatory site at present known . . . seems to offer advantages equal to those found at Wilson's Peak.'[33]

During Hale's absence in Pasadena, Barnard made a notable discovery with the 40-inch telescope. On June 15, 1903, on turning it toward Saturn, he noticed a 'decided bright spot' in the planet's northern hemisphere. Despite having looked at Saturn frequently over many years, Barnard had never before seen anything of the sort – he was, as we have seen, a severe skeptic of the claims of spots supposedly seen by observers with small telescopes. Soon after Barnard began his observation, the sky clouded over, and he did not have a further chance at the planet until June 24, when he recovered the spot and found that it was preceded by a smaller one, 'separated from the main spot by a small dusky patch.'[34] The rotation period was 10 hr, 39 min. Barnard continued to observe these and various other small spots – fragments of the original disturbance – for the next two months. They signaled one of the major eruptions seen from time to time on the planet, of which the only previous example had been the equatorial white spot observed by Asaph Hall with the 26-inch refractor at Washington in December 1876, though others have occurred since – in 1933, 1960, and 1990.

Hale continued to press the Carnegie Institution for the funds for a high-altitude solar observatory. At first he was turned down. Disappointed and without a clear sense of his future plans, he returned briefly to Yerkes, then decided, in December 1903, to return to California. Though he announced, 'I am practically on an expedition from the Yerkes Observatory,' he was actually paying for the expedition out of his own pocket. Leaving in charge at Yerkes the Dartmouth-trained spectroscopist Edwin Brant Frost, who had joined the staff in 1898, he arrived in Pasadena on December 20, and asked to have sent after him the $61\frac{1}{2}$ foot coelostat.

Following Hale's more optimistic reports from California, Barnard began hoping for the chance to bring the Bruce telescope to Mt. Wilson in order to photograph the southern Milky Way. Frost made an attempt to secure $500 from Harper for the purpose, but without success. At this time, Barnard was still trying to put the finishing touches on the Bruce Photographic Observatory, though the harsh Wisconsin winter made it difficult. Just after Christmas 1903, he informed Hale, 'It is very cold and disagreeable here. It has been as low as 20° below zero. It remains cold all the time with tremendous winds . . . We got the tinner to come and finish the shutters on the Bruce dome, but he has yet to come and do all the soldering. I hope to get him here in the January thaw. It is too cold now. The tin needs painting, but it cannot be painted until the soldering is done. The plastering is all dry but the final coat cannot be put on because of the cold.'[35]

Meanwhile, Hale, the phenomenal promoter, succeeded where Frost had failed. Soon after arriving in Pasadena, he obtained an introduction, through Campbell, to Joseph D. Hooker, a Los Angeles millionaire who had made a fortune in the hardware and steel-pipe business and was interested in astronomy. Hale met Hooker for lunch on February 3, and was able to secure from him a grant of $1000 for the 'Hooker expedition' to bring Barnard and the Bruce telescope out west.

Two weeks later, the Bruce telescope arrived from Warner and Swasey, and Barnard set it up temporarily in the corridor of the Yerkes Observatory's main building. Now that Hooker had promised the funds for Barnard to take the telescope to California, he hesitated to go to all the trouble of setting it up permanently on its brick pier, and confided to Hale that he hoped to have it out west by early spring.

Hale had now begun to worry that it would be too difficult to haul the Bruce telescope up Mt. Wilson, where there was only a trail to the summit, and that Barnard ought to set up instead on the more accessible Mt. Lowe. Mt. Lowe could be reached by a railway that extended from Altadena up Rubio Canyon to Echo Mountain, and then four miles further into the San Gabriel range to Mount Lowe. Astronomical observations had already been carried out at Echo Mountain; Barnard's old friend Lewis Swift had, at the invitation of Thaddeus Lowe, a many-sided entrepreneur whose careers had included Civil War balloonist, manufacturer of dry ice, and California real-estate speculator, brought the 16-inch Clark refractor from the Warner

Observatory to Echo Mountain in 1893, after Warner's fortunes had collapsed in the financial panic of that year. There Swift had picked up three comets – 1895 II, 1896 III, and 1899 I, the last at the age of almost seventy – before his eyesight began to fail. Finally, in 1900, he returned to Marathon, New York, where blind, deaf, and alone he died in 1913.

Ever eager, Barnard wanted to ship the Bruce telescope out to Mt. Lowe as soon as possible, and to arrive himself by early May. But Hale advised him 'to come out here as late as possible, since the weather is frequently not settled until after the middle of May ... I advise you not to come before June 1, if this will serve your purpose.'[36] Hale himself was by this time operating on a slender budget. Ellerman had hurried out to join him in California, and had at once adjusted to Western ways by donning a ten-gallon hat, mountain boots, and a pistol and cartridge belt. Hale was also hoping to bring out young Walter Sydney Adams. Born at Antioch, Syria, the son of Congregationalist missionaries, Adams had studied with Frost at Dartmouth, and had come to Yerkes with him in 1898 in order to work under Hale. Unfortunately, as usual, there were money problems – Hale was in the embarrassing situation of finding that he did not have the funds to pay the astronomers' salaries. He had appealed in vain to Yerkes, who had just left Chicago for good and was sailing to England full of plans to develop the London underground, as well to John D. Rockefeller, the benefactor of the University of Chicago, who had already given it millions but was unwilling to give any more. Hale nevertheless remained in good spirits: 'I feel more and more confident of the outcome of the solar observatory scheme,' he confided to Adams, 'and even if it fails, I intend to have a station where you and Mr. Ellerman and I can work to advantage on Mt. Wilson.'[37]

Barnard, who at the time was in bed with a bad case of bronchitis, as he almost always was by the end of the winter in Williams Bay, tried to commiserate with Hale's financial struggles, assuring him that he ought not to let his own wishes for the Bruce telescope interfere in any way. But he promptly added:

> It is needless to say that I am extremely anxious to get a chance at the southern part of the Milky Way from Mt. Lowe. It would be a wonderful chance for the Bruce, and I believe that we should get something extremely valuable. I looked out the other morning about 3 o'clock while sick, at the Milky Way and that glorious region in Scorpio and Sagittarius was coming up in the low south next, and it made me feel that it would be a great thing to get at it with the Bruce from Mt. Lowe where the altitude would be 9° or more greater, and the atmosphere clear and transparent: with the further fact that one could get at it every night when the moon did not interfere.
>
> I am satisfied that you will do all in your power to get the Bruce out there and I hope you will succeed, but as I say if it interferes with your plans for your other work, why let it go.[38]

Apparently deciding that the chances for the expedition to Mt. Lowe were too uncertain, Barnard decided to go ahead and mount the Bruce telescope on its brick pier at Yerkes, so that he could begin work there. The building which housed it, designed on

Dome of Bruce photographic telescope on grounds of Yerkes Observatory, 1904. Yerkes Observatory

cleancut classical lines, was located about midway between his residence on Lake Geneva and the 40-inch dome, and was capped by a 15-foot dome with an unusually wide shutter (almost 100°) to accomodate the large field of the Bruce telescope. The site enjoyed a clear horizon except, Barnard noted, 'for an unimportant part cut out by the 40-inch dome.'[39]

Unfortunately, though predictably, the telescope proved more difficult to get into working order than expected. Moreover, Barnard, forever unsatisfied, was already angling for a new lens, to be made of the recently developed Jena ultraviolet glass (glass with unusually high transmission of the ultraviolet part of the spectrum). 'I am wild to have a lens of this glass,' he exclaimed to Hale. 'There is no question but this glass will turn things upside down so far as portrait lenses for the nebulae are concerned, and I'd like to get an early chance.'[40] He proposed to Hale substituting a 6-inch lens made of Jena glass for the Bruce's $6\frac{1}{4}$-inch lens, and until the California plans were solidified intended to give the Bruce telescope a thorough working over at Yerkes.

As of early June, he had had only a few nights in which to make tests, and groused to

Campbell: 'When the moon is away, it is cloudy all the time here – since getting the Bruce telescope up, I have only a few exposures from this cause and none of long exposure.'[41] By then, however, Hale was also grousing about the conditions in California. The dust that he had earlier ignored was now causing serious problems. A disappointed Barnard summed up the situation for Campbell: 'It seems the dust of the desert comes up and covers the sky with a whitish haze that would hamper the work badly – so that it has been decided not to take the Bruce telescope there. I am very much disappointed but it is something that could not be helped.'[42] For a moment it seemed that there would be no point in bringing the Bruce telescope out west, then or ever, but the setback proved to be only temporary. By fall the trip was on again. Though the Jena lens had still not arrived – in the end, Barnard was to be disappointed in it anyway[43] – Hale was having better luck at Mt. Wilson, and the financial situation had also improved. Moreover, a new trail had been opened up Mt. Wilson, which meant that Barnard would be able to bring the Bruce telescope there after all. Hoping to avoid another dreadful Midwestern winter, Barnard was more eager than ever to come out. In early November he informed Hale that he had obtained an eight hour exposure on the Pleiades and another of nine hours on the Milky Way in Cassiopeia, but his results were little better than he had obtained with the Willard lens on Mt. Hamilton. 'The night was not transparent,' he explained. 'We have had very little clear weather this fall until the end of last week which has kept up till now and it is singularly perfect – it is Indian summer in appearance – but the sky is thickish at night and on faint nebulosities it is very poor.'[44] A few days later the weather at Williams Bay turned 'cold and raw and cloudy, with a miserable north wind,'[45] and after another attempt at the Pleiades, he announced to Hale: 'I am . . . disappointed and shall not try again. I get the brighter parts [of the exterior nebulosities] all right . . . but many of the details do not come out for want of a more transparent sky. I have decided not to delay on their account – and I may be on time to get at them at Mt. Wilson. I will begin to pack the telescope at once and will ship as soon as I get directions from you.'[46]

7

Hale gave Barnard the go ahead, and December 4, 1904, the Bruce photographic telescope was shipped to California – all except the 10-inch lens, which Barnard planned to carry with him personally as 'hand luggage.' Rhoda, meanwhile, was planning during his absence to travel to England to visit relatives there. However, she was suddenly taken ill, and Barnard had to delay – though not cancel – his own departure. 'I am very anxious to see her off from Chicago,' he told Hale. 'She is not very well – she has trouble with her heart again, a weakness and pain . . . She expects to leave Chicago Jan. 2. We have a friend in N[ew] Y[ork] who will look after her until the evening of the 6th when she will go aboard as the vessel sails at about 7 p.m. I thought I might get away earlier and leave her to look after her own departure – however, under the circumstances I had better not do it.'[47]

Barnard and Rhoda at home at Yerkes Observatory, photographed by Miss Gertrude Bacon, 1900. Letter Archives, British Astronomical Association (courtesy Richard McKim and Council of the British Astronomical Association)

While Barnard was waiting to see Rhoda off, he learned that the Executive Committee of the Carnegie Institution had given Hale a grant of $150 000 for each of the next two years and had authorized his plans for the Mt. Wilson Solar Observatory. Moreover, intending that the Mt. Wilson Observatory should be a separate institution from the University of Chicago, Hale now announced his intention to resign the directorship of the Yerkes Observatory. Already full of anxiety about his wife and about getting himself and the Bruce telescope out to California while the Pleiades were still in reach, Barnard reacted with shock. 'It seems strange,' he told Hale,

> after having labored so long to create a great observatory like this that you should cut free from it where it has attained to the high position it now holds. You however will know best. I suppose there will be no question of the Carnegie for continuing indefinitely the work on Mt. Wilson – if that is assured, I can see a life work before you of tremendous importance and one that will claim all your time. Nevertheless I wish you could have held on somewhat longer to make sure of everything.[48]

Expressing his confidence in Frost who was to take over the directorship at Yerkes – 'I know that [he] will deal justly with everyone here,' Barnard continued; 'there is a satisfaction in this for it would be extremely unsatisfactory to fall under someone with cranky ideas as sometimes occurs' – he expressed his personal regret at Hale's decision to sever his connection with Yerkes. 'I for one shall miss you greatly here for it was always a great pleasure and satisfaction to feel the uplifting and cheering influence that you always exerted over me.'[49] Within two weeks of penning this letter, Barnard saw his wife off – perhaps wondering whether he would ever see her alive again – and left Chicago for the west coast.

Barnard reached San Francisco on January 6, 1905, stayed the next night in San Jose after paying a brief visit to Lick, then took the train to Pasadena. At 1 p.m. on the afternoon of January 10 he started out on the so called 'new' trail for Mt. Wilson. Built by the Pasadena and Mount Wilson Toll Road Company, it was a steep and precipitous path about nine miles long, beginning six miles from Pasadena at the mouth of Eaton's Canyon in Altadena and zigzagging up the rugged southern face of the mountain range. Barnard reached the summit after a climb of five hours, most of it on foot, and noted that 'the 10 in. lenses were fastened on one side of a horse or mule and ballanced on the other side. They arrived at the summit ("Monastery") all safe. The last part of the journey up the mountain was in fog and rain.'[50]

The 'Monastery' was the name Hale had given to the newly erected living quarters on the mountain. The name was inspired by his reading about the monasteries of the Levant, 'perched on rocky promontories, looking out on distant peaks.'[51] Moreover, it expressed a resolve, born of his experience with the personality conflicts among the women of the isolated community at Williams Bay, 'who had no stars to watch, no figures to conjure with, no spectra to measure, and no other absorbing occupations,'[52] that if he ever founded another observatory, the astronomers and their families would not live on the observatory grounds. The building was intended as a place where male astronomers could live while observing on the mountain, and was located on a dramatic ridge which fell away in sheer precipices on three sides, from whence it looked out on the valleys, canyons and distant peaks beyond. The valleys were often filled with 'a vast ocean of clouds,' and the peaks – Old Baldy, San Bernadino, and San Jacinto – were distinctly visible on clear days.

By the time Barnard arrived, Hale, like a magnet, had drawn after him several of the leading members of the Yerkes staff – Adams, Ellerman, and Ritchey. Still another

Mt. Wilson group, 1905. Seated in front is F. Ellerman; from left to right are Construction Superintendent H. L. Miller, C. G. Abbot, G. E. Hale, L. R. Ingersoll, W. S. Adams, E. E. Barnard, and C. Backus. Archives of Mt. Wilson Observatory, Huntington Library

transplant from Yerkes was the Snow solar telescope, a horizontal telescope of coelostat design that had originally been set up by Hale and Ritchey at Yerkes in the autumn of 1903; it was named for its benefactress, Helen Snow. With the creation of the Mt. Wilson Solar Observatory as a separate institution, these astronomers and the Snow telescope would all remain with Hale at Mt. Wilson. The loss of the core of its staff was a serious blow to the Yerkes Observatory, and could not help but lead to somewhat bitter feelings on the part of those who were left behind.

Others who came out to Mt. Wilson on a temporary basis during the time Barnard was there were Charles G. Abbot of the Smithsonian Institution, his assistant, L. G. Ingersoll, and Henry Gale, a former football player who was now a University of Chicago physicist interested in sunspot spectra. Adams later warmly recalled those early days at the Monastery:

> Hale had introduced Abbot to the Oriental stories . . . on the monasteries of the Levant, and our evenings usually started off with a dramatic rendering by Abbot of the tale of the Jew of Constantinople and Solomon's Seal which he knew by heart.

Occasionally the Smithsonian challenged all comers to a game of duplicate whist, but more often the group would gather around the fireplace for discussions of plans of work or of the state of the world in general. Hale's amazing breadth of interests, his great personal charm, and his stories of important figures in science and national and international affairs make these evenings stand out in memory.[53]

8

Barnard decided to erect the Bruce telescope on a small hillock on the trail midway between the Monastery and the observatory shop. By the end of his first month on Mt. Wilson, he had mounted together on the cement pier on this site four telescopes – the 10-inch and $6\frac{1}{4}$ inch photographic telescopes, and a $3\frac{1}{2}$-inch doublet and a lantern lens.

Barnard made his first exposures on the night of January 27, and over the next seven months set himself a feverish pace of work. For many years he had been used to getting along with less than four hours of sleep a night. On Mt. Wilson he often gave up sleep altogether. As Adams later recalled:

> Barnard's hours of work would have horrified any medical man. Sleep he considered a sheer waste of time, and for long intervals would forget it altogether. After observing until midnight, he would drink a large quantity of coffee, work the remainder of the night, develop his photographs, and then join the solar observers at breakfast. The morning he would spend in washing his plates, which was done by successive changes of water, since running water was not yet available. On rare occasions he would take a nap in the afternoon, but usually he would spend the time around his telescope. He liked to sing, although far from gifted in the art, but reserved his singing for times when he was feeling particularly cheerful. Accordingly, when we at the Monastery heard various doleful sounds coming down the slope from the direction of the Bruce telescope, we knew that everything was going well and that the seeing was good.[54]

On first arriving on Mt. Wilson, when he was guiding the Bruce telescope alone in the darkness, Barnard sometimes experienced the terror that he had known at earlier times. The observatory was still in the construction stage, and sometimes he was the only one on the mountain. Moreover, he recalled, 'the Bruce Observatory was separated quite a little distance from the monastery, which was hidden by heavy foliaged spruce trees, so that while observing I was essentially isolated from the rest of the mountain':

> I must confess that at times, especially in the winter months, the loneliness of the night became oppressive, and the dead silence, broken only by the ghastly cry of some stray owl winging its way over the canyon, produced an uncanny terror in me, and I could not avoid the dread feeling that I might be prey any moment to a roving mountain lion. The sides of the observatory were about five feet high, so that it would have been an easy thing for a hungry mountain lion to jump over it and feed upon the astronomer. So lonely was I at first that when I entered the Bruce house

Bruce photographic telescope in its temporary shed on Mt. Wilson, August 1905. Archives of Mt. Wilson Observatory, Huntington Library

and shoved the roof back I locked the door and did not open it again until I was forced to go out.[55]

Fortunately, with the coming of spring, the loneliness and oppressiveness that he experienced during the winter months began to lift. A good part of this had to do with the reawakening of insect life, which 'began its notes in the chaparral':

the dread of the night soon passed away and the door was left open and it became a pleasure to sit and listen to the songs of nature while guiding the telescopes in long exposures, heedless of all beasts of prey. No one knows what a soothing effect these 'noises of the night' have on one's nerves in a lonely position like that on Mount Wilson.[56]

Once spring arrived, and the night terrors departed, Barnard felt an exhilaration that he had not known since the early days on Mt. Hamilton. Though his first two months had been hampered by heavy rains, by May he was getting beautiful photographs and was 'delighted with the conditions on the mountain.'[57] Hale wrote to Frost:

> [Barnard] is getting magnificent photographs that will be a great credit to the Hooker Expedition from the Y[erkes] O[bservatory]. I never saw him in such high spirits as at present, since the summer weather began. Even through the bad weather he was in a far more cheerful state of mind than he would have been under similar circumstances at home. I think his general health has improved considerably.[58]

Adams also noticed the improvement in Barnard's spirits. 'He at once fell in love with the mountain and everything connected with it,' Adams remembered afterwards.

> He was fascinated by the views, studied the birds, measured the growth of yucca stalks, and treasured the sight of a deer. I remember his excitement one winter morning when he came in to breakfast and announced that he had just seen a wildcat walking through the snow outside his bedroom window . . . Barnard's devotion to the mountain may be judged by the fact that during four months of his stay he made but one trip to the valley. This was to Sierra Madre to see a notary and to have his hair cut, after which he turned around and started back up the trail. His health was excellent at this time, and to those who knew him in later years it will be a surprise to learn that he once clambered down the steep walls of the ridge below the Monastery, crossed the deep canyon, and climbed the side of Mount Harvard to the toll road, perhaps as difficult a trip as any around the mountain top.[59]

Only Frost was worried that Barnard might overdo it, and wrote to him: 'It is a pleasure to know that you are having such fine weather for work, but I hope you will not overdo, and you will give up some clear nights when you need sleep.'[60] As usual, Barnard ignored this advice.

Since there was no running water on the summit, water, including that Barnard used for developing his plates, had to be packed up the mountain by burro from Strain's Camp, named after the first pioneer who had settled on the mountain. It was located on the north side of the mountain, where there were springs. The old burro which dutifully performed this laborious task was named Pinto. Barnard became especially fond of him when he discovered that his hair was much finer than a human's – thus exceptionally well suited for making crosswires for a guiding telescope.

Barnard's adventures on Mt. Wilson also included a rather close scrape with a rattlesnake. The floor of the small wooden building which housed the Bruce telescope was about three feet above the ground, and a trap door allowed access to the weights of the driving clock which were suspended by cable. When guiding the telescope, Barnard often opened the trap door and sat on the floor with his legs dangling through the opening. One summer morning at breakfast he casually mentioned that he had been hearing strange rustling sounds for several nights, and that on further investigation he had found that a rattlesnake had been making its home beneath the floor. 'Whether the snake had ever attempted to investigate the intruding legs we never knew,' wrote Adams, 'but Barnard took the episode quite calmly and on the following night [after the snake had been killed] the trap door was open as usual.'[61] Barnard later quipped that 'it must have been a friendly snake and was there for the purpose of warming the observer's feet.'[62]

Generally, Barnard found the skies at Mt. Wilson much more transparent than those at Yerkes – after the drenching rains early in the season, the dust which had been a matter of concern had completely settled out, and Barnard was obtaining magnificently deep plates of the southern Milky Way. In all, he exposed some five hundred plates with his three telescopes and his lantern lens. He photographed the vacant regions of ρ Ophiuchi and θ Ophiuchi, which had been discovered with the Willard lens, on the larger scale of the Bruce telescope, and in the search for diffused nebulosities Barnard exposed a large number of plates on the upper part of the Scorpion.[63] Many of the plates, by chance, recorded the trails of faint asteroids, and on three plates exposed on July 22, 1905, he recorded the trail of an unknown comet. However, he did not notice the trail for over a year. 'The trail was rather conspicuous,' he then wrote, 'and how it was overlooked at Mount Wilson is a mystery, unless it was from the wearied condition of the observer at the time, for a sharp lookout was generally kept for just such objects.'[64] Unfortunately, he had not recorded it on any other plates; a search by E. C. Pickering's assistant, Henrietta Leavitt, of Harvard plates also failed to turn it up, and the comet had to be given up for lost.

There is no question that Barnard would have loved to have stayed on Mt. Wilson with Hale, Adams, Ellerman, and Ritchey, rather than return to the depressing winters of the Midwest. But Frost would never have agreed to give him up. Moreover, he was a skillful practitioner in the old methods of astronomy rather than a pioneer of the new approaches of astrophysics, and it was on the latter that Hale had staked the future of the new observatory. So in mid-September 1905, after a stay of only eight months, Barnard packed up the Bruce telescope and his precious plates and, escorted down the mountain early one morning by Adams, returned east.

9

On returning to Yerkes, Barnard remounted the Bruce telescope in its small dome, where it remained until the telescope was moved and the dome torn down in the early 1960s.[65] He returned to Wisconsin in time for another miserable winter. 'This has been an awful bad winter for observing,' he summed up to Wesley in 1906; 'no clear weather at all hardly.'[66]

The oppression of that winter was lightened somewhat, however, by the arrival of his niece, Mary Rhoda Calvert (one of Ebenezer's daughters), who came from Nashville to join the Barnard household. A self-effacing woman, she devoted herself entirely to

Pipe Nebula, photographed with Bruce photographic telescope at Mt. Wilson; the plate above shows the region of θ Ophiuchi and eastward, and is a 4 hr 45 min exposure taken on June 30, 1905; the plate opposite follows the stem of the 'pipe' through Ophiuchus into Scorpio and is a 4 hr 33 min exposure taken on June 28, 1905. Yerkes Observatory

Barnard and his work, helping him in the office with correspondence and computations, and after his death remaining at Yerkes where she served as chief computer and photographic technician for many years. She was one of the unsung women of astronomy during an era when discrimination against women was flagrant, as demonstrated by the following letter written by Frost to a woman applying for a computer position at the observatory: 'We give preference to men for this work, when we are able to get them, because they can assist in the observing with the telescope, which is too heavy for a woman.'[67]

After the weather began to improve in the spring of 1906, Barnard used the 40-inch for a series of visual observations of Phoebe, the ninth satellite of Saturn which had been discovered photographically by W.H. Pickering in 1898, and of the fifth and sixth satellites of Jupiter, the latter discovered by Perrine at Lick in 1905.

In the summer of 1907, Mars came to a good opposition, but because of its far southerly declination Barnard was unable to get satisfactory observations with the 40-inch. Typical notes from his observing books read: 'The outlines of the "seas" are strong and well defined. The air is like running water in front of the planet.' And: 'Have taken off the diaphragm. Full aperture now – though the image is not so well defined, I can once in a while really see it better than with 15 inches . . . The planet is very low and the air is moving across in waves.'[68]

On the other hand, his observations of the phenomena of Saturn's passages through

Region north of θ Ophiuchi; a 3 hr 30 min exposure by Barnard with the Bruce photographic telescope at Mt. Wilson, May 8, 1905. The S-shaped marking north of θ Ophiuchi is the 'snake nebula,' B72 in Barnard's catalog of these objects, shown also in the previous plate. Yerkes Observatory

the ring-plane that year are classic. There had not been an opportunity to make such observations since 1891, when Barnard had noted the disappearance of the rings with the 36-inch refractor on Mt. Hamilton. The first of three passages of the Earth through the ring-plane occurred in April 1907, but Saturn was then too close to the Sun to be visible. Between then and the end of July, the 'dark' side of the rings was on view from Earth. Barnard was prevented by unusually bad weather from observing with the 40-inch in the early summer, and when he finally got underway, on July 2, he noticed

Barnard, Rhoda, Mary R. Calvert and two of her sisters at Yerkes Observatory in 1909.
Courtesy of Professor Robert T. Lagemann

that 'the entire surface of the ring was easily seen, though the Sun was not then shining on its visible surface. Where it was projected against the sky, the ring appeared as a greyish hazy or nebulous strip.' In addition he described two 'nebulous condensations' of greater brightness on the ring on each side of the planet, which were of a pale grey color.[69] Barnard deduced at once that what he was seeing was not the actual sunlit edge of the ring, but the oblique surface of the ring shining by sunlight 'percolating' through the particles making it up. The condensations, however, continued to perplex him for some time.

The Earth made its second passage through the ring-plane on October 4, 1907. With precautions such as using a hexagonal diaphragm over the object glass to collect the stray light into six rays, leaving clearer sky in between, and an occulter in the eyepiece to block out the planet, he was able to make out 'very feeble traces of the ring.'[70] He continued to observe with the 40-inch refractor throughout the rest of the fall, keeping up observations on the nearly edgewise ring's aspect almost daily and from hour to hour. During this interval the appearance of the ring and the condensations remained more or less unchanged until December 25, when Barnard noted that the thread-like ring appeared much thinner than it had looked two weeks earlier. By then southern Wisconsin was experiencing its usual harsh winter weather. Barnard, though he had 'not been well' for some time, as he told Wesley,[71] had not refrained from his usual demanding observing schedule, but now he became seriously ill just as these phenomena were passing through another critical phase – the Earth was due to make its final passage through the ring-plane on January 7, 1908. On January 2, he got up from his sick bed long enough to note that the ring was still visible in the 40-inch telescope, though 'very thin' and with a satellite at each end. 'Without occultation it was almost impossible to see any trace of the ring on the sky,' he noted 'The condensations were feebly seen as slightly brighter parts of the ring.' On January 5, the ring was much fainter, and no trace of it could be seen without the occulter. On January 6, with poor seeing, it was completely invisible, and Barnard concluded that the Earth must have passed through the ring-plane that night.[72]

In sending these critical observations to T. Lewis of the Greenwich Observatory, Barnard wrote: 'I have been sick in bed . . . and am up for a bit to day to get this off but shall have to go back to bed. I managed to get the observations of Saturn by taking big risks and wrapping up good to go into [the] big dome.'[73]

Having obtained the observations he sought, Barnard now collapsed into his sick bed, though he continued to work, whenever he felt strong enough, on the report of his observations for the *Monthly Notices of the Royal Astronomical Society* (the archives of the RAS show that he sent constant revisions and corrections which must have come close to driving the poor secretary, W. H. Wesley, mad). Finally he was able to clear up the mystery of the so-called condensations.[74] On comparing their positions with his earlier measures of the ring system, he discovered that they lined up exactly with the crape ring and the Cassini division, from which he hazarded the guess – correct, as we now know – that the Cassini division is not entirely devoid of particles. 'If . . . the Cassini division were filled with particles as closely clustered as they are in the crape ring,' he suggested,

> a satisfactory explanation of the condensations would be that they were simply due to the sunlight shining through and illuminating the particles in the crape ring for the inner condensations, and a similar effect of the Sun shining through the Cassini division and illuminating the particles in it would produce the outer condensations. The fact that the inner and outer condensations were essentially of the same intensity would require that the particles should be as closely clustered in the Cassini division as in the crape ring.[75]

This was a marvelous deduction, and has now been completely verified by the Voyager spacecraft images.

1   E. E. Barnard, 'The Total Eclipse of the Sun in Sumatra,' *PA*, **9** (1901), 528–44:534–5

2   J. M. Bacon, 'Wadesborough, North Carolina,' in E. Walter Maunder, ed., *The Total Solar Eclipse 1900: Report of the Expeditions Organized by the British Astronomical Association to Observe the Total Solar Eclipse of 1900, May 28*, (London, Knowledge Office, 1901), pp. 16–17

3   R. H. Tucker to his mother, February 17, 1901; SLO

4   EEB to GD, March 26, 1900; BL
5   EEB, undated notes; VUA
6   ibid.
7   ibid.
8   ibid.
9   Barnard, 'The Total Eclipse,' p. 529
10   ibid., p. 543
11   EEB, Sumatra notebooks in possession of Robert Lagemann
12   ibid.
13   Barnard, 'The Total Eclipse,' p. 540
14   ibid.
15   ibid.
16   S. A. Mitchell, 'With Barnard at Yerkes Observatory and at the Sumatra Eclipse,'
     JTAS, 3:1 (1928), 27
17   ibid., p. 26
18   ibid., p. 27
19   EEB to W. H. Wesley, July 29, 1901; RAS
20   EEB, unpublished lecture notes; VUA
21   E. E. Barnard, 'Peculiarity of Focal Observations of the Planetary Nebulae and Visual
     Observations of Nova Persei with the Forty-Inch Yerkes Telescope,' *Ap J*, 14 (1901),
     151–7
22   The discovery was actually made by Antoniadi, who complained to Wesley of 'M.
     Flammarion wedding his name to mine . . . merely as Director of the Juvisy
     Observatory.' EMA to W. H. Wesley, November 15, 1901; RAS
23   See James E. Felten, 'Light Echoes of Nova Persei 1901,' *Sky and Telescope*, 81 (1991),
     153–7
24   Isaac Roberts, 'Herschel's Nebulous Regions,' *Ap J*, 17 (1903), 72–6
25   ibid.; Roberts's description of Herschel's field no. 25
26   E. E. Barnard, 'Diffused Nebulosities in the Heavens,' *Ap J*, 17 (1903), 77–80:78
27   ibid., 78. For an insightful discussion of Roberts vs. Barnard on this question, see
     David Malin, 'In the Shadow of the Horsehead,' *Sky and Telescope*, 74 (1987), 253–7
28   GEH to WWC, April 1, 1903; SLO
29   EEB to WWC, April 15, 1903; SLO
30   ibid.
31   Quoted in Helen Wright, *Explorer of the Universe: A Biography of George Ellery Hale*
     (New York, E. P. Dutton & Co., 1966), p. 161
32   Quoted in ibid., p. 165
33   GEH to J. S. Billings, July 6, 1903; HHL
34   E. E. Barnard, 'White Spot on Saturn,' *AJ*, 23 (1903), 143–4:143
35   EEB to GEH, December 29, 1903; HHL
36   GEH to EEB, March 9, 1904; HHL
37   GEH to WSA, March 7, 1904; HHL
38   EEB to GEH, March 23, 1904; HHL
39   E. E. Barnard, 'The Bruce Photographic Telescope of the Yerkes Observatory,' *Ap J*,
     21 (1905), 35–48: 45
40   EEB to GEH, March 24, 1904; HHL

41   EEB to WWC, June 4, 1904; SLO
42   ibid.
43   EBF to GEH, March 26, 1906; HHL
44   EEB to GEH, November 4, 1904; HHL
45   EEB to GEH, November 10, 1904; HHL
46   EEB to GEH, November 24, 1904; HHL
47   EEB to GEH, December 22, 1904; HHL
48   EEB to GEH, December 29, 1904; HHL
49   ibid.
50   EEB, observing notebook; YOA
51   Wright, *Explorer* p. 188
52   ibid., p. 123
53   W. S. Adams, 'Early Days at Mt. Wilson–II' *PA*, **58** (1950), 97–115: 99–100
54   ibid., pp. 97–8
55   EEB, unpublished MS draft; YOA
56   ibid.
57   GEH to WWC, May 15, 1905; SLO
58   GEH to EBF, May 16, 1905; HHL
59   Adams, 'Early Days,' p. 97
60   EBF to EEB, July 27, 1905; VUA
61   Adams, 'Early Days,' p. 98
62   Philip Fox, 'Edward Emerson Barnard,' *PA*, **31** (1923), 195–200:199
63   E. E. Barnard, *Atlas of Selected Regions of the Milky Way*, E. B. Frost and M. R.
     Calvert, eds. (Washington, DC, Carnegie Institution, 1927), vol. 1, p. 7
64   E. E. Barnard, 'Photographic observations of an unknown comet on 1905 July 22 (1905
     f),' *AN*, **174** (1907), 3–8:4
65   The Bruce photographic telescope is now at the Athens Observatory in Greece.
66   EEB to W. H. Wesley, April 4, 1906; RAS
67   EBF to Mary Lilly, February 19, 1912; YOA
68   EEB, observing notebook; YOA
69   E. E. Barnard, 'Observations of Saturn's Ring at the time of its Disappearance in 1907,
     made with the 40-in. refractor of the Yerkes Observatory,' *MNRAS*, **68** (1908),
     346–60:346–7
70   ibid., p. 350
71   EEB to W. H. Wesley, November 30, 1907; RAS
72   E. E. Barnard, 'Additional Observations of the Disappearances and Reappearances of
     the Rings of Saturn in 1907–08, made with the 40-in. refractor of the Yerkes
     Observatory,' *MNRAS*, **68** (1908), 360–6:363–4
73   EEB to T. Lewis, January 11, 1908; RAS
74   Earlier astronomers had noted them, but had given incorrect explanations. Thus
     William Cranch Bond of Harvard had surmised that they were the edges of rings A and
     B seen through the Cassini division. Others had supposed that they were clumps of
     material in the rings, but Barnard had disproved this on January 6 by showing that they
     disappeared when the ring was exactly edgewise.
75   Barnard, 'Additional Observations,' pp. 365–6

20

The comet and Milky Way photographs

After Keeler died in 1900, W. W. Campbell became director of the Lick Observatory. One of his first actions as director was to find out what progress Barnard had made toward the long dormant project of publishing his Milky Way and comet photographs which he had taken at Mt. Hamilton in the 1890s. Though Barnard was then busy preparing for the Sumatra expedition, he had Hale do some experiments for him back East, which he hoped would lead to adequate reproductions of his photographs.[1] On returning from Sumatra, he was at first too discouraged to do anything further about the project. Nevertheless, Campbell queried him again in July 1902:

> I am anxious to help you in every possible way in the publication of your excellent photographs. If your funds are insufficient . . . it will give me great pleasure to make efforts for the securing of the funds. I really am very anxious that your volume should be issued as soon as possible. The photographs, in my opinion, are extremely valuable; and, in justice to yourself and to the Lick Observatory, – and to the whole profession, – the issue should take place as promptly as possible. I hope you will feel that I am actuated only by the most appreciative feelings for your successful work on the photographs.[2]

This generous and diplomatically phrased offer of help succeeded in stirring Barnard back to action. Though Hale's experiments had come to naught, Barnard decided to put his negatives in the hands of a Chicago firm specializing in the halftone process.[3] Briefly he was encouraged, and for a while in the winter of 1902–3 it looked as if the project would soon be moving forward again. Hale, too, was doing everything he could to encourage Barnard, and assured Campbell:

> I shall do all I can to induce Barnard to rush the reproduction of his Lick photographs. In fact, I have spoken to him periodically on this subject ever since he dropped the work after his exceedingly disappointing experience in Chicago. He now has strong hopes of successful results with the half tone process . . . I fully agree with you that further delay in publication would be extremely unfortunate.[4]

However, the perfectionist was almost impossible to satisfy. He decided that the halftone process would never do, and went back to the collotype and photogravure processes again. Still he was unable to get satisfactory results, and early in 1907, with the project continuing to be stalled after more than a decade of effort, he told Campbell:

> You have doubtless had experiences in trying to get work of this kind done in a
> satisfactory manner, and can in part understand the sad disappointments that come
> from such efforts. To me the whole thing has been a most bitter disappointment. It
> has caused many an illness from worry over it.[5]

He promised Campbell that he would either have the work completed 'before the end of
the present year,' or 'every cent of the money, dollar for dollar . . . will be returned to
California, and I shall once more be free from worry in that direction.'[6]

Campbell, whose own frustration with the endless delays was by now mounting,
made every effort to shore up Barnard's flagging spirits: 'I am glad to have your
statement of the condition of the reproductions and of further experiments that are in
progress. It has always been a great personal pleasure, and at the same time I feel it to be
my genuine duty as Director, to promote work on your volume in every possible
manner . . . You will understand from my several letters that I have not meant to urge
haste at the expense of quality.' Presumably to underscore the point about his patience,
he quoted from his own letter of July 1902, 'I am anxious to help you in every possible
way,' etc.[7] Barnard replied somewhat defensively: 'I fully appreciate the interest you
have shown in the matter. While I clearly understand that one can not hope for
perfection in these reproductions I was fully justified in stopping the work where I did
because of the introduction of errors that would have made the work unreliable.'[8]

A few months later, no further ahead than he had been and his spirits drooping
because of the illness which had afflicted him all fall and had worsened with the winter,
Barnard worried that he might die without having the matter resolved. On the day after
Christmas 1907 – as he was in the midst of his crucial series of observations of Saturn's
edgewise rings – he wrote almost hysterically to Schaeberle, who in 1898 had left Mt.
Hamilton in disappointment after being passed over for the directorship after Holden's
departure and was now back at Ann Arbor. Schaeberle, of course, had been one of the
earliest supporters of Barnard's plans to publish his Milky Way and comet
photographs, and Barnard wanted to explain to his former colleague why the
publication had not yet appeared. 'It has been a heart rending affair,' he wrote, 'and I
have finally given it up. The pictures already made are many of them full of errors and I
don't propose to do anything with them':

> Life is short and uncertain, and I cant stand the strain any longer, I had hoped this
> year to make another effort to get the work out but disappointment again came to
> me. I would rather die than to have a faulty work go out. I have therefore decided to
> give up any more efforts, and to put the money out of my hands, as I do not want to
> die with anything in my possession that does not belong to me . . .
>
>   To get this dreadful responsibility off my hands, I have decided to place the
> entire sum $2225 that was given me, into the possession of the Lick Observatory as a
> fund for some purpose or other. I have already spent in the work, over one thousand
> dollars – this shall be my loss . . . The whole thing has been a very sad affair to me
> and has caused me many heart aches and I want to straighten it out before I die. I
> had set my heart on these pictures, but I would infinitely rather lose personally what
> I have spent on them than to have the work go out full of errors.[9]

Soon afterwards, Barnard sent Campbell a check payable to the Lick Observatory for $2225, which was the entire amount he had collected in 1895, together with the printed sheets of the Milky Way reproductions that were already completed. Barnard would have preferred to return the money directly to the contributors, but this was impossible since many of them were now dead. He wrote to Campbell a complete history of the sorry affair, and concluded:

> My sole desire in the matter has been to put in the hands of astronomers a trustworthy set of reproductions of these Milky Way and comet photographs. I have no desire to get out a volume simply for the sake of the volume. Recognizing at last the hopelessness of bringing out these pictures to my satisfaction, and feeling of late the uncertainty of life, I have finally decided, while it is within my power (but not without much pain and disappointment) to close up the matter and to abandon the work to its fate.[10]

Barnard also sent a set of his reproductions to Wesley at the Royal Astronomical Society, describing them bitterly as 'the wreck of the Milky Way and comet volume.'[11]

The subject was closed as far as Barnard was concerned. However, Campbell, who just then returned from an eclipse expedition to Flint Island in the central Pacific Ocean, had no intention of allowing the matter to drop. The Lick director pointedly asked Barnard to indicate which of the reproductions sent to him were not satisfactory.[12] He was aware that Barnard had been ill, but he was also suspicious that Barnard's lapse of interest was partly a result of his having received a grant from Carnegie to publish his photographs taken with the Bruce telescope at Mt. Wilson. As he confided to University of California president Benjamin Ide Wheeler:

> For many years following 1895, Mr. Barnard's ambition appears to have centered in securing a considerably larger photographic telescope than the one he used here, – a 10-inch, whereas ours is a 6-inch. With this instrument he obtained another series of Milky Way photographs on Mount Wilson, California . . . [and] the Carnegie Institution was considering the reproduction of this later series . . . Mr. Barnard has not offered further information as to a reproduction of his later series . . . I shall be surprised if my duty in this direction does not lead to bad feeling in certain quarters, but I shall get full information before expressing my sentiments and taking action.[13]

Barnard responded to Campbell's further queries with annoyance. 'I thought my letter indicated that the thing is closed for me. I am sick of it.'[14] However, he gave Campbell the information requested, and thus Campbell succeeded in keeping the project, however precariously, alive. Though in late February 1908 Barnard was complaining to Wesley that he still did not yet have his strength back, he was feeling somewhat better, and wrote optimistically, 'I suppose I will be all right by the beginning of the spring and summer.'[15] Indeed, he was feeling well enough to resume his work at the telescope, which he had given up ever since he had completed his observations of Saturn's edgewise rings in early January. Now he was no longer talking about dying, and he was again looking at the problem of the Milky Way and comet photographs 'in a more hopeful way.'[16] Campbell suggested that he ought to carefully consider the implications of 'the prior, or even simultaneous, reproduction of a later series of photographs,' and added that 'I do not agree that the returning of the money to the donors, even if this impossibility could be carried out, would be doing justice to the donors, to yourself, or to the Lick Observatory.'[17]

Barnard did reconsider. In another three months, he was proceeding with further experiments in reproducing his plates with a firm in Boston, and was negotiating once more with A. B. Brunk of the Chicago Photogravure Company, the same firm with which he had broken off the contract more than ten years before.[18] Eventually he decided that Brunk was the best man for the job after all, but still he continued to drag his feet, prompting another round of irritated letters from Campbell. 'It is somewhat over a year since you wrote me concerning the Chicago proofs of the Milky Way photographs that "they are quite as good as one can expect," ' Campbell scored him in August 1909. 'In this matter which concerns us it can scarcely be denied that I have been reasonably patient.'[19]

The work continued for another two years. Finally Barnard pronounced himself satisfied with the one hundred and nine plates to be included in the volume, and paid tribute to the efforts of Brunk and the Chicago Photogravure Company. '[They] have done everything in their power to get the best possible results,' he told Campbell. 'They have disregarded expense. The manager, Mr. A. B. Brunk, has given his personal attention and a remarkable devotion to the faithful reproduction of the pictures':

> I think you will be pleased with the results – especially with the Milky Way pictures. The comet pictures were more difficult. The scientific accuracy of these (comet plates) has, however, been retained at the expense of looks in some cases . . .[20]

On finally receiving the proofs, Campbell was as relieved as Barnard to see an end to the project, and explained the unprecedented delays in completing it to the Comptroller of the University of California:

> Astronomical subjects are extremely difficult to reproduce satisfactorily, and Barnard's temperament is such that discouragements led him to put the subject entirely aside for two or three years at a time. Shortly after I became Director I insisted that he go on with the work, partly because the photographs were the first great successes in their lines, and partly to show our good faith with private contributors. Barnard's entire freedom from business ability has made the administrative questions difficult, but the scientific merits of the subject have been sufficient to preserve my patience.[21]

Barnard also provided an introduction and descriptions to go along with the plates, to which Campbell suggested a few editorial changes. He felt that Barnard ought to say something about Holden's role in acquiring the telescope – 'I feel sure that you will want always to remember having done him full justice' – and he also thought that Barnard would want to give credit to Colonel Crocker who had paid for the lens and the equatorial mounting.[22] Barnard agreed to make these changes, and also, regretting that no reference had been made in the originally prepared text to the work done by him in the summer of 1889, when he obtained his first photographs of the Milky Way, added such a reference to the description of one of the plates. He explained to Campbell that 'I have avoided any reference to the cause of my not continuing the work with the lens when first started, but some reference to the fact that the lens then passed out of my hands for some time is necessary. It has been put, however, in a form that can reflect on no one.'[23] The long and bitter struggle with Holden was thus sanitized.

After a delay of nearly two decades, the book was finally ready for the press. In all, a thousand copies of volume 11 of the *Publications of the Lick Observatory*, 'Photographs of the Milky Way and Comets,' were printed (another two hundred copies of each plate were left over in California, which Barnard wanted sent to him by freight).[24] Among those to whom Barnard sent copies were many of the great astronomers and institutions of the day. In addition, as an afterthought, he sent one to his brother, Charles, with whom he had hardly had any communication since the early days in Nashville. Charles's reply is touching:

> Just received your Book all O.K. I think it is just fine and it is highlay apreciated by me and I thank you many times for it and Shall keep is a allwas as a Prize.[25]

Barnard, of course, was not finished with the Milky Way – by the time 'Photographs of the Milky Way and Comets' appeared, in September 1914, he was already working on reproducing the photographs taken with the Bruce telescope for what eventually became his *Atlas of Selected Regions of the Milky Way*. But his photographs with the Bruce, wonderful as they are, do not decrease the significance of those taken with the Willard lens, whose value, Campbell justly wrote, 'from my point of view, is immensely increased by the fact that they represent the first great pioneer successes.'[26]

1 WWC to EEB, February 1, 1901; EEB to WWC, February 6, 1901; SLO
2 WWC to EEB, July 10, 1902; SLO
3 EEB to WWC, February 26, 1903; SLO
4 GEH to WWC, April 15, 1903; SLO
5 EEB to WWC, January 8, 1907; SLO
6 ibid.
7 WWC to EEB, January 19, 1907; SLO
8 EEB to WWC, February 8, 1907; SLO
9 EEB to JMS, December 26, 1907; SLO
10 EEB to WWC January 27, 1908; SLO
11 EEB to W. H. Wesley, January 28, 1908; RAS
12 WWC to EEB, February 6, 1908; SLO
13 WWC to Benjamin Ide Wheeler, February 8, 1908; SLO
14 EEB to WWC, February 11, 1908; SLO
15 EEB to W. H. Wesley, February 25, 1908; RAS
16 EEB to WWC, March 1, 1908; WWC to EEB., March 6, 1908; SLO
17 WWC to EEB, March 6, 1908; SLO
18 EEB to WWC, June 3, 1908; SLO

19   WWC to EEB, August 19, 1909; SLO
20   EEB to WWC, May 15, 1911; SLO
21   WWC to Ralph P. Merritt, Comptroller of the University of California, July 26, 1912; SLO
22   WWC to EEB, March 18, 1913; SLO
23   EEB to WWC, February 7, 1914; SLO
24   EEB to RGA, September 1, 1914; SLO
25   C. H. Barnard to EEB, September 16, 1914; VUA
26   WWC to EEB, May 23, 1911; SLO

21

Comet tales

1

Partly from early associations, comets had for Barnard an irresistible allure. He had first established his reputation as a prolific discoverer of comets, and even after going on to a 'higher class of work,' he continued to be fascinated with their behavior on the photographic plate – especially by what he called the 'freaks,' that is, the unexpected caprices, of their tails.

After bringing the Bruce telescope back from Mt. Wilson and setting it up permanently at Yerkes Observatory in late 1905, Barnard could almost always be found in the dome whenever there was a comet in the sky and the Moon out of it. 'Nothing else so aroused Mr. Barnard's enthusiasm or gave him more pleasure,' wrote his niece Mary Calvert, 'than the announcement of the discovery of a new comet. If it was a bright one, or if it showed any interesting or unusual features, he would make a photograph of it every night. Some very active comets . . . he photographed from hour to hour to try to get a record of the rapid changes that took place in them.'[1] The next morning he would always pore eagerly over the freshly developed plates, and Calvert remembered that 'he would lay them down with a deep sigh, exclaiming, "O dear, I wish I knew something about comets!" '[2]

Yet though still mysterious, comets had begun to yield a few of their secrets. The orbital motions of comets had occupied astronomers ever since Edmund Halley had drawn up a list of twenty-four comets that had appeared between 1337 and 1698, and had laboriously computed their orbits around the Sun; from these computations he discerned that the comets of 1531, 1607, and 1682 followed nearly the same orbit – thus that it was most likely the same comet returning with a period of seventy-six years. He predicted that it would return again in 1758, and so it did. At its next return, in 1835, it was favorably placed for observation, and telescopic observers described 'an outrush of luminous matter, resembling in shape a partially opened fan,'[3] which seemed to be shot forward from the nucleus in the direction of the Sun then bent sharply backward as if by a strong repulsive force. Among those who observed these remarkable phenomena was the German astronomer Friedrich Wilhelm Bessel, who compared the appearance of the comet at this time to that of a blazing rocket. Bessel, developing the ideas of another German, Heinrich Olbers, first worked out in mathematical detail what became known as the 'fountain model' – in the same way that water from a fountain curves gracefully back to the ground owing to gravity, the particles emanating from a comet's nucleus are swept back along parabola-like paths by a repulsive force from the Sun, which Bessel estimated to be about twice as powerful as the attractive force of gravity. Bessel believed that electrical charges on the particles in the comet were of opposite charge to the charges on the Sun, and declared that the 'emission of the tail was a purely electrical phenomenon.'[4]

The next influential student of comets was the Russian, Feodor A. Bredichin, who noted that tails come in several varieties, and in the 1870s went so far as to attempt to calculate from Bessel's fountain theory the repulsive force exerted in the formation of each type. What he called Type I tails were long, straight, and faintly blue and pointed almost perfectly straight back from the Sun. His Type III tails were short, curved and faintly yellowish, while his Type II tails were intermediate between the other two (astronomers today still use his categories, though they generally lump Bredichin's Type II and III tails together as Type II). Bredichin pointed out that on being ejected from a comet and driven by a repulsive force away from the Sun, a particle would move in a hyperbolic path, thus falling behind the comet in its orbital motion. The greater the repulsive force, the straighter its path away from the comet; a lesser force would cause the particle to lag more quickly behind, so as to give a short but small tail. Bredichin surmised that a comet's multiple tails were composed of different kinds of matter. As Agnes Clerke summed up his reasoning:

> Hydrogen, as the lightest known element – that is, the least under the influence of gravity – was naturally selected as that which yielded most readily to the counter-persuasions of electricity. Hydro-carbons had been shown by the spectroscope to be present in comets, and were fitted by their specific weight, as compared with that of hydrogen, to form tails of the second type; while the atoms of iron were just heavy enough to compose those of the third, and, from the plentifulness of their presence in meteorites, might be presumed to enter, in no inconsiderable proportion, into the mass of comets.[5]

Bredichin's theories continued to dominate thinking about comets right up until his death in 1904. However, there were modifications. For one thing, light pressure, instead of some kind of electrical force, was increasingly invoked as the repulsive force that drove comet tails backward from the Sun. Predicted by James Clerk Maxwell from his theory of electromagnetism, light pressure was first demonstrated in the laboratory in 1900 by Peter N. Lebedev of Russia and G. F. Hull and E. F. Nichols in the United States, and Barnard later explained how it was supposed to work in the case of comet tails:

> If the particle is so small that its surface is relatively large compared with its mass, then the sunlight pressing against it will overcome the pull of gravity and drive the particles away into space. The smaller the particles the greater will be their velocity. From this it will be seen that when the conditions are right the smaller particles will be sifted out from the more massive ones and will flow away from the comet, thus forming the tail, which will point away from the Sun.[6]

Nevertheless, as Barnard fully realized, light pressure alone was incapable of accounting for all of the phenomena of tails, and he added: 'The abrupt change of direction of the tail and the sudden transformation from a straight to a violently curved tail demand some other explanation . . . Doubtless the tail is produced by light pressure, but some other agency is responsible for its distortions, deflections, etc.'[7]

The first comet Barnard studied in detail from Yerkes was Comet Borrelly (1903 IV), which, as the Bruce photographic telescopic was not yet finished, he photographed with the small 'magic lantern' lens attached to the 12-inch refractor. In July, when the comet made its closest approach to both the Earth and Sun, it appeared as a hazy star between 2nd and 3rd magnitudes, with only feeble traces of a tail, but the lantern-lens photographs showed that the tail actually extended for some 17°. On July 24, one of his plates recorded that 'some two or three degrees back of the head the tail was apparently broken off and the outer portion shifted bodily toward the direction from which the comet was receding and in a line parallel with the remaining portion of the tail – as if some force had suddenly broken off a large section of the tail and pushed it to one side of its former position.'[8] Another exposure was begun at the end of Barnard's by Yerkes assistant R. J. Wallace, and confirmed the apparent separation, 'as if the fragment were being left behind by the comet in its flight through space.'[9]

From the comet's orbital elements, Barnard calculated that the fragment of tail was separating from the head at a rate of 29 miles/sec. As the comet was approaching the Sun at 22 miles/sec., the actual velocity of motion of particles of which the tail was composed away from the Sun was 7 miles/sec. Drawing on the fountain model, Barnard concluded that 'the appearance in question would seem to be due to a sudden change in the direction of emission of the particles . . . This theory may be illustrated by a jet of steam issuing from a nozzle. If the nozzle is suddenly slightly changed in direction, the old stream will recede for a few moments before it dissipates, and a new one will follow it closely in a slightly different direction.'[10]

The first bright comets photographed with the Bruce telescope were Giacobini's Comet (1906 I), which began to show strong activity as it approached perihelion in January 1906 but was unfavorably positioned low in the east before dawn, and Daniel's Comet (1907 IV). The latter, in particular, was expected to be a good performer. As Barnard noted, 'it had apparently every advantage in its favor – a relatively near approach to the Sun and a naked eye visibility that extended over at least two months.'[11] Barnard photographed it extensively; however, 'the changes in the tail – which was long and splendid – were relatively few, and when the comet finally reached perihelion it had settled down to a steady flow of tail producing particles with essentially no changes. Altogether it seemed to be a well-behaved body, without any desire for spectacular display.'[12]

2

By contrast to the tame display of Comet Daniel, Comet Morehouse (1908 III), though visually a relatively inconspicuous object, would supply him with 'all that was wanting in the comet of last year in the way of new phenomena,' and unexpectedly become 'the most bizarre comet that we have had to deal with since photography began to register the freaks of the tails of comets.'[13]

Its discoverer, Daniel Walter Morehouse, was born in Mankato, Minnesota, in 1876, and had since 1900 been professor of astronomy at Drake University in Des Moines, Iowa (and later became its president). The years 1908–9 he spent at Yerkes Observatory, studying and photographing comets with Barnard. Morehouse relieved Barnard at the Bruce telescope whenever Barnard had work to do on the 40-inch, and also continued his series of observations of comets and auroras during Barnard's occasional absences from the observatory. One night he came to the dome at about midnight and started an exposure with the 10-inch Bruce refractor. Barnard, meanwhile, had stepped outside, and at once noticed the gegenschein, which was then unusually bright. 'He stood there for a half-hour describing it to me,' Morehouse recalled. 'What torture, with a desire generated through many years of search for this unusual phenomenon, to keep my eyes glued to the telescope. What agony, I say, that I must keep to my post and listen to the words of this master as he described the unusually elusive phenomenon.'[14]

Morehouse discovered his only comet with the Bruce photographic telescope on September 2, 1908. At the time of its discovery it was a 9th magnitude object near the Camelopardalis–Cassiopeia border; it was picked up independently the next night by Borrelly at Marseilles. Visually, the comet was faint, with little or no tail. Morehouse's discovery plate showed, however, that the comet was really a very active one. It showed a long and conspicuous tail, and Barnard made every effort to follow its development photographically.

During September, as the comet's distance from both the Earth and Sun was decreasing steadily, the atmosphere at Yerkes was filled with dense haze and smoke, so that for some days the Sun itself was scarcely able to shine through with a feeble yellow light, and at night only the brighter stars could be seen. Under these conditions, Barnard found it difficult to guide on the comet because of its faintness. Nevertheless, when he developed his plates he found that the effort had been worthwhile. On one of them, exposed through the thick sky on September 16, he found the tail violently curved at a large angle to its normal direction (away from the Sun). Even more remarkable was the series of plates he obtained on the night of September 30. His first exposure on the comet showed a rather small head, from which a thick tail 'ran out in a straggling manner with a faint sheeting of matter having a sharp edge on the south side.'[15] In the next exposure, the whole tail had moved out bodily from the head. A violent change was taking place in the comet throughout the night:

> The tail [became] cyclonic in form and was attached to the head (which was small and almost star-like) by a very slender curved tapering neck which was all but detached, and must within an hour or two have become completely separated from the head. That this separation took place was evident from the photographs of the succeeding night (October 1), where the wreck of the tail was shown as a very large long mass some two degrees from the comet, and apparently attached to it by one or two slender threads or streams of matter. This mass which had formed the tail on

September 30, was visible on the photographs for several days, farther away and fainter until it finally dissipated into space.[16]

Barnard suspected another disturbance on October 6, when the tail presented a similar appearance – 'a straight narrow tail for one degree . . . joined on to great masses like clouds of smoke.' Unfortunately, Barnard's observations that night were hindered by moonlight. On October 15, Barnard exposed two plates, then immediately developed them to see if the comet was in any way abnormal. As soon as he saw the remarkable condition of the tail, he commenced another set of exposures. The comet's high northern declination, $+76°$, permitted exposures to be made throughout the night, and Barnard continued the series from dusk to dawn, changing the plates at hourly intervals. The early photographs showed that the tail was made up, for the first half degree or so, of a slender stream joined to 'great cloudlike masses.' Successive photographs showed that, as before, these masses were hurtling away from the comet, and seemed to have a lateral motion in the direction of the comet's flight, but apparently moving faster than the comet itself.

Afterwards Barnard came up with the idea of combining several pairs of plates for viewing with a stereoscope. The results of this experiment were striking, giving a picture of the comet 'in beautiful relief suspended in space, as we know it is in reality, with the various parts of the tail in individual perspective. There is a wonderful effect of reality in these pictures, and the filmy, breath-like character of the comet is shown as no single picture can even hope to show it.'[17] He admitted that these pictures had to be viewed with caution; some of the perspective effects were undoubtedly 'delusive' – what he called pseudo-stereoscopic effects due to changes in the forms of the masses. Nevertheless, they were in his opinion of great value in understanding the features of comets, if only because no other method could even begin to suggest what a comet really looked like in space. In the case of his combinations of pictures of Morehouse's Comet,

> They clearly show the gradual transformation of the near end of the old tail. At first it was twisted or cyclonic in form, as if it had received some twisting motion when it left the head. It slowly formed into a thickish fragment of a ring, from all parts of which streams of particles swept back to form the old tail, giving it the appearance of part of an open sack, or a partly opened scroll, with irregular sides. Without the aid of the stereoscope one would never have guessed the real form of the tail. It seems that immediately after the separation of the tail from the head a new and slender tail was shot out from the head at a different angle from that of the receding one. In the stereoscope this new tail is seen to pass behind the old one – away from us and toward the background of stars. It was moving out much faster than the rear portion of the old tail – a peculiarity that seems to be always present in the general process of forming a new tail . . .[18]

In all, Barnard obtained 350 photographs of the comet, and was able to record further disturbances in the tail on or about October 30, November 15 and December 11. He later commented that 'no other comet has approached [Comet Morehouse] in interest and importance since photography has been applied to these bodies . . . The photographs showed the extraordinary rapidity with which a comet can alter its appearance when in one of its changing moods.'[19]

The sudden freaks in the comet's tail were especially puzzling to Barnard. 'I have tried to think up some cause for these extraordinary conditions,' he wrote to W. J. Humphreys at the US Naval Observatory, 'which apparently defy our ordinary means of reasoning.'[20] He continued to study his photographs, and finally the solution dawned on him. Ever since coming to Yerkes, he had kept careful records of the auroral displays, which were often brilliant from that comparatively northern latitude. In late 1908, the Sun was approaching sunspot maximum; auroras were frequent, and Barnard noted that there were displays on most of the dates when he observed disturbances in Comet Morehouse's tail. On September 30, the night of the greatest disturbance, he had recorded in his observing book, for example: 'Cloudy until 10h. The clouds, disappearing, revealed a strong auroral glow . . . The sky was very luminous all night – not haze, but like moonlight. I could read my watch in the Bruce dome at 14h 0m without artificial light.'[21] From such correlations, Barnard came to suspect that the changes in the comet's tail were due to the effects of solar disturbances similar to those that produce auroras and magnetic storms, and asked Humphreys, an expert on geomagnetic phenomena, 'if you can conveniently tell me if there were unusual magnetic disturbances, say about September 30, October 6, 15 and 30, and about November 15 and December 11. These are some of the dates on which abnormal phenomena were noted in the tail of the Morehouse comet.' However, he added: 'A coincidence of magnetic disturbances on earth with disturbances in the tail of the comet, is not really necessary to establish the connection, as I believe these disturbances, produced by solar storms, are not simultaneous in direction throughout the solar system.'[22]

He first published this highly original idea in the *Astrophysical Journal* for January 1909,[23] and later worked it out in somewhat greater detail. 'If we accept the fact,' he wrote, 'and I think it is a fact, that the tail of a comet is sometimes broken or changed by some force or influence not directly due to light-pressure or the forces present in the comet itself, it becomes a very important problem to determine what this unknown cause may be.' He had once thought that the changes he had recorded in the tail of Brooks's Comet of 1893 were due to its encounter with a swarm of matter. However, he no longer regarded this as probable. 'Swarms of meteors,' he continued, ' . . . could not account for some of the changes: that, for instance, where the tail is accelerated at right angles to the radius vector . . . apparently in defiance of the laws of gravitation. No ordinary matter thus encountered could possibly produce this effect. We must therefore seek for some other cause':

> Throughout my work in photographing this comet [Morehouse], I was impressed with the frequent displays of the Aurora . . . It finally occurred to me that a possible explanation of this peculiar effect in comets' tails might be found in a similar cause to that which makes the aurora [and] magnetic storms on the earth. These disturbing influences – which we will call them for want of a better name – must be going out from the sun in various directions at more or less frequent intervals. The

effect of this upon the attenuated matter of a comet's tail would likely be great, especially as it might change the electrical conditions of the particles forming the tail. We know that the speed of such an influence must be vastly greater than that which could possibly be produced by gravity. It would not appear that the effect of this disturbing influence should necessarily be radial with respect to the sun's center . . .[24]

We now know that Barnard was basically correct. Unfortunately, his ideas were forgotten, more or less, for sixty years. Not until the late 1960s were the details of the disconnection events worked out, by John C. Brandt and his colleagues, who only afterwards discovered that Barnard had already been far along the same path.[25]

As we now know, the coma of a comet consists in part of gases ionized by sunlight and glowing by fluorescence. As the comet moves through its orbit, the ions of the coma are buffeted by the solar wind, the stream of ionized gas (plasma), having a temperature of a million degrees, which is continuously emitted from the Sun. The solar wind captures the cometary ions and repels them backward with forces up to a thousand times that of solar gravitation, thereby forming the Type I or ion tail. The 'freakish' tails of comets which so fascinated Barnard are of this type – they are blown back like windsocks in the solar wind, and twist and bend in its eddies (by contrast, the more sedate Type II tails consist of dust, and shine by reflected sunlight). High-speed streams in the solar wind are associated with magnetic fields, whose field lines alternate direction (towards or away from the Sun) every few days, the reversals being quite sudden. As a comet crosses from a sector having one magnetic polarity to another, separation of the ion tail occurs – a disconnection event – followed immediately by the formation of a new tail. Not only was Barnard prescient in thinking that the freaks of comet tails might be related to solar weather, records of terrestrial magnetism have made it possible to do as he hoped to do, and to associate some of the events in the tail of Comet Morehouse with magnetic disturbances.

3

While Comet Morehouse was registering its changes on the photographic plate, Barnard was already preparing for the return of Halley's Comet in 1910. As early as October 1908, he had begun sweeping the region of the sky in which the ephemerides placed the comet with the 40-inch refractor. By the end of the year he had begun a photographic search with the Bruce photographic telescope. However, the comet was not discovered until September 11, 1909, when its image was weakly registered by his old rival, Max Wolf, at Heidelburg. Morehouse, who was still at the observatory at the time, recalled the scene:

> For weeks every time the telephone would ring, the observers would step into the hall and listen for the news. One afternoon . . . about three o'clock the telephone rang. As was our custom, we stepped to the door. Dr. Frost called out, 'Hoo-hoo, it

is found.' I shall never forget Dr. Barnard's white face as he stepped to the hall. 'Who found it?' he said. 'Dr. Max Wolf.' Dr. Barnard closed his eyes for a moment. He asked for the position, and without saying a word, he turned and walked back to his office, picked up a photographic plate of two nights before, and by the aid of the [blink] comparator found he had the object on a Bruce plate, but not one word or complaint that would detract in the slightest from the glory of Dr. Wolf's discovery was uttered. Dr. Barnard wanted his own, but he was equally insistent on giving everybody else due and just credit.[26]

The comet was seen visually by Burnham with the 40-inch refractor on September 15, and two nights later by Barnard, who described it as a 15th magnitude 'fleck of light surrounded by a faint nebulosity.'[27] Barnard followed it with the great refractor until February 1910, when, still a rather inconspicuous object, it disappeared into the solar glare. It passed perihelion in late April, when it emerged from behind the Sun into the early morning sky, and thereafter began brightening dramatically with decreasing distance from the Earth. Unfortunately, Barnard noted, the timing of its reappearance 'could not have been under more unfortunate circumstances at the Yerkes Observatory. That part of the year is always unpropitious here, and it seemed as if everything combined, on this particular occasion, to hide from us the growth of the comet and its approach to the earth.'[28] Vast forest fires were then raging in northern Wisconsin, producing a smoky pallor which 'cut off with a thick yellow veil all but a glimpse of the bright head.'[29] Only on the mornings of May 3 and 4 did the sky become clear and reasonably transparent. Barnard then found the comet a beautiful object, with a 2nd magnitude head and a tail traceable, before moonrise, out to 18°, and he succeeded in obtaining some excellent photographs with the Bruce telescope, despite the fact that its light was somewhat dulled by smoke from the observatory powerhouse which was blowing directly across it during the exposures.

Throughout the next two weeks the sky was again hazy or covered with clouds, allowing only occasional glimpses of the comet, but at midnight on May 17 there was a sudden clearing – just in the nick of time for Barnard to observe the phenomena of the comet's closest approach to the Earth (within only 0.15 AU). Much of the world had awaited this date with dread. It had been predicted that the Earth, on or around the date of the comet's closest approach, would brush through the comet's tail. From Sir William Huggins's earlier findings with his spectroscope that the coma and tail of a comet contain hydrocarbons, including cyanogen, a deadly poisonous gas, the French astronomer Camille Flammarion predicted that the cyanogen gas in the tail of Halley's comet would impregnate the Earth's atmosphere, possibly snuffing out all life on the planet. Flammarion should have known better – a comet's tail is so tenuous that even a brush through it would be without any noticeable effect on Earth, but the result of his announcement was, predictably, widespread hysteria. Barnard, who still remembered the superstitious dread with which his Southern neighbors had looked at a comet, probably the Great Comet of 1861 through whose tail the Earth had also passed with no apparent ill-effects, wrote that 'there was one fact which was brought forth by the comet with startling vividness. It showed that the superstitious terror formerly attending the

Drawing by Barnard showing the naked-eye appearance of the tail of Halley's Comet, on the night of the Earth's passage through the tail, May 18–19, 1910. Yerkes Observatory

appearance of a great comet is by no means dead in the human breast. Cases of this kind developed all over the country and abroad – from the stopping-up of keyholes and cracks in doors and windows in Chicago (according to the daily papers) to keep out the deadly comet gases, to the manufacture and sale, among the negroes of the South, of "comet pills," which were supposed to ward off the evil effects of the comet.[30]

The expected passage of the Earth through the tail of Halley's Comet on May 18–19 kept Barnard on the *qui vive*, and he carefully watched the sky throughout the day and night. On the night of May 18, with a gibbous Moon above the horizon, Barnard noted that the 'sky had a feeble misty look everywhere, which was not due to ordinary haze, for apparently the sky was very clear.' After midnight (May 19), he found the illumination growing over toward the eastern horizon; this illumination later proved to be the comet's tail. Two hours later, the tail was quite noticeable even in the moonlight, cutting a broad swath through the lower part of the Great Square of Pegasus – in its brightest portion, it appeared as bright as the Milky Way. Altogether, he was able to map it visually through a length of 120°.

After the passage of the Earth through the tail – needless to say, without ill effects – the comet's head rose later and later in the evening sky after sunset, and in the last week of May the comet put on a brave show. From Yerkes, Barnard saw it as strongly bluish-white, with the nucleus resembling a 1st magnitude star wrapped in haze, and the tail extending for some 25°. Though in cities smoke and electric lights robbed the comet of some of its glory and caused it to be regarded as a disappointment (as many a comet since has been), Barnard censured the glowing and sensational newspaper accounts which had 'raised expectation beyond all reason' and pronounced himself satisfied with its performance. 'It is unfortunate,' he reflected,

> that the newspapers and the general public were so greatly disappointed in the comet . . . It would have been a gratification to know that everyone who saw this wonderful object saw it with the same spirit of elation and wonder – one would almost say veneration – with which the average astronomer regarded it. This was, at least, the feeling of the present writer when he looked at this beautiful and mysterious object stretching its wonderful stream of light across the sky . . .[31]

On June 6, the comet discarded one of its tails and developed a new one – the event was unusually well documented, as photographs of it in different stages were obtained at Yerkes, Cordoba, Mt. Hamilton, Honolulu and Beirut. The comet was now quickly receding from the Earth, and had faded, as Barnard remarked, to a 'ghost of its former self.' A few days later, on June 11, it dropped below the threshold of naked-eye visibility. Nevertheless, Barnard continued to follow it visually with the 40-inch refractor for another eleven months, until May 23, 1911, when it had receded to just over 5 AU from the Sun and was, he noted, 'excessively faint and difficult.'[32] He was the last person to see it visually until January 1985.[33]

4

In the aftermath of the Halley's Comet frenzy, most astronomers were only too eager to turn their attention from comets to other objects. Barnard, however, continued to keep up a photographic record of every comet visible from Yerkes for as long as he had strength to observe. The most notable of these was Brooks's Comet (1911V), discovered by his old rival, William Robert Brooks. Though Barnard had long since given up the search for new comets, Brooks remained as diligent in their pursuit as ever. To the eleven comets he had found with his reflectors at the old Red House Observatory in Phelps, New York, he had gone on to add fourteen more at the observatory of Hobart and William Smith Colleges in Geneva, New York, between 1888 and 1905, most of them being found with its 10-inch refractor. His record is the more remarkable considering that he had to carry on his comet-seeking 'in the few intervals between other duties, among which is the entertainment of visitors, the Observatory being freely open to the public on every clear night. This explains why most of my Geneva comets have been discovered in the morning sky.'[34] His last comet, which he captured as a 10th magnitude object in Pegasus in July 1911, proved to be his best. It brightened rapidly as it approached the Sun and Earth, and in mid-October loomed in the northwestern sky after evening twilight, reaching 2nd magnitude with a bluish-white tail extending 30°. 'At its best,' Barnard noted, 'it was but little inferior to Halley's comet.'[35] As spectacular as it was to the naked eye, photographically it was, until the end of October, a disappointment; Barnard wrote that his photographs were 'very much alike night after night and it looked as if the comet would not be of any special importance.'[36] Only after it passed perihelion on October 28 and looped into the morning sky did the comet become more active. The head began to shrink rapidly in size (due, Barnard opined, to the increased pressure of the sunlight), and the tail widened into a great bundle of threadlike streamers, at times appearing wavy in form, 'as if the tail had been combed with a coarse comb with an undulatory motion.'[37]

Among later comets that appeared during Barnard's lifetime, only Delavan's comet (1914 V) was a striking object as seen with the naked eye – it reached 3rd magnitude at its best, and was circumpolar for a time in August 1914, allowing Barnard to photograph it throughout the night. Of other work on comets, mention should also be made of his recovery of the periodic Comet Encke at its expected return in 1914 and of Pons–Winnecke in 1921, though Barnard himself, as we have noted earlier, thought that there was little credit to such discoveries, since they naturally tended to fall to those observers who were equipped with the largest instruments.

His real credit, as he knew, was his photographs, some 1400 in all, which revolutionized the study of comets. He was always most fascinated by the remarkable 'freaks' in the tails of some of them, which had gone completely unsuspected in centuries of visual observation. His photographic record of these dramatic events, combined with the opportunity that moving to Yerkes had given him to study the auroras, was what led him to realize in early 1909 that these phenomena were caused by magnetic disturbances from the Sun. Among his greatest insights, it remains a landmark in the study of these wild and mysterious visitors from the icy fringes of the Solar System.

1   Mary R. Calvert, 'Some Personal Reminiscences,' *JTAS*, 3:1 (1928), 29
2   ibid.
3   A. M. Clerke, *History of Astronomy During the Nineteenth Century*, (London, Adam & Charles Black, 3rd ed., 1893), p. 126
4   ibid., p. 127
5   ibid., p. 420

6   E. E. Barnard, 'Photographic Observations of a Very Remarkable Comet,' *PA*, **16**
     (1908), 591–6: 594–5
7   ibid., p. 595
8   E. E. Barnard, 'Photographic Observations of Borrelly's Comet and Explanation of the
     Phenomenon of the Tail on July 24, 1903,' *Ap J*, **18** (1903), 210–17: 211
9   ibid.
10   ibid., pp. 213–14
11   Barnard, 'Photographic Observations of a Very Remarkable Comet,' p. 591
12   ibid.
13   ibid.
14   D. W. Morehouse, 'Reminiscences of Edward Emerson Barnard,' *JTAS*, 3:1 (1928),
     21
15   E. E. Barnard, 'Comet c 1908 (Morehouse),' *Ap J*, **28** (1908), 292–9:293
16   Barnard, 'Photographic Observations of a Very Remarkable Comet,' pp. 592–3
17   E. E. Barnard, 'On the Erroneous Results of a Stereoscopic Combination of
     Photographs of a Comet,' *MNRAS*, **69** (1909), 624–6:624. Barnard sent a reprint of
     this article to Joseph Jastrow, a professor of psychology at the University of Wisconsin,
     who agreed with him that 'it is quite possible that the variations of the two selected
     plates may combine to suggest false depth relations,' though he felt that the 'general
     stereoscopic appearance is legitimate for most of the purposes involved.' Joseph
     Jastrow to EEB, November 13, 1909; VUA
18   E. E. Barnard, 'On Comet 1919b and on the Rejection of a Comet's Tail,' *Ap J*, **51**
     (1919), 102–6:106
19   Barnard, 'Comet c 1908 (Morehouse),' pp. 292, 299
20   EEB to W. J. Humphreys, March 19, 1909; VUA
21   EEB, observing notebook; YOA
22   EEB to W. J. Humphreys, March 19, 1909; VUA
23   E. E. Barnard, 'Comet c 1908 (Morehouse) – Third Paper,' *Ap J*, **29** (1909), 65–71
24   EEB, 'On a Possible Cause of the Extraordinary Changes in the Tails of Comets,' MS
     dated March 11, 1909; VUA
25   J. C. Brandt to W. Sheehan, personal communication, June 28, 1994
26   Morehouse, 'Reminiscences,' p. 21
27   EEB, observing notebook; YOA
28   E. E. Barnard, 'Visual Observations of Halley's Comet in 1910,' *Ap J*, **39** (1914),
     373–404:381–2
29   ibid., p. 382
30   Barnard, 'Visual Observations of Halley's Comet,' p. 375
31   ibid., p. 374; pp. 375–6.
32   EEB, observing notebook; YOA
33   S. J. O'Meara, 'The Visual Recovery of Halley's Comet,' *Sky and Telescope*, **69** (1985),
     376–7
34   W. R. Brooks, 'Dr. Brooks's Discovery of his Twenty-fourth Comet,' *MNRAS*, **64**
     (1904), 840–3:843
35   E. E. Barnard, 'Photographic Observations of Comet 1911 c (Brooks),' *Ap J*, **36** (1912),
     1–13: 1
36   ibid., p. 3
37   ibid., p. 6

22

Observer of all that shines – or obscures

1

Apart from probing the nature of comets, the other scientific question that absorbed Barnard's closest attention during his later career was the nature of the dark markings of the Milky Way. Some of these dark markings he had recognized visually as early as the 1880s, when he was canvassing for comets with his 5-inch refractor, and ever since, he wrote, 'they have always appealed to me with an interest scarcely less than that of any other natural feature of the sky.'[1] Their proper study began only with Barnard's pioneering investigations with wide-angle lenses – first with the Willard portrait lens, and later with the Bruce photographic telescope. His photographs showed that many of the large diffused nebulae were associated with vacant regions – the best example being the remarkable nebula of ρ Ophiuchi, situated 'in apparently a large hole in the Milky Way.'[2] This meant that the existence of the nebulae in these regions was in some way – though he was not yet sure in what way – the cause of the scarcity of stars.

He had made a beginning, but he still had a long way to go. Indeed, as late as December 1904, on the eve of departing with the Bruce telescope for Mt. Wilson, he was still espousing the view that these features were in most cases real vacancies among the stars. 'In reference to these dark lanes and holes,' he wrote,

> there seems to be a growing tendency to consider them dark masses nearer to us than the Milky Way and the nebulae that intercept the light from these objects. This idea was originally put forward by Mr. A. C. Ranyard. Though this may in a few cases be true – for some of them look very much that way – I think they can be more readily explained on the assumption that they are real vacancies. In most cases the evidence points palpably in this direction. In the few cases where the appearance would rather suggest the other idea – and this is mostly in reference to the nebulae – the evidence is still not very strong.[3]

Barnard obtained hundreds of plates of the Milky Way from Mt. Wilson, from which he later quarried most of the material used in his *Atlas of Selected Regions of the Milky Way*. But nothing in them caused him to change his mind about the dark markings. In a paper, 'On the Vacant Regions of the Sky,' read before the Astronomical and Physical Society in December 1905, only two months after his return, he repeated his still strong belief that 'most of these blank regions . . . impress one as being actual holes.'[4]

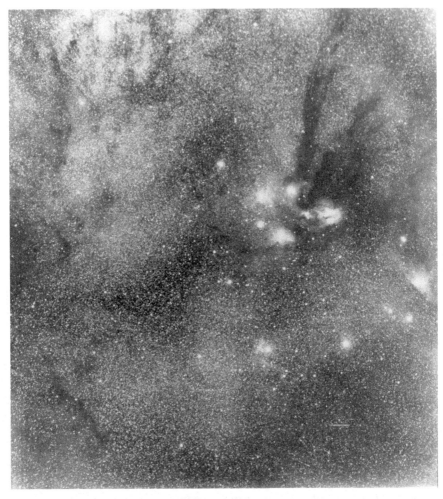

Region of the great nebula of ρ Ophiuchi, 4 hr 30 min exposure by Barnard with the Bruce photographic telescope at Mt. Wilson, April 5, 1905. Yerkes Observatory

Nevertheless, he had to admit that there were a few parts of the sky where this did not seem to be the correct explanation. In particular, he was deeply troubled by the peculiar features in Ophiuchus and Scorpio – what he called 'the most puzzling region that I know of in the sky.' 'Here occur vacancies within vacancies,' he wrote:

> [T]here are vast regions almost entirely free from stars, in a surrounding region thick with small stars. These regions seem veiled over with some sort of material in which occur blacker spaces, as if all this part of the sky were involved in a thin faint nebulous sub-stratum which partly veils the blackness of space beyond. In this, apparently, occur rifts and openings giving a clearer view of space.[5]

Here was the nearly starless chasm beginning near θ Ophiuchi, which turned west in shattered form, then strengthened into a definite lane extending for another 15° and finally connecting with the 'remarkable vacancy in the dense sheeting of small stars'[5] near ρ Ophiuchi. These features were more complicated than the simple holes in sheets of stars that he thought he could discern so clearly elsewhere in the Milky Way:

> The blending of this great nebula into the surrounding region, where it seems to mingle with the material of the vacancies, makes it hard to tell where the nebula leaves off . . . There is a slight suspicion that certain outlying whirls of the nebulosity have become dark and that they are the cause of the obliteration of the small stars near . . . [6]

No sooner had he put forward his 'slight suspicion' – that nebulae might fade out in parts, and eventually they become dark and obscure – than he felt the need to draw back:

> I think this is fanciful however, for the irregular vacancy in which it lies connects readily with the vacant lane running east to the region of θ Ophiuchi. No one would suspect for a moment that this lane is anything but an actual vacancy among the stars.[7]

The only way to decide the whole question was to obtain more and better observations, so he returned to photographing the sky with the Bruce telescope. The question of the nebulous sub-stratum which seemed to exist in Ophiuchus and Scorpio nagged at him, but he also worried that the vacant lanes in the Milky Way might be entirely subjective – due to scarcity of stars alone – rather than 'channels in a bed-work of nebulous substratum.' If the stars were removed, he asked, would the lanes still exist? Early in the frigid month of January 1907, he trained the Bruce telescope on the sky north and east of the Pleiades, where earlier photographs with the Willard lens had caught dim suggestions of dark lanes extending far to the east of the cluster. He exposed the plates for five and a half hours. When the plates emerged from the developing tray, the lanes showed up unmistakably. They were not only devoid of stars but darker than the surrounding sky, satisfying Barnard that they would still be visible even if the stars were all removed. In addition to the dark lanes, the plates showed a large nebula apparently in a hole almost devoid of stars, from which one of the lanes straggled away to the southeast for several degrees. He later described these remarkable features in his paper, 'On a Nebulous Groundwork in the Constellation Taurus' (1907):

> The pictures seem to show that the brighter part of this nebula is only a small portion of it, and that the nebula is feebly luminous over most of the vacancy . . . The feebler portions of the nebula would almost suggest the idea that a large nebula exists here, but that the major portion of it is dead or non-luminous, and that it actually causes the apparent vacancy by cutting out the light from the stars, while the few stars visible are perhaps on this side of the nebula. I give this simply as what the picture would suggest to one, and not as what may really be the truth.[8]

Barnard with Bruce photographic telescope. Yerkes Observatory

He had returned to the idea he had suggested then dropped a year earlier: why might not a luminous nebula – the cause of whose shining was not, admittedly, known – fade out and eventually die, just as the stars themselves did?

> The dying-out of nebulae . . . is a probability fully as warranted as the belief and certainty that the stars must die out. What would be the condition of a nebula that no longer emitted light[?] [I]t is likely that we should simply have a dark nebula which would not be visible in the blackness of space unless its presence were made known by its absorption of the light of the stars beyond it . . .[9]

Dark lanes in Taurus, 5 hr 29 min by Barnard with Bruce photographic telescope at Yerkes Observatory, January 9, 1907. Barnard later wrote that this photograph was 'one of the most important of the collection, and bears the strongest proof of the existence of obscuring matter in space.' Yerkes Observatory

At last Barnard was knocking at the very door of truth:

> This idea of the absorption of the light of the stars by a dead nebula or other absorbing matter has been used by some astronomers as an explanation of the dark or starless regions of the sky. Though this has not in general appealed to me as the true explanation – an apparently simpler one being that there are perhaps no stars at these places – there is yet considerable to commend it in some of the photographs . . . I have been slow in accepting the idea of an obscuring body to account for these vacancies; yet this particular case [that of the dark lanes in Taurus] almost forces the idea upon one as a fact.[10]

Even in 1907, he was not prepared to embrace the idea, however; once again he drew back. 'The idea of the dying-out of a nebula,' he suggested timidly,' . . . is not strengthened by the presence of the lanes, for we do not find in general any great streams of nebulosity extending away from the nebulae.'[11]

2

All this time, Barnard had been almost alone in struggling with the meaning of the dark markings revealed in his plates. Curiously, other astronomers seem to have taken scant interest. One exception was Max Wolf, who since the 1890s had been photographing the Milky Way from the Königstuhl Observatory in Heidelberg. Wolf also noted the curious relationship which had impressed Barnard – the fact that extended nebulae were almost always situated within larger regions that contained only a very small number of faint stars. His North America nebula, for instance, was at the edge of a great, nearly starless region. Another astronomer who was working on the problem – though from an entirely different direction – was Jacobus Cornelis Kapteyn, of the Gröningen Observatory, in the Netherlands. While a research associate at Mt. Wilson in 1909, Kapteyn published a paper in the *Astrophysical Journal*, 'On the Absorption of Light in Space.' Kapteyn suggested that the 'enormous mass of meteoric matter' which filled space would undoubtedly intercept some part of the starlight; this, in turn, would produce a dimming of the stars as one looked farther into space and would cause astronomers to believe that the more remote stars were dimmer than they really were, thus leading to an exaggerated estimate of their distances.[12]

Barnard undoubtedly read Kapteyn's paper, and it is quite probable that it influenced his own choice of a title for his next important paper on the subject of the dark markings of the Milky Way, also published in the *Astrophysical Journal* – 'On a Great Nebulous Region and on the Question of Absorbing Matter in Space and the Transparency of the Nebulae' (1910). Barnard concentrated on the large straggling nebula around ν Scorpii which he had first discovered with the small lantern lens in 1893 and whose extensions seemed to reach to, and in a feeble manner connect with, the great nebula of ρ Ophiuchi. 'The greatest interest in this nebula,' he wrote, 'lies in the fact that it seems to show a veiling of the stars in certain of its portions . . . The line of demarcation between the rich and poor portions of the sky here is too definitely and suddenly drawn by the edges of the nebula to assume the appearance due to an actual thinning out of stars':

> It looks, where this part of the nebula spreads out, as if the fainter stars were lost, and the brightness of the others reduced by at least a magnitude or more . . . In the region of ρ Ophiuchi there is every appearance of a blotting-out of the stars by the fainter portions of the nebula, but from its complicated and irregular form the hiding of the stars is not so clearly evident as is the case of the ν Scorpii nebula. At present we have no means of determining whether a nebula is transparent or not. The assumption has always been that they are transparent like the comets . . . I think in the present case . . . that the nebula of ν Scorpii is shown to be at least partially transparent, but the absorption of the light of the stars behind it must be

considerable. The picture is quite conclusive evidence that the nebula is nearer to us than the general background of stars at this point.[13]

Though clearly on the threshold of accepting the view that the dark markings were opaque matter, he remained on the fence. 'If these dark spaces of the sky are due to absorbing matter between us and the stars – and I must confess that their looks tempt one to this belief – such matter must, in many cases, be perfectly opaque, for in certain parts of the sky the stars are apparently blotted out,' he wrote. But he added his customary disclaimer: 'It is hard to believe in the existence of such matter on such a tremendous scale as is implied by the photographs.'[14]

3

For some time Barnard had been edging closer and closer to accepting the opaque nebula idea, but he was at heart a conservative. He wanted to be certain before changing his mind. The decisive turning point for him finally came 'one beautiful transparent moonless night' in the summer of 1913. He was photographing the southern Milky Way with the Bruce telescope at Yerkes Observatory:

> I was struck with the presence of a group of tiny cumulous clouds scattered over the rich star-clouds of Sagittarius. They were remarkable for their smallness and definite outlines – some not being larger than the moon. Against the bright background they appeared as conspicuous and black as drops of ink. They were in every way like the black spots shown on photographs of the Milky Way, some of which I was at that moment photographing. The phenomenon was impressive and full of suggestion. One could not resist the impression that many of the small spots in the Milky Way are due to a cause similar to that of the small black clouds mentioned above – that is, to more or less opaque masses between us and the Milky Way. I have never seen this peculiarity so strongly marked from clouds at night, because the clouds have always been too large to produce the effect.[15]

In 1913 there were still few city lights, and Chicago light dome that now interferes with sensitive observations at Yerkes did not yet exist. Under these conditions, the clouds appeared perfectly black and darker than the background sky. Nowadays, they would appear brighter, due to reflection of artifical lights, and the effect that made such a strong impression on Barnard would be lost.

After his moment of revelation, Barnard began thinking back to earlier visual observations of some of the dark markings in the Milky Way. The most striking of these markings, because of their smallness and definite form, were in Sagittarius. One was the small object, 'like a drop of black ink on the background of the Milky Way,' which he had found with his 5-inch refractor in Nashville (it would be entered in his catalog of these objects as B86). With the 36-inch Lick refractor in 1895, it had nearly filled the field of view. The western half was well defined, the eastern half more diffused, and considerable nebulosity was associated with it.[16] The other, B92, was even more

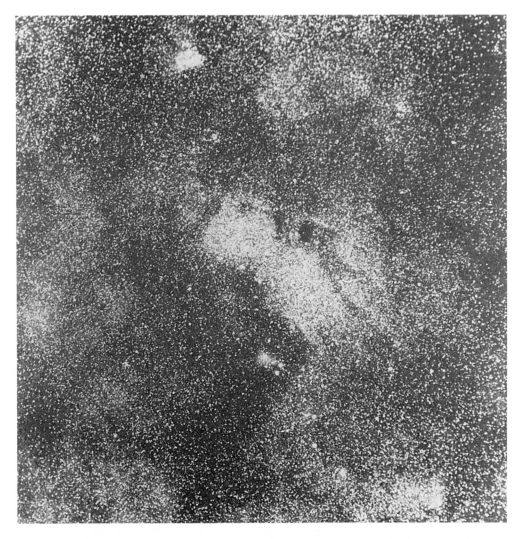

Small star cloud in Sagittarius. The conspicuous dark spot in the upper part of the cloud is B92. Exposure of 4 hr 30 min by Barnard with the Bruce photographic telescope at Mt. Wilson, July 31, 1905. Yerkes Observatory

striking. In previous observations with the Lick 36-inch and Yerkes 40-inch refractors, he had sometimes had the distinct suspicion that there 'was an actual object at this point,' but he had never been quite sure. On July 27, 1913, conditions were unusually favorable both for transparency and steadiness, and Barnard carefully examined this dark 'hole or spot' with the 40-inch refractor:

> With its following edge cutting across the middle of the field, which is some three times smaller than the spot, it was quite distinctly seen that the preceding half of the field, in which there were no stars, was very feebly luminous, while the following

side showed a rich, dark sky with the few small stars on it. From the view, one would not question for a moment that a real object – dusky looking, but very feebly brighter than the sky – occupies the place of the spot. It would appear, therefore, that the object may be not a vacancy among the stars, but a more or less opaque body.[17]

Though Barnard continued to caution against accepting *all* the dark regions as consisting of dark matter, insisting that some must be dark 'purely from the fact that there are no stars there,'[18] he never again doubted that this was the true explanation for most of them. However, he still fretted about how such objects – which in many cases appeared to be totally opaque – could be seen at all. They were silhouettes – but what was the nature of the luminous background against which they were silhouetted? 'If I have proved that there are dark objects in the heavens that are shown on photographs through being projected on a luminous ground,' he wrote, 'I have opened the way to prove something else. [The] very fact of there being a luminous background may prove of the greatest value to us in our solution of the problems of space, because one form of this background suggests a feeble luminosity through interstellar regions and perhaps beyond.'[19] We now know that the luminous background he was photographing consisted, in the Milky Way, merely of the innumerable faint background stars. In the Milky Way itself, Barnard himself guessed that this was the correct explanation, but in regions far away from the Milky Way he continued to believe in a 'widespread and undoubtedly universal (so far, at least, as our stellar universe is concerned) . . . feeble illumination of distant space.'[20] He did not know that here, too, there were faint galaxies in numbers beyond belief.

4

In his quest to demonstrate conclusively the obscuration of light in space, Barnard called attention to the spectacular object which carries the catalog designation B33 but is better known as the 'Horsehead Nebula,' near ξ Orionis. This object was discovered by Isaac Roberts on a photograph taken in 1900. Roberts himself described it as an 'embayment.' Barnard declared that 'This object has not received the attention it deserves. It seems to be looked upon as a rift or hole in the nebulosity, as implied . . . from Dr. Roberts' paper.' However, so too would Barnard have looked upon it, until now. Now that he had seen such dark objects in a new light, he had a different interpretation. His own photographs of it from February 1913 revealed 'instead of an indentation, the almost complete outline of a dark object . . . projected against the bright nebulosity,' leading Barnard to conclude that 'it is clearly a dark body projected against, and breaking the continuity of, the brighter nebulosity.'[21] On the night of November 4, 1913, he examined it visually with the 40-inch, using magnification of × 460:

> The outlines of the spot – so sharp and clear in photographs of this region – could not be made out with any definiteness. The view showed that the spot is certainly

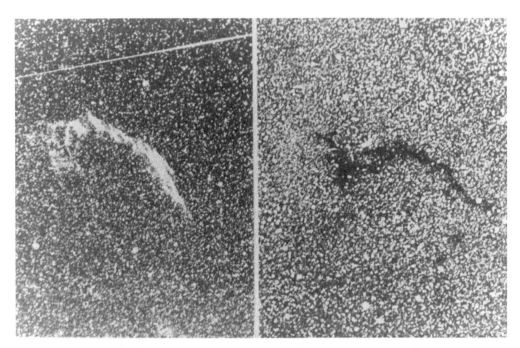

Photographs published by Barnard in the *Astrophysical Journal* for January 1916, showing analogous shapes of bright and dark nebulae which he hoped would strengthen his case for the existence of dark nebulae. The bright nebula on the left is NGC 6995 in Cygnus, shown in this exposure of 5 hr 43 min taken by Barnard on July 15, 1909; the dark nebula on the right is B150 in Cepheus, exposure 6 hr 2 min by Barnard on October 1, 1910. Barnard enhanced the weak image of the latter by making multiple printings with the position shifted slightly each time, which gives the illusion of a dense star field. Yerkes Observatory

not clear sky, for the field was dull, apparently indicating the presence of some material substance at this point. To me the observation would confirm the supposition of an obscuring medium.[22]

   Though now certain of their existence, Barnard was cautious in expressing any idea as to the nature of these opaque bodies. 'What their nature is we do not know,' he wrote in October 1915, 'and the spectroscope cannot help us because the objects are devoid of light, or nearly so. But there is strong evidence that they are of the nature of the nebulae – that is, that they are dark nebulae.'[23] He believed that they were nebulae that had lost their light, or had never been luminous – 'it is possible,' he suggested, 'that the original condition of a nebula is dark.'[24] Somehow he hoped to prove the connection, but there was no obvious way. 'Perhaps,' he suggested, 'if we show a close resemblance in form and size of one of these [dark opaque objects] to one of the well-known nebulae, it may aid us in connecting the two kinds of objects.'[25] He thus produced two photographs on the same scale, one showing part of his dark nebula in Cepheus and the other the luminous Veil nebula (NGC 6995) in Cygnus. The resemblance was striking and suggested analogy:

There is a striking resemblance in the forms of these two objects; but one is a luminous nebula and the other a dark – what? One can readily see that if the nebula were to lose its light, it would, if dense enough, still be shown against the sky and would strongly resemble the dark object. For this and many other reasons I am constrained to believe that the dark object is really a non-luminous nebula seen against a luminous background.[26]

In fact, we now know that in this case Barnard was misled by the similarity in shapes. The Veil nebula has a gaseous spectrum, and represents the far-flung remnants of an exploded star – a supernova – that continues to be faintly luminous after having been violently hurled outward from the explosion. The dark nebula Barnard had photographed had no spectroscopic signature, but we now know that it is made up of a different kind of matter.

Indeed, only three years later, in his 1919 paper, 'On the Dark Markings of the Sky,' in which he published his famous catalog of 182 of these objects, Barnard retreated from the notion that they were dead nebulae. What influenced him to do so was the spectroscopic work of V. M. Slipher, Percival Lowell's assistant at Flagstaff, who had shown in 1912 that the nebulosity which surrounded Merope, in the Pleiades, had a star-like rather than a gaseous spectrum.[27] It was a reflection nebula, consisting of dust that reflected the light from the nearby star, and in 1914, Slipher obtained the same result for the ρ Ophiuchi nebula. Barnard was one of few astronomers who appreciated the significance of these results at the time. 'To me [there is] conclusive evidence that masses of obscuring matter exist in space and are readily shown on photographs with the ordinary portrait lens,' he wrote. 'What the nature of this matter may be is quite another thing. Slipher has shown spectroscopically that the great nebula about ρ Ophiuchi is probably not gaseous . . . The word "nebula," nevertheless, remains unchanged by this fact, so that we are free to speak of these objects as nebulae. For our purpose it is immaterial whether they are gaseous or non-gaseous, as we are dealing only with the question of obscuration.'[28]

5

As important as it was, Slipher's discovery of reflection nebulae was completely overshadowed by his far-reaching work on spiral nebulae. Directed to the problem by Lowell, who believed as most astronomers of the day had that the spiral nebulae were planetary systems in formation, Slipher had to overcome great difficulties in obtaining spectrograms of these faint objects. He began with the brightest spiral, M31 in Andromeda, but the spectrum was still so faint that in order to capture it with the means available at the time, he had to use very long exposures – one of his plates, obtained at the end of 1912, required exposures over three consecutive nights. The results were astonishing, to say the least. The spectral lines were abnormally shifted toward the violet end, indicating an unusual velocity of approach. The velocity worked out to 190 miles/sec, which was greater than had been measured for any other object up to that time. 'It looks as if you had made a great discovery,' Lowell told him. 'Try some more spiral nebulae for confirmation.'[29] This Slipher did, starting with NGC 4594, the spindle-shaped 'Sombrero Hat' nebula in Virgo, which showed an even greater displacement in its spectral lines – this time toward the red instead of the violet, from which he worked out a velocity of recession of 600 miles/sec. Over the next three years he obtained spectrograms of twenty-two more spirals, all shifted toward the red and all having velocities of recession of the same order. His work on NGC 4954 led, moreover, to another important discovery. On placing the slit of the spectroscope parallel to the long axis of this nebula, he found that the spectral lines were not only red-shifted but tilted, indicating a measurable rotation. A few months later he obtained similar evidence for the rotation of the Andromeda nebula.

Though Slipher himself did not at first grasp the full significance of what he had discovered – for a while he continued to believe that the spirals were planetary systems of some sort, 'composed of matter from dust-clouds to suns in size and development' – he later changed his mind, possibly owing to Lowell's influence, who in a November 1915 lecture cited Slipher's spectrograms as showing that 'the spiral nebulae are not the prototype of our system, but of something larger and quite different, other galaxies of stars.'[30] There was, by then, other evidence pointing in the same direction. From 1909 on, Heber D. Curtis took hundreds of direct photographs of spirals with the Crossley reflector. From these photographs, he estimated that the number of spirals within reach of this telescope was much greater than Keeler's estimate of 120 000, and more like 700 000 or even 1 000 000.[31] Moreover, in reaching his conclusion that these objects were 'inconceivably distant, galaxies of stars or separate stellar universes so remote that an entire galaxy becomes but an unresolved haze of light,'[32] Curtis referred not only to Slipher's spectrograms but also to the fact that some of the spirals which were seen edge-on, such as NGC 891 in Andromeda and NGC 4594 in Virgo, showed dark bands, which he believed must be due to 'occulting matter' similar to the 'dark nebulae' and 'coal sacks' which were already well known from Barnard's photographs of the Milky Way.[33] Barnard himself had written in 1915 of the grand edge-on spiral NGC 4565, in Coma Berenices:

> Another beautiful example of this kind is shown in photographs of [this] very elongated nebula, . . . which seems to be an object similar to the great nebula of Andromeda, with its edge toward us, where the darker outer periphery of the nebula is seen cutting across the brighter central region as a black irregular streak.[34]

However, he fell short of concluding, as Curtis did, that NGC 4565 and others like it were themselves galaxies seen edge-on, and that the dark nebulae he was photographing in the plane of our own Galaxy belonged to a similar ring of obscuring matter.[35]

Curtis had not yet published his ideas about the spiral nebulae when, in July 1917, George W. Ritchey, using the 60-inch reflector at Mt. Wilson, announced the discovery of a 14th magnitude nova in the spiral NGC 6946.[36] Soon other novae were located in plates of other spirals – they were ordinary novae, like those frequently observed in our own Milky Way, unlike the outburst in Andromeda in 1885, which had been an intrinsically much brighter object, what we now call a supernova. The apparent faintness of these ordinary novae showed, moreover, that the objects in which they were located must be at vast distances. Thus by 1917, the view that the spirals were other galaxies of stars was rapidly gaining ground, and Barnard, who followed Ritchey's nova visually with the 40-inch refractor between July 28 and August 28, wrote to Ritchey that he was now 'beginning to believe that the spirals really are outside universes.'[37]

6

Not everyone was willing to accept this conclusion. The leading skeptic was Harlow Shapley, a native Missourian, who had begun studying variable stars as a student of Henry Norris Russell at Princeton, and while there had become interested in one particular type of variable stars – Cepheids, known after the prototype δ Cephei, whose light variations had been discovered by John Goodricke in 1784. In 1912, Henrietta Leavitt of Harvard showed that if one plotted the periods of the Cepheids which she had identified in the Small Magellanic Cloud against their apparent brightnesses, the resulting graph was linear. Since the distances of these variables could, to a first approximation, be considered the same, their apparent brightnesses indicated their true luminosities. This meant, as the Danish astronomer Ejnar Hertzsprung realized the following year, that if only one could reliably measure the distance to one of them, the intrinsic brightness of all the other Cepheid variables could be worked out from the period alone, and one could use them as a powerful measuring stick across space. Hertzsprung mentioned his idea to Russell, who in turn mentioned it to Shapley, who was still Russell's doctoral student. Shapley went on to show that the Cepheid variables were large, intrinsically bright stars. Moreover, they were not binaries but true pulsating stars. After he left Princeton for Mt. Wilson in 1914, Shapley discovered Cepheid variables in the globular clusters, and using a calibration worked out from a handful of Cepheids in the Milky Way, tackled the problem of finding the distances to them. In 1918, he announced that the typical globular was on the order of 50 000 light years distant. Moreover, since most of the globulars were concentrated in the direction of Sagittarius, he assumed that the nucleus of the Galaxy was located in the center of this halo of globulars. Whereas the English astronomer Arthur S. Eddington had just a few years earlier estimated the extent of the Galaxy as on the order of only 15 000 light years, Shapley recalculated its breadth at 300 000 light years, and put the Sun in the remote outskirts far from the galactic center.[38] Because he failed to take into account dimming by obscuring matter, Shapley overestimated the distances to the Cepheids – the Galaxy is only about a third as large as he estimated, but his figure was certainly on the right order. It was so large that he could not bring himself to believe that the spirals could be outside it, and he was still troubled by the nova in the Andromeda nebula in 1885. In 1920, he and Curtis were invited to give lectures at the National Academy of Science in Washington, what later became known as the 'Great Debate.' Robert G. Aitken, the double star observer at Lick Observatory, wrote to Barnard at the time:

I would like to hear the debate between Curtis and Shapley. I have read Curtis'
paper – a very good one – and have had long talks with Shapley also, and each one
has many very good arguments to present. For my own part, I am still 'on the fence'
on the question. I very greatly doubt the visibility of half-a-million or more 'island
universes' on the one hand, and, on the other, I am not ready to accept Shapley's
conclusions *on the basis of his measuring-rod*. It seems to me that its value is not yet
sufficiently demonstrated. I am open to conviction.[39]

Shapley talked about the scale of the universe, while Curtis, who did not accept
Shapley's view of the scale of the Milky Way system, presented his arguments in favor
of regarding the spirals as island universes. Both astronomers were partly right and
partly wrong, though Curtis is generally regarded as having 'won.'[40] In any case, by
early 1924 Edwin P. Hubble, who had been an assistant on the Yerkes staff in 1914 and
came to Mt. Wilson just after the War, would use the 100-inch reflector on Mt. Wilson
to identify a Cepheid variable in the Andromeda nebula – it proved to be excessively
faint, and using Shapley's methods ('He never acknowledged my priority,' Shapley
wrote ruefully long afterwards, 'but there are people like that'),[41] Hubble was able to
work out the distance – well over 1 000 000 light years. Thus there could no longer be
any doubt that the spiral nebulae were indeed vast star systems far beyond the confines
of the Milky Way.

7

Shapley's globular cluster results came as no surprise to Barnard. He had himself been
carefully measuring stars in some of them since 1898 – one of the most prodigious
pieces of work he undertook with the 40-inch refractor. In M13 alone, he measured and
remeasured no less than 247 individual stars; in all he obtained positions of 1363
individual stars in eighteen clusters. At first, as he later recalled, 'I had formed what I
now believe was an entirely erroneous idea of their dimensions and of the sizes of the
stars that compose them. They appeared to me as compressed groups of small suns that
did not in any sense rank with the ordinary stars in the sky. Their distances from us,
though great, were thought comparable with ordinary stellar distances. From these
considerations I had reasonable hopes of detecting some relative motion of the
individual stars in a few years' time from accurate micrometer measures.'[42] However, to
his 'great regret and disappointment,' his measures repeated at ten years showed no
changes. When, at twenty years, he continued to face the same negative results, he was
finally ready to face the fact that these clusters 'were at vaster distances from us and on a
more magnificent scale than their apparent insignificance might imply.'[43] His
photographs of the Milky Way showed that some of them were superimposed on the
great star clouds and therefore had to be nearer than the star clouds themselves.[44]
Indirectly, his measures lent strong support to the very great distances which Shapley
was claiming for these objects.

8

The heroic period between 1912, when Henrietta Leavitt identified the period–luminosity relationship of Cepheids in the Small Magellanic Cloud and V.M. Slipher began measuring the red shifts of spirals, and 1929, when Edwin P. Hubble plotted the distances of galaxies against their red shifts and discovered the expansion of the universe, was one of unprecedented change in astronomy; the classical methods gave way to those of astrophysics, and the modern view of the universe took shape. The globular clusters were identified as systems of stars on a vastly greater scale than had been hitherto supposed, forming a framework around the galactic nucleus which is itself situated far away in the direction of the constellation Sagittarius. The galactic nucleus is hidden from direct observation because, in silhouette between us and the galactic center, are clouds of obscuring matter, similar to the lanes of obscuration which were found in edge-on spirals. Barnard had photographed and struggled to understand these dark clouds over most of his scientific career, and though he eventually realized that they consisted of obscuring matter of some kind, he still had no real idea as to the nature of this matter. Just before Barnard's death, Henry Norris Russell proposed, though he could not yet prove, the correct idea – that this obscuring matter consisted of fine dust.[45] During the 1920s, this idea gained ground, and was finally proved by Robert J. Trumpler in his 1930 paper, 'Absorption of Light in the Galactic System.'[46] Interestingly, in this paper Trumpler took the existence of dark nebulae for granted. By 1930, writes Gerrit L. Verschuur, 'this was taken to be so obvious that no reference to any specific work was given, which paints Barnard's struggle in a sobering light.'[47] Thus do the great insights of one generation fade into the light of common day for the next and become accepted commonplaces.

Still later, in 1947, Bart J. Bok and Edith Reilly proposed that small dark clouds are sites of star formation (they are known today as 'Bok globules,' though Bok himself disliked the name. 'They should be called Barnard globules,' he protested; 'he discovered them').[48] Indeed, the dark nebulae that Barnard first photographed in Taurus and around ρ Ophiuchi are now known to be teeming with young stars. Protostars form in the dense cores of these dark clouds, and when they begin to generate enough heat blow off their cocoons of interstellar dust. At that point they may become visible in optical telescopes as T Tauri stars (named after the prototypical star associated with Hind's variable nebula and located among the vast dark clouds in Taurus which Barnard first photographed in 1907). Though the details are complex, the broad outlines of the process of star formation are being worked out.[49] These clouds, whose ominous darkness fascinated Barnard, are not, as he once speculated, the remnants of dead nebulae, they are the birth places of suns.

1   E. E. Barnard, 'On the Vacant Regions of the Sky,' *PA*, **14** (1906), 579–83:579
2   E. E. Barnard, 'The Bruce Photographic Telescope of the Yerkes Observatory,' *Ap J*, **21** (1905), 35–48:46

3   ibid., pp. 46–7

4   Barnard, 'On the Vacant Regions of the Sky,' p. 582

5   ibid., pp. 580–1

6   ibid., p. 581

7   ibid., p. 581

8   E. E. Barnard, 'On a Nebulous Groundwork in the Constellation Taurus,' *Ap J*, **25** (1907), 218–25: 220

9   ibid., p. 219

10   ibid., p. 221

11   ibid., p. 222

12   J. C. Kapteyn, 'On the Absorption of Light in Space,' *Ap J*, **29** (1909), 46–54. Within a few years, Kapteyn had revised his thinking, concluding that the absorption of starlight was actually negligible. Thus he arrived at the erroneous belief that the Sun was near the center of the Milky Way.

13   E. E. Barnard, 'On a Great Nebulous Region and on the Question of Absorbing Matter in Space and the Transparency of the Nebulae,' *Ap J*, **31** (1910), 8–14:8–9

14   ibid., p. 13

15   E. E. Barnard, 'Some of the Dark Markings on the Sky and What They Suggest,' *Ap J*, **43** (1916), 1–8: 4

16   He used a power of 350, which produced a field of view some 6' of arc across. EEB, observing notebook; LO

17   E. E. Barnard, 'Dark Regions in the Sky Suggesting an Obscuration of Light,' *Ap J*, **38** (1913), 496–501:497

18   EEB to Harlow Shapley, May 12, 1919; VUA

19   Barnard, 'Some of the Dark Markings,' p. 6

20   E. E. Barnard, 'On the Dark Markings of the Sky with a Catalogue of 182 such Objects,' *Ap J*, **49** (1919), 1–23:4

21   Barnard, 'Dark Regions,' p. 500

22   ibid., p. 501

23   Barnard, 'Some of the Dark Markings,' p. 3

24   ibid., p. 4

25   ibid.

26   ibid.

27   V. M. Slipher, 'On the spectrum of the nebula in the Pleiades,' *Lowell Observatory Bulletin*, no. 55 (December 20, 1912)

28   Barnard, 'On the Dark Markings of the Sky,' p. 2

29   P. Lowell to V. M. Slipher, February 8, 1913; LOA

30   P. Lowell, 'Nebular Motion,' text dated November 23, 1915 of lecture to Boston's Melrose Club; LOA

31   Barnard was skeptical of these high numbers, telling J. L. E. Dreyer of the Armagh Observatory that even Keeler's more modest estimate was probably 'much exaggerated.' J. L. E. Dreyer to EEB, February 28, 1909; VUA

32   H. D. Curtis, 'Report 1913, July 1–1914, May 15,' draft; SLO.

33   H. D. Curtis, 'A Study of the Absorption Effects in the Spiral Nebulae,' *PASP*, **29** (1917), 145; 'A Study of Occulting Matter in the Spiral Nebulae,' *PASP*, **13** (1918), 45

34   Barnard, 'Dark Markings on the Sky,' pp. 6–7

35 As soon as he realized that the Milky Way had a ring of occulting matter, Curtis understood why the spirals were found exclusively outside the Milky Way's 'Zone of Avoidance' – they could not be seen there through the obscuration.

36 Curtis had already been on the same track. Already in March 1917, he had found a new star of 14th magnitude on a plate exposed two years earlier on the spiral NGC 4527. Finding that the star had since disappeared, he suspected that it might be a nova. He then compared other plates that he had taken in 1915 with earlier plates of another spiral, NGC 4321, and found two more new stars. However, he delayed publishing his results, wanting to be certain that these were not variable stars, and before he was finished, Ritchey had made his discovery.

37 EEB to G. W. Ritchey, December 6, 1917; HHL

38 Harlow Shapley, 'Globular Clusters and the Structure of the Galactic System,' *PASP*, **30** (1918), 50

39 RGA to EEB, April 27, 1920; YOA

40 For details, see Michael Hoskin, '"The Great Debate": What Really Happened,' *JHA*, 7 (1976), 169–82 and Robert W. Smith, *The Expanding Universe: Astronomy's 'Great Debate,' 1900–1931* (Cambridge, Cambridge University Press, 1982)

41 Harlow Shapley, *Through Rugged Ways to the Stars* (New York, Charles Scribner's Sons, 1969), p. 57

42 E. E. Barnard, 'Micrometric Measures of Star Clusters,' E. B. Frost, G. Van Biesbroeck, and M. R. Calvert, eds., *PYO*, 6 (1931), 3

43 ibid., p. 1

44 E. E. Barnard, 'On the Comparative Distances of Certain Globular Clusters and the Star Clouds of the Milky Way,' *AJ*, 33 (1920), 86

45 Henry Norris Russell, 'Dark Nebulae,' *Proceedings of the National Academy of Sciences*, **8** (1922), 115

46 R. J. Trumpler, 'Absorption of Light in the Galactic System,' *PASP*, **42** (1930), 214–27. There were several steps in Trumpler's reasoning. His main interest was in star clusters, specifically open clusters like the Pleiades. Most of them lie close to the galactic plane, and thus are called 'galactic clusters.' He obtained spectra and measured the apparent brightnesses of many of the individual stars in these clusters. From the Hertzsprung–Russell diagram that had recently been worked out by Henry Norris Russell and Ejnar Hertzsprung, which gives the relationship between spectral type and intrinsic luminosity of stars, he was able to determine the true brightnesses of these stars, and comparing this result with their apparent brightnesses was able to estimate their distances. His final step was to plot the apparent diameters of these clusters (their angular diameters in the sky) against their distances. When he did so he found that the sizes of the clusters showed a steady increase with their distance from the Sun. Since he could not accept that the Sun was at the center of the Galaxy, he reasoned that the more remote stars did not appear fainter due to distance alone – they were also being dimmed by some kind of intervening matter. When Trumpler carried out the same analysis of globular clusters, of which the most distant lie far from the galactic plane, he found that the absorbing effects were much less. Thus he was finally able to prove what had been strongly hinted at ever since Barnard began photographing the dark clouds of the Milky Way – the interstellar matter is concentrated mainly in the galactic plane.

47 Gerrit L. Verschuur, *Interstellar Matters* (New York, Springer-Verlag, 1989), p. 102

48   Bruce Medal acceptance speech, 1977; quoted in Joseph Tenn, 'E. E. Barnard: The Fourteenth Bruce Medalist,' *Mercury*, **21** (1992), 164–6:166.

49   For a brief popular account, see Charles Lada, 'Deciphering the Mysteries of Stellar Origins,' *Sky and Telescope*, **85** (1993), 18–24

23

Eclipse and decline

1

Barnard's career straddled the remarkable period in which the methods of classical astronomy increasingly yielded to those of the modern astrophysics. He witnessed but did not contribute to this transformation. While the younger generation forged ahead, he continued to rely, as he had always relied, on the tried and true methods of eye and micrometer. Despite his passion for precise measurements, Barnard regarded his splendid photographs as pictures, not as subjects for measurement. Thus the principal work he set himself with the great refractor, measuring stars in globular clusters, he carried out visually for over twenty years with the filar micrometer – even though as early as 1903, his younger colleague Frank Schlesinger, who later went on to a distinguished career at Allegheny and Yale, was showing that it was not difficult to measure the stars on Ritchey's plates of globular clusters with only a third the probable error of Barnard's visual micrometer measurements. 'I do not know which of us was the more surprised,' Schlesinger wrote; 'perhaps it was I, for Barnard had a very modest notion of all his attainments, and I realized better than he did how high a standard his micrometer work had set. To reach a high plane of accuracy with the micrometer is a difficult matter requiring in any case years of patient practice, and even then it is not attained by all':

> On the other hand it is characteristic of the photographic method . . . that an observer acquires in a few months, or even a few weeks, a standard of accuracy which years of experience do not enable him to surpass . . .
>
> The economy of the photographic method is equally surprising . . . [T]he time spent at the telescope to secure the same result is at least one hundredfold greater with the micrometer, and the measurement on photographs of one hundred stars in each of four globular clusters is a task that can be carried out in a few weeks at most.[1]

Nevertheless, the conservative Barnard went on with his work with the filar micrometer. It is more than probable that he invested more time and effort in this work than in any other he ever undertook, but the negative results – though eventually he drew the correct inference from them, namely, that the globulars were much more remote than had been hitherto thought – were undoubtedly a great personal disappointment to him.

2

At Yerkes, Barnard had always been a most inspiring figure to all of his colleagues there. His sheer diligence, and the range of his interests, made him so. His director, Edwin Brant Frost, wrote that to Barnard

> a night at the great telescope was almost a sacred rite – an opportunity to search for truth in celestial places. Rarely has a priest gone up into the temple with a deeper feeling of responsibility and of service than did this untiring astronomer go up into the great dome. He was usually ready before the sun had set, and impatiently waiting until the darkness should be sufficient for him to 'get the parallel' for the thread of the micrometer before he could observe faint objects. During the day preceding one of his nights, his associates in the observatory were generally conscious of his keen anxiety for a clear sky, as evidenced by a frequently repeated nervous cough, which was always worse if the prospects for the night were unfavorable.[2]

Mary Calvert, too, emphasized the great pleasure her uncle took in his work with the great telescope:

> He looked forward to each night . . . with an eagerness that seemed never to be dulled. This work came ahead of everything else, and nothing was allowed to interfere with it. When his night at the telescope came he would have his early supper, be dressed for the night and at the observatory, often before the sun was down. There was no last-minute rush to the dome.[3]

It was not only when it came to observing that Barnard was obsessed with punctuality. 'He was always ready for any engagement he might have just on time or, more likely, a little ahead of time,' Calvert continued. 'He could never understand and was always irritated by a guest who, invited for a certain time, would come a little (or perhaps much) later than the time named. Probably his long work at the telescope was partly responsible for this habit of extreme punctuality. One cannot be late in making an observation of an eclipse or any such celestial phenomena.'[4]

In his chosen career of astronomer, it helped that Barnard was able to get along with only four hours (and often less) of sleep a night. After a night of clear skies and good seeing, he was often merry – 'laughing and joking, or he might be heard singing in the dark room . . . From his dark room also would come the sound of his voice reciting poetry and sometimes a song improvisation of such things as "The Burial of Sir John Moore." '[5] He did not allow clouds to deter him from his vigils; when the night was cloudy, he found it difficult to relax, and was constantly on the lookout for a possible clearing of the sky. Nevertheless, he could usually be counted upon to appear in his office by seven o'clock the next morning.

He willingly endured the greatest extremes of temperature in the pursuit of his work. Visitors to the observatory were amazed that he continued to work when the temperature in the dome was ten or fifteen degrees below zero Fahrenheit, and when they asked him how he kept warm, he replied: 'We don't!' One profoundly miserable

Barnard, 1915. Yerkes Observatory

winter night, when the temperature plunged to $-26°$ Fahrenheit, Barnard was observing with the 40-inch refractor; S. A. Mitchell, who was at that time an assistant at the observatory, was working at the 12-inch refractor. At 2 a.m., a haze came over the sky. 'We each left [our respective] dome to go down stairs to thaw out,' Mitchell recalled. 'Inwardly, I must confess that I hoped it had clouded for good. If he felt the same he did not say so. The haze was only the last traces of moisture being frozen out of the atmosphere for it cleared off and we both went back to work until seven o'clock . . . [O]h! the torture of working so long at a stretch at such temperature with one's vitality at so low an ebb!'[6] On another equally bitter night, Barnard did stop work and close the dome, although the stars were still shining brightly. The next morning, he explained that he had done so from worry that the telescope might break or be injured in the extreme cold. Frost, who knew that Barnard would never abandon work solely with a thought to his own comfort, agreed that this precaution was prudent, and suggested that henceforth whenever the temperature in the dome dropped below $-25°$ Fahrenheit work should stop – for the sake of the telescope![7]

Probably no one, with the possible exception of William Herschel, was ever more diligent in taking advantage of every fragment of clear sky. Otto Struve, who at the end of World War I came to Yerkes from Russia where he had been fighting Bolsheviks, and later succeeded Frost as director, learned from Barnard 'the importance of catching the rare moments of good seeing . . . that to accomplish something in observational astronomy we must make use of all clear sky: there is not enough of it to waste by starting late or by being too choosy about the conditions':

> Barnard was always on duty and no sooner would the sky clear up after a snow storm than we would hear him opening the dome. It was a familiar sight for us when Mr. Barnard would walk up in the afternoon to the barograph in the library and would sigh deeply and disconsolately if the pressure was going down [foretelling the imminent clouding up of the sky].[8]

Barnard's personality impressed many of those who knew him as simple, straightforward – indeed almost childlike. Heber D. Curtis, who first became acquainted with him on the eclipse expedition to Sumatra, thought of him as 'an astronomical Peter Pan who never grew old, and whose every waking thought was concerned with his beloved science.'[9] Frost, perhaps with a hint of envy given his own absorption in business as director of a great observatory, claimed that part of Barnard's charm lay in his apparent 'detachment from the affairs of the mundane world,'[10] and added: '[He] was not entirely of this world. His interests and his thoughts had been so long in the heavens that many of the affairs of the earth had little interest for him, and often escaped him.'[11]

Though his interests were mainly astronomical, and he had little leisure for anything else, Barnard did take pride in the orchard he planted near his home on Lake Geneva, and the notes in his observing book reflect a sensitive awareness of natural phenomena generally. He often jotted down in the midst of observations of specifically astronomical interest: 'Autumn foliage very beautiful today,' 'The lake froze over last night,' 'Heard the first whipporwill of the season tonight just after dark,' or 'Great numbers of wild geese on the lake making a great commotion.' He also enjoyed entertaining visiting astronomers at his home. 'Away from his observatory and from his home,' Frost recalled, 'Mr. Barnard was always rather shy and restive, but in his office, and

Yerkes Observatory ski club, about 1914. Yerkes Observatory

particularly in his home, he was a most delightful host and entertained his guests with charming hospitality, in which Mrs. Barnard fully shared.'[12] However, they were always made aware of the fact that he considered it his duty to excuse himself if the sky was clear.

He never went out of his way for exercise, though for that matter pushing the 40-inch refractor around to its various pointings during a night in the pure out-door air of the dome was itself more than enough exercise for most mortals. Occasionally – and then under protest – he joined in a round of golf on the course adjoining the observatory grounds. In the winter he was a member of the Yerkes Observatory cross-country ski club (skiing Scandinavian style – with only one pole), and in the summer was an occasional swimmer in Lake Geneva. A younger colleague at Yerkes, the double star observer Philip Fox, wrote that Barnard was

> a powerful swimmer having developed his stroke in the swiftly flowing Cumberland River. One day of the first summer that Professor Frank Jordan spent at the Yerkes Observatory, Professor Barnard joined us for such a swimming party. Some of us had told Professor Jordan that Barnard was not over-skillful but a very venturesome swimmer, and that should he venture into deep water he would need careful watching. We have often joked over Jordan's evident distress, and subsequent immediate perplexity and chagrin, – when Barnard dove without hesitation into

Barnard swimming in Lake Geneva, around 1914. Yerkes Observatory

deep water at the end of the pier and came up blowing water from his moustache
before starting to swim with strong swift strokes.[13]

3

Though a conservative when it came to measuring off of photographic plates, Barnard
gradually modified his skepticism about the value of reflectors. During the nineteenth
century, reflectors had largely deserved the ill-repute from which they suffered. When
it first arrived on Mt. Hamilton, Barnard had described the Crossley reflector as 'no
good old junk,'[14] but by 1900 he was forced to admit that Keeler's nebular photographs
taken with the Crossley were 'the finest that have ever been made,'[15] and by then
Ritchey was starting to do important work with the 24-inch reflector in the southeast
dome at Yerkes. Like Barnard, Ritchey was 'an artist at heart,' and took endless pains
with his exposures. In order to bring out the faintest details, he sometimes used such
highly diluted developer that the treatment of the plates in the darkroom took many
hours.[16] In this way he obtained photographs of clusters and nebulae which set the
standard for decades. After leaving Yerkes with the rest of Hale's first team for Mt.
Wilson, Ritchey turned to building an even larger reflector with the 60-inch mirror that
Hale's father had bought soon after the Yerkes Observatory opened. Ritchey had
already begun figuring it at Yerkes, but at the time there had been insufficient funds to

complete the telescope. At Mt. Wilson, Ritchey finished the mirror, and Hale succeeded in obtaining a grant from the Carnegie Institution to have it mounted. The first observations with this marvelous instrument were made in December 1908, and Hale wrote that visually, 'the star images are very small and sharp . . . Such an object as the Great Nebula in Orion shows a bewildering variety of detail.'[17]

Though Hale's chief aim with the 60-inch reflector had always been to use it to obtain spectrograms of the brighter stars, it was ready in time for the favorable opposition of Mars in 1909. Interest in Mars continued at a very high pitch, with the general public still as fascinated as ever by the personality of Percival Lowell and the sensationalism of his Martian canal theories. At the same time, the view first proposed by E. Walter Maunder in 1894, that the 'canals' were an optical illusion in which irregular, discontinuous details were joined together by the eye into continuous lines, was gaining ground. Maunder had explained to Barnard in 1907:

> We see with our eyes and consequently . . . our vision is subject to all the imperfection of those eyes. It is the principle that Mr. Samuel Weller enunciated when he was explaining why he could not see through two pair of stairs and a deal door[.] 'Being only heyes, you see my wision is limited.' I fear that Mr. Percival Lowell has sometimes forgotten this important truth.[18]

Barnard, who had seen Mars better perhaps than anyone up to that time, shared the skepticism of most professional astronomers toward the canals, telling Lowell's nemesis W. W. Campbell, 'I perfectly agree with you as to the great harm Lowell is doing.'[19] However, he was careful to avoid making any fulminant public pronouncements on the subject, and as a result remained on good personal terms with Lowell, often meeting him in Chicago whenever Lowell was en route between Flagstaff and Boston where he lived for about half of each year.

Instead of visually observing Mars in 1909, Barnard concentrated on photographing it with the 40-inch refractor. Barnard noted that 'better conditions are required for successful work in this direction than for visual observations. One can do much visually under conditions where the best definition is only momentary; but for these enlarged photographs any break in the definition for even a single second during the exposure means injury or total ruin to the image.'[20] Only on September 28 did he enjoy what he considered to be 'fairly satisfactory conditions' for this work. Observers elsewhere entertained the highest expectations of his results. Thus E. M. Antoniadi, who had formerly been Flammarion's assistant at Juvisy and was now, at the invitation of director Henri Deslandres, 'astronome volontaire' at Meudon Observatory, wrote to him: 'We all felt . . . that the very favorable apparition of Mars now over would greatly increase our knowledge of the planet; and the eyes of the astronomical world were turned to its foremost observer . . . In fact, all our hopes hung on you; we were awaiting with feverish anxiety your results.'[21] Antoniadi himself was keenly interested in the 'canal' question. His own results in 1909 proved to be far more sensational than Barnard's, if only because they were more provocatively framed. He had previously had the opportunity to observe Mars only with smallish telescopes. As soon as he had the chance to peer at the planet with the 33-inch Henry refractor at Meudon on September 20, he immediately realized the crushing superiority of such large instruments. That night the disk was perfectly calm, and Antoniadi – enjoying views like those Barnard had with the Lick refractor in 1894 – was astonished by the bewildering mass of intricate details on view. Being one of the most skilled draughtsmen ever to sketch the planet, he was able to capture much of this detail. 'After my excitement . . . was over,' he wrote, 'I sat down and drew correctly both with regard to form and intensity all the markings visible.'[22] But for all the detail visible, there was not a straight line anywhere on the disk, and it was obvious to him that the geometrical network of single and double canals was a gross illusion. Henceforth he resolved to 'work day and night to demolish these whimsical provocations of truth.'[23]

With the Mt. Wilson 60-inch telescope, diaphragmed to 44 inches, Hale had much the same experience. 'I was able to see a vast amount of intricate detail – much more than has been shown on any drawings with which I am acquainted,' he told Antoniadi. 'In spite of the very fine seeing on certain occasions . . . no trace of narrow straight lines, or geometrical structure, was observed. A few of the larger "canals" of Schiaparelli were seen, but these were neither narrow nor straight . . . I am thus inclined to agree with you in your opinion . . . that the so-called "canals" of Schiaparelli are made up of small irregular dark regions.'[24] Hale added that he was joined in his conclusions about Mars by the other astronomers who were on Mt. Wilson at the time, including Adams, Ellerman, and A. E. Douglass, Percival Lowell's former assistant who was now a critic of the canals. 'All . . . agree with me regarding the character of the details shown,' Hale wrote.[25] Barnard, of course, had had this view of the planet ever since 1894, and he too wrote to Antoniadi: 'I am particularly glad that you have had the same view of the planet, and have drawn it with the great Meudon refractor. I note that in your drawings the canals have become broader, more diffuse and more irregular than most people show them. This is in accord with my own observations . . . made with large instruments.'[26]

Reacting to the increasingly vituperative tone of the debate about Mars, Robert G. Aitken of Lick Observatory proposed that some of the leading observers such as Barnard and Antoniadi go to Flagstaff in order to witness for themselves what Lowell and his assistants were claiming they could see.[27] Antoniadi, however, felt that the effort would be in vain, and told Barnard: 'I believe that . . . your views . . . will [not] change at all . . . Having seen Mars better than anybody else in 1892 and 1894, I am sure that you have nothing to learn, as to the true structure of the minor details of the planet, by studying it anew with an inferior, or crippled telescope' (alluding to Lowell's penchant for stopping down his 24-inch object glass to 12 or 15 inches).[28] He wrote in similar vein in the conclusion of one of his observing reports for the British Astronomical Association: 'The frail testimony of small refractors has vanished before the decisive evidence of giant instruments.'[29]

Though Antoniadi accorded Schiaparelli's canals at least a limited basis in reality – they were, he thought, 'the optical products of very complex and irregular natural [markings], sporadically scattered all over the Martian surface,' he regarded Lowell's network as entirely non-existent. He gave credit to Barnard's observations of 1894 as a 'rebuke from which the spider's webs were never to recover,'[30] but he was himself the first to do what Barnard had despaired of doing – sketch accurately the intricate details of the Martian surface. After comparing Antoniadi's drawing of the Syrtis Major region with one of Hale's photographs, Barnard congratulated him, 'The very remarkable agreement . . . proves to me that your drawings are precise.'[31] Sharing Barnard's deep sense of duty in making his observations as accurate as possible, Antoniadi wrote in another letter to Barnard: 'My only ambition is to defend the truth and write nothing susceptible of being overthrown. When we feel sure that our work will remain, that our representations of the heavenly bodies are accurate . . . then may we quit this world with the satisfaction of accomplished duty.'[32]

4

Hale's observations of Mars with the 60-inch telescope were an incidental affair; the telescope had been designed chiefly for spectrographic work on the stars. Barnard was eager to see it in action, and had his first chance in late August 1910, when he went to Mt. Wilson for the third meeting of the International Union for Cooperation in Solar Research, an organization which Hale had brought into existence at the World's Fair in St. Louis in 1904. About a hundred of the world's most distinguished astronomers and physicists attended the meeting on Mt. Wilson, though most of them were astronomers who studied stars and nebulae rather than the Sun. From Pasadena, the delegates traveled the first part of the way in horse-drawn carriages, but had to take mules and burros along the Sierra Madre trail leading to the summit. In Barnard's carriage were Sir Joseph Larmor, Frank Dyson (later Sir Frank and Astronomer Royal of England), and W. S. Adams. Halfway up the road, a whiffletree struck a projecting rock and was broken, and Adams, who was the best mountaineer on the Mt. Wilson staff, ran ahead in order to stop the next carriage, while Barnard and the others followed at their leisure. Adams later recalled with amusement: 'A short distance up the road on an inside curve I stopped to see how the others were getting along, and have always regretted the absence of a camera to record the expressions of these famous scientists. All three had shed their coats and had settled down into serious business, plodding along, Barnard with a look of placid determination, and Sir Joseph and Dyson with the grim resolution and doggedness characteristic of their race.'[33]

Hale was at the meeting only briefly, making a one-day appearance before he and his wife Evelina dashed off for Chicago, en route for New York and Europe. Though still comparatively a young man at forty-two, he had been under great strain for years, and was now suffering from severe bouts of 'nervousness.' After suffering what he later called 'a preliminary nervous attack' in 1908, from which he sought, unsuccessfully, relief through a European sojourn, he returned to the observatory still feeling under tremendous strain. Always grasping, he was full of plans for a 100-inch telescope, for which $45 000 had been donated by the hardware magnate Joseph D. Hooker, who had

Barnard at third meeting of the International Union for Cooperation in Solar Research on
Mt. Wilson, August 1910. Yerkes Observatory

earlier funded Barnard's photographic expedition to Mt. Wilson. The St. Gobain glass
works in Paris had contracted to produce a suitable 100-inch glass disk; but repeated
attempts had thus far succeeded in producing one that was badly flawed, and the aging
Hooker was running out of patience and threatening to withdraw further financial
support from the project. Moreover, a rift had grown up between Hale and Ritchey.
Ritchey was full of visionary ideas for telescope designs that were far ahead of their
time. In 1909, he floated to Barnard plans to mount the 100-inch telescope high above
ground, in the open air, with no dome but a shelter on wheels to be rolled over it for
protection during the daytime and in inclement weather, and to be pulled away at
night.[34] Barnard, who had suffered greatly from 'dome seeing' at Yerkes and had once
told Antoniadi that 'he hoped the great Yerkes dome could be mounted on a rail, so that
he could push it away and work in the open air with perfect images,' was presumably
sympathetic.[35] Nothing came of this particular plan, but a year later Ritchey proposed
to Hooker himself that instead of using the flawed 100-inch disk, the telescope ought to
be fashioned with a built-up mirror, consisting of three thin 100-inch disks separated
by spacers or ribs of the same type of glass cemented between them. Air circulated
between the three disks by fans would keep them at uniform temperature. As if this
were not bold enough, he suggested that the 100-inch ought to incorporate an
innovative optical system which he and Henri Chrétien had just developed in order to
reduce distortions in star images at the edge of the telescope's field. Hale was opposed to

the built-up mirror idea, though he was willing to look further into the Ritchey–Chrétien optical concept. Nevertheless, he saw Ritchey's attempts to lobby Hooker on behalf of his ideas as disloyal, and henceforth became increasingly negative toward his assistant.[36]

These were a few of the worries that weighed Hale down in the weeks just before the International Solar Union meeting. His wife saw him on the verge of a complete breakdown, and urged him to check in to a sanatorium (as she herself had done in 1905). However, saying that he could not bear to be away from her for even a brief period, he refused, and instead of preparing for the meeting went on a fishing vacation to Lake Tahoe and Oregon. He proposed, moreover, to leave his trusted assistant W. S. Adams in charge of the observatory while the two of them sailed again for Europe, to 'travel about as we might choose, to new places and old, avoiding all scientific men and institutions, and renewing our youth in a second wedding journey.'[37] In the event, Hale would be out of astronomy for sixteen months. Moreover, the same theme would recur again in later years – Hale taking off for lengthy travels or stays in sanatoria trying to shake what he called his 'head problems.'

Even without Hale's participation, the International Solar Union meeting was an enormous success. On each of the three nights when the meeting was in session, Ritchey showed off the 60-inch to the assembled dignitaries, and on one of them Barnard made a brief observation of Saturn. He did not expect much; though he had long been aware of the reflector's value in photographic work, for some reason – perhaps due to an early observation with a reflecting telescope under unfavorable conditions, as he later explained to Hale – he had formed the idea that a reflector would not be satisfactory for visual work. However, even this brief view was enough to show him that he had been wrong.

After Barnard had returned to Williams Bay, Hale, on the way through the south of France and on to Italy, with his eventual destination being Egypt, learned that Andrew Carnegie was donating $10 million more to the Carnegie Institution. Thus, even without Hooker's continued support, there would be money enough to finish the 100-inch telescope. At the same time, Hale came across an article by Barnard in the *Monthly Notices of the Royal Astronomical Society*, in which Barnard had included the best of his photographs of Mars from 1909. Perhaps he intended to show up Ritchey, who had thus far failed to obtain any good planetary photographs with the 60-inch. In any case, he suggested that Adams ought to invite Barnard to come to Mt. Wilson for the purpose of phtographing Mars. Frost agreed to part with Barnard for a month, and Barnard, eager to use the 60-inch reflector, arrived in Pasadena on November 6, 1911, and stayed through December 8. The season was not the best, with bad conditions setting in with the winter rains, but the timing was dictated by the November 24 opposition of Mars (much less favorable than that of 1909). Most of the time Barnard spent photographing Mars at the 100-foot Cassegrain focus, though because of Mars's greater distance, the results were disappointing and did not measure up to expectation. He was more successful with Saturn, which was well placed for observation at the time, and Hale published some of his best exposures in the Mt. Wilson Observatory *Year Book*.

Even more interesting than Barnard's photographic results, however, were his visual observations at the telescope's 25-foot Newtonian focus (owing to the time required to change the telescope from the Cassegrain to the Newtonian arrangement, it was not possible to use both arrangements on a single night; thus only a few visual observations were made, but even so they were very suggestive). Compared with the images of Saturn and Mars in the 60-inch, Barnard noted, those in refractors now seemed to him to have a 'muddy or dirty look.' The images of bright objects in refractors were always more or less degraded by chromatic aberration – though their focal lengths were more or less constant in the yellow and green spectral region to which the eye is most sensitive, there was considerable dispersion in the blue and violet regions, and thus a bright object was always surrounded with blue haze (this could be eliminated through the use of a yellow filter, but at the expense of losing the 'natural' colors of a star or planet). The reflecting telescope's parabolic mirror brought a star to a perfect focus without any chromatic aberration whatsoever. Barnard found the absence of this secondary spectrum 'very remarkable,' and added: 'The planet looks as if cut out of paper and pasted on [the] background of sky. It is perfectly hard and sharp with no softening of edges. The outline and general deffinition are much superior to that of a refracting telescope.' The disk of Mars appeared 'very feeble salmon – almost free of color'; the dark markings (Syrtis Major and Mare Tyrrhenum) were 'light grey.'[38]

Even with the aperture stopped down to only 12-inches because of indifferent seeing, Barnard saw Mars, on November 23, better than he had ever seen it before. 'The Syrtis Major was broken up into a great number of wispy masses,' he wrote.

> The momentary best seeing gave the impression that the broken masses were still further shattered, so that the whole mass would be a flock of wisps with no continuity of form . . . whatever. Certainly the true nature of this remarkable region has never been so clearly seen by me before. Even in 1894 with the 36 [inch] of L[ick] O[bservatory] it was not so well seen. The impression I now get of the Syrtis [Major] is . . . in being more thoroughly broken up . . . No trace of any thing resembling a canal either in the dark or the bright regions could be seen. I think this is perfectly decisive.[39]

A few nights later he was able to use the full aperture of the 60-inch reflector on Mars, to much the same effect: '[The Syrtis Major] is broken up in to cloud like masses with wispy details . . . The great mass . . . preceding the trunk of the Syrtis is almost as conspicuous as the Syrtis [and] seems to be made up of a shredded appearance . . . It is not possible to draw the details because they are so complex.'[40]

After such revelations, Barnard did not hesitate to pronounce that, given an equal facility for handling the large reflector, he would prefer it for visual work on the planets to either the Lick 36-inch or Yerkes 40-inch refractors.[41] His views of the stars were equally impressive. On turning the telescope toward the Milky Way, Barnard told Hale, he found that 'the stars looked like jewels on black velvet. The sky was rich and dark, and every star was a glowing, living point of light.'[42]

Barnard, though belatedly, had seen the light about reflectors. Indeed, all the great telescopes of the future would be reflectors – the Yerkes refractor was the last instrument of its kind, and will probably never be surpassed, if only because there is no reason for anyone to bother to do so. The Mt. Wilson 100-inch reflector, with mirror finished by Ritchey and mounting designed by Pease, went into operation in June 1919. Shortly afterward, Hale dismissed Ritchey from the Mt. Wilson staff, citing among other things, in a confidential letter to Robert S. Woodward of the Carnegie Institution, his long-time assistant's 'unreasonable attitude with regard to the 100-inch telescope.'[43] For his part, Hale continued to have 'head problems,' relapsing repeatedly into depression and inertia, and for reasons of health reluctantly stepped down as director of Mt. Wilson in Adams's favor in 1923 (he remained as 'honorary director'). Increasingly in his later years, he endured life only by avoiding people and spending much of his time in the seclusion of his private solar observatory, an improved version of the Kenwood Physical Observatory, which he built at his home in Pasadena in the attempt to recapture his youth. He confided to Frost that he was 'revert[ing] in part to an earlier state.'[44] In 1928, after a stay at Austen Fox Riggs's sanatorium for (wealthy) 'psychoneurotic' patients in Stockbridge, Massachussetts, he was able to pull himself together and emerge from seclusion long enough to secure funds from the Rockefeller foundation for another large telescope, the 200-inch Palomar giant, though it was not finished until long after Hale died in 1938.

His last visit to Yerkes was in 1932. According to staff astronomer George Van Biesbroeck, 'He came in, in a hurry as always,' rushed in to see Frost, stayed there for about twenty minutes, and after exchanging a few remarks with Van Biesbroeck in the hall was about to head out the front door. Then suddenly he turned and said, 'I must have a look at the 40-inch.' He dashed up the stairs to the dome, opened the iron door and entered. He then took off his hat and stood there in silence, looking up at the venerable telescope. Finally he remarked, 'Noble instrument.' Then he left, never to return.[45]

5

Barnard remained in generally good health well into his fifties – except for the usual bouts with colds and bronchitis from which he suffered almost every winter, of which the worst case was in 1907–8 when he had to give up observing for several weeks. Ellerman attempted to console him, and urged him to leave Wisconsin during the winters:

> I learned the other day that you have been having a most miserable time with the cold this winter, and laid up most of the time. I feel awefully sorry for you as I know how painful and distressing your attacks are . . . What you ought really to do is to pack up your material as the autumn approaches and head out into some mild climate and work up your summer observations, and not try to do any observing in winter, but get away from that abominable climate from December to April.[46]

As he had always done, Barnard rejected this sensible plea, and soon was back to pushing himself as hard as ever. Nevertheless, his age was starting to tell on him. He was putting on more weight. This development may have helped him push around the great telescope, but it was not, alas, without serious complications. In February 1914 he noticed that he was beginning to tire more easily; he was thirsty much of the time and passing copious amounts of urine, which made life difficult for someone who was often required to guide a telescope while exposing a photographic plate for six or eight hours without a break. He continued to work, but on March 8, 1914, he took to his bed – the nearly total eclipse of the Moon three nights later he observed from his bedroom window.[47] His physicians were called in to examine him, and at once diagnosed diabetes mellitus. Alas, though the diagnosis was easy, the treatment was obscure – in those days before insulin, the outlook for the diabetic patient was bleak. Barnard's physicians agreed with the prescription recommended by Sir William Osler in his influential textbook of medicine (and recommended, but to little avail, by Barnard's friends ever since Bishop McTyeire had long before encouraged the young Vanderbilt fellow to 'Rest, *rest*.') 'Sources of worry should be avoided,' Osler had directed, 'and [the patient] should lead an even, quiet life, if possible in an equable climate' (hardly attainable in Wisconsin).[48] As part of the proposed rest cure, Barnard's doctors ordered him not to observe with the 40-inch telescope. For him, this was the greatest imaginable privation, but as Frost observed, he bore it 'manfully.'[49] Still, the director hastened to add that it was 'almost impossible for Mr. Barnard to keep away from the Bruce photographic telescope when the sky was clear and the moon did not interfere.'[50]

6

By giving up the great telescope for a year and escaping to California for three months during the worst of winter (between December 19, 1914 and March 22, 1915), Barnard recovered his health to the point that he was once more able to resume observations with the 40-inch. However, for the rest of his life he would continue his struggle with the vitality-sapping, wasting disease.[51]

Generally speaking, he no longer had the energy to begin new programs of research. It was enough to carry on and prepare for publication work that was already underway. He kept up his observations of globular clusters and steadily added to his monumental catalog of dark nebulae, which reached 182 objects by 1919 and 349 by the time of his death. Above all he worried over the prints, and later the text, of his *Atlas of Selected Regions of the Milky Way*, which he had been thinking about almost constantly ever since 1907, when he had received funds from Carnegie to publish the photographs he had taken with the Bruce telescope. Frost, hoping to spare him years of frustration like those he had experienced in trying to reproduce the photographs with the Willard lens, convinced him that photographic prints would most faithfully reproduce the details of the original negatives. Thus Barnard selected fifty of his best negatives, and then painstakingly made a second negative from each from which the prints could be made.

In this way, the contrast in faint regions was increased and details brought out which otherwise would have been lost. A Chicago firm of commercial photographers, A. Copelin & Son, was hired to make seven hundred prints from each of Barnard's second negatives, and the prints were then carefully mounted on muslin. From May 1915 onward, Barnard made frequent trips to Chicago so that he could personally supervise the work in progress and hold Copelin and his son Bert, who did most of the prints, to his high standards.

But the story of the 'Milky Way and Comet photographs' repeated itself. Oppressed by his huge self-demand, Barnard had an almost constitutional inability to finish such monumental works. Every one of the 35 700 prints to be used in making up the edition of seven hundred copies of the *Atlas* had to be personally inspected and passed by Barnard. Hundreds and even thousands were rejected, but even so he was still not completely satisfied. 'There were many cases,' he later wrote, 'where a rejection was unfair. This was when the print was slightly too dark or too light, but the difference was not large. Such a print must be passed, though the desire was great to throw it out. In other words, it seems impossible for the manipulator to attain to perfection in this work.'[52] In the end, it was not the plates but the text that delayed the volume's appearance, for Barnard was still struggling to understand the features that his photographs had revealed. In 1919, he wrote a letter in which he attempted to account for the long delay to Robert S. Woodward, president of the Carnegie Institution:

> I have suffered greatly because of my inability to get the volume out. The fault lies entirely with me, but I am blameless so far as any intentional neglect is concerned. For some years I have been badly handicapped with the inability, or very great difficulty, to get anything like this volume finished. It has caused me great sorrow and endless worry.
>
> The prints arc all finished and are here at the Yerkes Observatory . . . The only thing lacking is to complete the descriptive matter . . . I have endeavored to complete this descriptive part but have each time been disappointed with it, for my brain gets fagged quickly. I will go at it again and perhaps I can complete it. The Bruce photographic telescope is out of commision now (at Warner and Swasey's for changes) and perhaps this will give me less night work and maybe I can soon finish the MS.[53]

But it was not to be.

7

In all, the total number of plates Barnard exposed on the Milky Way and other sky fields at Yerkes amounted to 3500. These were in addition to the 500 he had exposed at Mt. Wilson in 1905. These plates contained a rich repository of potential information concerning stellar proper motions, variable stars, and even an unknown planet, Pluto, which was drifting across the field of a pair of plates he exposed in January 1921, though it remained hidden from astronomers for another nine years. Barnard himself found time to examine only a few of pairs of his plates on the 'blink comparator,' an ingenious instrument which had been devised by the Zeiss firm, in Jena, and allowed two plates of the same part of the sky taken at different times to be examined side by side, in quick succession, so that any changes leapt out immediately to the eye.

On thus confronting a plate exposed on Ophiuchus with the 6-inch lens of the Bruce photographic telescope in May 1916 with another of the same region that had been exposed with the 6-inch Willard lens in August 1894, Barnard found a 10th magnitude star on the more recent plate that seemed to have left no trace on the earlier one. At first he suspected that this star might be a faint nova. Investigating further, he located another star 4' of arc to the south which was bright on the 1894 plate but faint on the 1916 one. He looked up yet another plate of the same region from June 1904, and this one showed a star that was bright on that plate and faint on the two others. At first glance, there appeared to be several variable stars in this region, but Barnard was suspicious – the images all lay upon a straight line. It then dawned on him that 'the bright images must all belong to the same object which on two of the dates accidentally fell on a faint star image and gave the impression of several variable stars.'[54] He confirmed this by examining yet another plate of the region, taken in April 1907. The same star was later identified on Harvard College Observatory plates from 1888 and 1890.

This rapidly moving star was at first referred to by the Yerkes staff as 'Gilpin,' after a character whose horse runs away with him in a poem by William Cowper.[55] It is now known as 'Barnard's Star' or 'Barnard's Runaway Star.' It is a red dwarf, and has the distinction of having the greatest apparent motion (proper motion) of any star known, its annual motion of 10.29" being so rapid that with the large image scale of the 40-inch refractor, plates taken only a week apart show a measurable motion. After its discovery, the star was at once put on the parallax list at Yerkes, Lick, and other observatories, so that its distance could be worked out. It proved to be only 6.0 light years away, making it the nearest star to the Earth after the α Centauri system.

8

The years of Barnard's decline were years of decline generally for the Yerkes Observatory. Burnham's health was failing. At about the same time that Barnard fell ill from diabetes, Burnham gave up his own work on the 40-inch refractor and stopped making his weekend visits to the observatory. Frost, moreover, was not the mover and shaker that the dynamic Hale had been, and as a result many of the most gifted researchers left Yerkes for Mt. Wilson and elsewhere. To make matters worse, the observatory was increasingly depressed financially. When Yerkes died in England at the end of 1905, he left a $100 000 endowment for the care and maintenance of the observatory that had been named for him. But it was not enough to keep the observatory healthy, especially as the Carnegie Institution was increasingly transferring its financial support from Yerkes to Mt. Wilson. This made it impossible for the observatory to upgrade its equipment – its Bruce spectrograph, for example, which had been a gift of Catherine Bruce in 1901, was equipped with badly-annealed prisms of 'a notoriously opaque and dense Jena glass,' and was unable to compete with the better spectrographs that were being launched at Mt. Wilson, Lick, and Allegheny. Compared with Hale, Frost was an ineffective fundraiser, and had never been a very creative research scientist. Then, in 1915, he began to go blind. He had been myopic as a boy, needing glasses at the age of only ten, and on a cold winter night (December 15, 1915) while he was working alone at the 40-inch telescope, he suffered a retinal tear in his right eye which within a year led to the total loss of its vision. The other eye was affected by a cataract, and several years later it too suffered a hemorrhage, leaving him unable to read and hardly able to see for ordinary purposes. 'Those who knew him in the years of his blindness will forever retain . . . a vision of the superb courage with which he faced his affliction,' Struve later recalled:. 'Never did he complain or show annoyance.'[56]

Blessed with a phenomenal memory, Frost retained much of what he had read in earlier years, but he was no longer able to read the current astronomical literature himself and was dependent on the help of others to read it to him. Even with the most valiant effort, he was unable to keep up. He naturally remained most interested in problems such as the spectrographic measurement of the radial velocities of stars which had been in the foreground in the early days of his own research, before he had lost his sight, but these were no longer the most important problems of astronomy. Nevertheless, without a pension to retire on, he had no choice but to hang on. Already by the end of World War I, when Struve joined the staff, Yerkes was 'hopelessly outdistanced in astrophysics.'[57] The later years of Frost's administration were even more lackluster.

9

During these doldrum years, when a frost indeed seemed to have laid its cold hand upon the Yerkes Observatory, Barnard worked compulsively on. The Yerkes staff celebrated his sixtieth birthday in 1917 with a surprise party in his honor, and for the occasion he received tributes from around the astronomical fraternity. Adams's was typical in wishing him

> the heartiest congratulations on your attainment of the years of discretion. I know of no one to whom the term sixty years young could apply more aptly for all of your friends know the fresh and youthful vigor of your heart and mind. Perhaps Methuselah at sixty might have equalled you but I can think of no other comparison.[58]

For the most part, Adams's comments were based on his recollections of the younger, more vigorous Barnard he had known at Mt. Wilson. However, in the fall of 1917, Barnard's health temporarily improved enough for him and Frost to make a

Barnard, 1917. Yerkes Observatory

strenuous two-week trip out west to scout three potential observing sites (Denver and
Matheson, Colorado, and Green River, Wyoming) along the path of totality of the
forthcoming solar eclipse of June 8, 1918. Though Frost acknowledged that 'it is not at
all fair to judge a station by observations covering a few days in any one year; and still
less at a different season of the year from that of the eclipse which is to come,'[59] he and
Barnard were most impressed with conditions at Green River, where totality would last
for ninety-eight seconds. This was a sparse settlement on the river of that name,
overlooked by picturesque buttes, of which the largest and most impressive was
700-foot high 'Castle Rock.' (It had been from this site, incidentally, where pioneer

Barnard at Green River, Wyoming, eclipse expedition, May 1918. Courtesy of Professor Robert Lagemann

explorer Major John Wesley Powell had started on his famous voyage to the Grand Canyon in 1869.)

The advance party of the Yerkes Observatory eclipse expedition, consisting of Barnard, Mary Calvert, the observatory carpenter, and the gardener, returned to Green River on May 4, 1918, five weeks before the eclipse. Barnard had stopped briefly in Chicago to see his old friend Burnham, who was now eighty and in poor health. He was 'very feeble,' with his left hand shaking badly, and Barnard guessed that he did not have much longer to live. 'I told him I was going to the eclipse,' Barnard jotted in his notebook. 'It deeply grieved me to see him so feeble.'[60]

At Green River, Barnard received help from prisoners from the local jail in transporting the equipment from the train. A borrowed Ford truck was used to transport it to the camp, which was too far out of town to be easily reached on foot. The equipment included the observatory's 12-inch 'Kenwood' refractor, with which Barnard hoped to obtain direct photographs of the Sun, and the coelostat apparatus which he had used at Wadesboro in 1900. Exposures with the latter instrument were to be made by Calvert, under Barnard's supervision.

During the first few weeks, while the camp was being set up, Barnard took infinite pains in adjusting the coelostat and assuring perfection in the focusing of all the cameras for which he had any responsibility. The Kenwood telescope was mounted on a temporary pier. The dry air and alkali dust cracked the lips of the members of the party, and at times the camp was visited by swarms of flies and 'very large and unrelenting' mosquitoes. But Barnard's greatest torment had to do with the weather, which for the most part proved very unfavorable and, after the ill-fated Sumatra expedition, almost wore him out with worry. As Calvert recalled: 'It was soon seen that this spot in the desert could show a great variety of weather conditions in a short period of time – beautiful sunshine with a deep blue sky, rain, snow, high wind, sand storms (from which driving clocks and telescope gears and bearings had to be protected), and even a small tornado which threatened the camp, but passed a few yards to one side of it.'[61] No matter how beautifully clear the sky was in the morning, the clouds 'had a wicked way of gathering during the day – a few fleecy white ones at first, then in greater numbers, until in the afternoon about the time the eclipse was to occur, the sky would be more or less overcast.'[62]

A week before the event, Barnard and Miss Calvert were joined by the other Yerkes expedition members, Frost, J. A. Parkhurst, and Storrs Barrett, who planned to obtain spectroscopic observations during the eclipse. They were joined also by L. A. H. Warren and H. R. Kingston of the University of Manitoba, and Hale, Ellerman, and other members of an expedition from Mt. Wilson, had also arrived, and had set up their camp not far from their old colleagues.

The day of the eclipse started out particularly fine, but then clouds came up after noon and reached the Sun by the time of first contact. All the astronomers could do was wait and hope that they would drift past the Sun before totality. Unfortunately, they did not – to Barnard's great disappointment, at the moment of totality the Sun stood behind a thin cloud that partially obscured the corona. Ironically, only two or three minutes after totality had ended, the clouds had drifted by, and the Sun emerged into a perfectly clear sky.

Depressed and feeling very pessimistic about the value of his photographs, which he was sure had been ruined by the clouds (they were not developed until after he returned to Yerkes, but in the event proved to be very successful, beautifully showing the prominences and inner corona), Barnard returned to the hotel in Green River where the members of the expedition were staying when not actually in camp. Late that day Barnard together with a few volunteer assistants from Green River took the Ford truck back out to the camp in order to pack up the plates he had exposed. On returning again to the hotel that night, his companions were mystified by his 'incoherent exclamations of surprise.'[63] Above Castle Rock toward the east, Barnard had noticed not far from Altair a brilliant new star. It was, he estimated, about half a magnitude brighter than Altair, and yellower than that whitish star. He returned to the hotel at 11:30 p.m, weary to the point almost of total exhaustion, but with the adrenalin flowing pointed out the new star to Hale and Frost, who were preparing for bed. At once Barnard, Frost, and four or five others drove back to camp to observe the nova, and Parkhurst that night succeeded in obtaining its spectrum.

The nova was the brightest of the twentieth century, and unsurpassed since Kepler's star of 1604. Because the distance to the star is 1200 light years, the explosion Barnard observed in 1918 had actually occurred before the birth of Charlemagne. Barnard was only one of many independent discoverers who were startled by its sudden appearance; by the time he saw it, it had already been noticed by observers in Europe and the Eastern United States, and at the same hour he first recognized it, it was discovered independently by seventeen-year old Leslie Peltier of Delphos, Ohio, who later went on to become a prolific discoverer of comets. Nevertheless, Barnard's independent discovery was one of the crowning moments of his later career. Because of poor health and the disappointment of his hopes in various lines of work, Barnard's later years were not particularly happy, but on that one night he became a child once again, full of wonder at the sky.

At the time of its discovery Nova Aquilae was still on its ascent to maximum brightness, which it reached two days later. Catching it on its rising above Castle Rock on June 9, Barnard found that it was 'very white and very much brighter than Vega. It twinkles very much – low altitude. It must be as bright as Sirius.'[64] On returning to Yerkes, Barnard followed the star with the 40-inch as it slowly faded. By fall, the star had assumed the appearance of a small yellowish planetary disk, as the shell of gas thrown off during the eruption expanded.[65] Gradually this shell of gas dissipated to the point of invisiblity. Today Nova Aquilae is a wholly unremarkable looking bluish object of the 12th magnitude.

10

The Green River adventure was to prove the last of Barnard's major scientific expeditions. He no longer had the strength left to do much travel, and had to refuse invitations even to return to California again.[66] Indeed, he would never see again what he had once described as 'God's Own Country.'[67]

After Copelin finished the Milky Way plates, Barnard no longer had much incentive to go to Chicago, and rarely lectured at the University during his later years. It took too much out of him to do so. He was 'always nervous for some days in advance of such an engagement,' Frost recalled.[68] Among the last lectures he gave was one on the Milky Way as part of an evening course of lectures at Mandel Hall on the University of Chicago campus in the summer of 1919.

As already mentioned, he did not take students in any formal sense, but was always free with advice to those who asked for it. His correspondence was vast – not only with established members of the astronomical community, but with amateurs and members of the interested public. He was asked by a newspaper reporter for his opinion of Frederick Leonard, a young Chicagoan with an interest in astronomy who later became one of Frost's advisees as a graduate student at the Yerkes Observatory. Barnard replied that Leonard was 'too young, of course, to have made his mark in the astronomical world, but there is much promise for one so intelligent and enthusiastic in any branch of science,' and went on: 'May I suggest that public notoriety, in a case of this kind, can do an infinite amount of harm . . . [O]ften scientific men are apt to judge inversely the qualities of a person by the amount of publicity he receives. Such, therefore, is directly harmful to him.'[69] Personally, he was now indifferent to the celebrity that had meant so much to him in earlier years, and he wrote in rejecting the request of a photographer who wished to have a portrait of him for publication in a magazine, 'I have made it a rule to avoid publicity as much as possible.'[70]

To a reporter who wanted to know whether a passage in the Book of Job, 'He stretcheth out the north over the empty place,' might refer to one of the vacant regions of the sky that Barnard had studied for so many years, he patiently explained: 'There are comparatively small regions [in the sky] where few stars are seen or are shown on photographs. The most remarkable of these are not in the northern part of the sky. They are found in the southern heavens, especially in the constellations Sagittarius and Ophiuchus. They have no bearing whatever on any statement made in the Book of Job.'[71] A lady asked him about the canals of Mars, because she regularly visited the planet (psychically) and went on to describe the craft on them. He replied that he had seen neither canals nor craft, but politely asked for more details.[72] However, in reality, he had little patience for spiritualism or spiritualists, and on the occasion of a visit to the United States by the famous English scientist and spiritualist Sir Oliver Lodge, told Joseph Jastrow, psychology professor at the University of Wisconsin: 'I have been much concerned about the present revival of spiritualism. When I was a boy it was rampant, especially after the close of the civil war. It has appeared to me that something should be done to combat its pernicious influence.' Regarding Lodge's activities, Barnard suggested that 'the very fact that a man is "scientific" is hardly a proof that he is better prepared to investigate such a thing as spiritualism than other men, especially if he has a leaning toward spiritualism. He who wants to be deceived is easily deceived.'[73]

A more momentous issue that occasionally surfaces in Barnard's correspondence is Einstein's theory of relativity – in particular, the test of the theory suggested by Einstein in 1911, involving the gravitational bending of a ray of light as it passed near a massive body like the Sun. The amount of this bending could be determined by measuring the positions of stars near the Sun at an eclipse, and from his special theory of relativity Einstein calculated that it would amount to just $0''.87$ for stars whose rays just passed the limb of the Sun. Soon after Einstein published these calculations, Erwin Findlay-Freundlich, an assistant astronomer at the Royal Observatory in Berlin, appealed to Barnard and others for eclipse plates in order to test the theory, but unfortunately none of these plates had recorded stars close enough to the Sun. It was clear that if the gravitational bending of light was to be detected, special observations would have to be made for that purpose. The first attempt to do so was by W. W. Campbell and Heber D. Curtis of the Lick Observatory, at the Russian eclipse of August 21, 1914, but they were completely clouded out. (Incidentally, Findlay-Freundlich was also in Russia to observe the eclipse. While he was there, Germany declared war on Russia, and he and the other members of the party were arrested,

Albert Einstein and Yerkes staff, May 1921. Yerkes Observatory

though they were later exchanged for Russian prisoners who had been captured in Germany). The Lick observers tried the experiment again at the June 8, 1918 eclipse, which they observed at Goldendale, Washington. By then Einstein had published his general theory of relativity, which led to a new prediction for the gravitational bending of light near the Sun which was exactly twice as large as the former value. Campbell and his associates succeeded in recording a few stars near the eclipsed Sun on their plates, but unfortunately the images were somewhat elongated due to inaccurate guiding. Nevertheless, despite the large probable errors, Curtis attempted to measure them. He found no gravitational deflection. But though this negative result was confided to various astronomers including Barnard, it was never published.

Barnard followed all these developments with interest. At first he was skeptical that the matter could be settled by photographing stars during an eclipse. On hearing of British plans to send expeditions to Brazil and South Africa to attempt to measure the Einstein effect at the May 29, 1919 eclipse, when the eclipsed Sun would stand in the midst of the rich Hyades star cluster, he wrote:

> I am afraid that this very test will give an uncertain result because the ray must . . .
> pass within a few seconds of arc of the sun's limb to get the maximum result,
> through the densest part of the corona, and the displacement is very small even close

to the sun's limb. As we know nothing of the refractive power of the corona, and there must be considerable refraction from it, this may entirely vitiate and leave in doubt any results thus obtained.[74]

The results of this expedition were announced by Arthur Eddington at a November 1919 meeting of the Royal Astronomical Society in London, and seemed to confirm the prediction from Einstein's general theory of relativity.

In June 1921, a month after Einstein paid a brief visit to the Yerkes Observatory – unfortunately, Barnard does not seem to have recorded his impressions of the great man – Barnard replied to Reverend W. E. Glanville, of Baltimore, who had queried him about an article on Einstein's theory, that he still considered it possible that the 'deflection of the light from a star near the totally eclipsed sun may have some other explanation, though certainly the results of the eclipse of May 29, 1919 seem greatly in [the theory's] favor.'[75] As for the theory itself, however, Barnard admitted that he had no more grasp of it than most laymen – he was completely mystified. 'If light is ponderable then it must be subject to acceleration, I should think,' he told Glanville. But he added: 'It has been stated that there were only twelve men in the world who understand Einstein. I am not the thirteenth.'[76]

1   F. Schlesinger, 'Historical Notes on Astro-Photography of Precision,' in Harlow Shapley, ed., *Source Book in Astronomy, 1900–1950* (Cambridge, Mass., Harvard University Press, 1960), p. 113
2   E. B. Frost, 'Edward Emerson Barnard,' *BMNAS*, **21** (1926), 10–11
3   Mary R. Calvert, 'Some Personal Reminiscences,' *JTAS*, **3**:1 (1928), 28
4   ibid.
5   Philip Fox, 'Edward Emerson Barnard,' *PA*, **31** (1923), 195–200:199
6   S. A. Mitchell, 'With Barnard at Yerkes Observatory and at the Sumatra Eclipse,' *JTAS*, **3**:1 (1928), 25
7   Calvert, 'Reminiscences,' p. 29. The rule was later modified, and at present work on the telescope is given up whenever the temperature sinks below −5° Fahrenheit.
8   O. Struve, 'The Story of an Observatory – I,' *PA*, **55** (1947), 227–244:240
9   H. D. Curtis, 'The Comet-Seeker Hoax,' *PA*, **46** (1938), 71–5:75
10   E. B. Frost, *An Astronomer's Life* (Boston and New York, Houghton, Mifflin Co., 1933), p. 143
11   E. B. Frost, unpublished MS; YOA
12   ibid.
13   Fox, 'Edward Emerson Barnard,' pp. 199–200
14   EEB to GD, May 26, 1897; BL
15   EEB to GD, March 26, 1900; BL
16   Struve, 'The Story of an Observatory,' p. 238
17   G. E. Hale, Mount Wilson Observatory *Year Book*, **8** (1909), 170
18   E. W. Maunder to EEB, June 8, 1907; VUA
19   EEB to WWC, July 11, 1908; SLO
20   E. E. Barnard, 'Photographs of the Planet Mars,' *MNRAS*, **69** (1909), 471–2

21   EMA to EEB, December 11, 1909; VUA

22   EMA to P. Lowell, October 9, 1909; LOA

23   EMA to EEB, January 16, 1911; VUA

24   GEH to EMA, January 3, 1910; quoted in E.M. Antoniadi, 'Sixth Interim Report for 1909,' *Journal of the British Astronomical Association*, **20** (1910), 189–92:191–2

25   ibid.

26   EEB to EMA, November 27, 1909; quoted in E. M. Antoniadi, 'The Planet Mars,' trans. Patrick Moore, in *Astronomy & Space*, vol. 2 (1973), p. 254

27   R. G. Aitken, ' "Geometrical" Canals on Mars? A Suggestion,' *Science*, New Series, **31** (1910), 114–15

28   EMA to EEB, February 6, 1910; VUA

29   E.M. Antoniadi, 'Sixth Interim Report,' p. 192

30   E.M. Antoniadi, 'Mars Section, Fifth Interim Report, 1909,' *Journal of the British Astronomical Association*, **20** (1909), 136–41:138

31   EEB to EMA, May 16, 1913, in Antoniadi, 'The Planet Mars,' p. 254. Antoniadi greatly appreciated the encouragement he received from Barnard, and later paid him this tribute. 'These encouraging words were characteristic of this astronomer – a man as great for his personal qualities as for his scientific discoveries.' ibid., p. 259.

32   EMA to EEB, May 28, 1913; VUA

33   W. S. Adams, 'Early Days at Mount Wilson – II,' *PA*, **58** (1950), 97–115:107–8

34   G. W. Ritchey to EEB, September 7, 1909; VUA

35   Antoniadi, 'The Planet Mars,' p. 168

36   See Donald E. Osterbrock, *Pauper and Prince: Ritchey, Hale, and the Big American Telescopes* (Tucson, University of Arizona Press, 1993), pp. 103–62 for a detailed analysis of Ritchey's troubles with Hale.

37   GEH to Evelina Hale, August 6, 1910; HHL

38   EEB, observing notebook; YOA

39   ibid., recording observations on November 23, 1911

40   ibid.; November 30, 1911

41   G. E. Hale, Mount Wilson Observatory *Year Book*, **11** (1912), 197–8

42   ibid., p. 198

43   GEH to Robert S. Woodward, August 19, 1919; CIF

44   GEH to EBF, October 11, November 22, December 27, 1923; YOA

45   Quoted in Helen Wright, *Explorer of the Universe: A Biography of George Ellery Hale* (New York, E. P. Dutton & Co., 1966), p. 201

46   F. Ellerman to EEB, March 12, 1908; VUA

47   EEB, observing notebook; YOA

48   Quoted in Frank N. Allan, 'Diabetes Before and After Insulin,' *Medical History*, **16** (1972), 266–73:268

49   EBF to RGA, February 19, 1913; YOA

50   Frost, 'Edward Emerson Barnard,' p. 10

51   EEB to WWC, January 20, 1917; SLO

52   E. E. Barnard, *Atlas of Selected Regions of the Milky Way*, E. B. Frost and M. R. Calvert, eds. (Washington, DC, Carnegie Institution, 1927), p. 12

53   EEB to R. S. Woodward, May 23, 1919; VUA

54   E. E. Barnard, 'A Small Star with Large Proper Motion,' *PA*, **24** (1916), 504–8: 505

55   Cowper's poem 'The Diverting History of John Gilpin.' Planning to celebrate his twentieth wedding anniversary at the Bell in Edmonton, a small village north of London, Gilpin, a linen-draper, sets out on horseback; his wife, her sister, and children are to follow in a horse and chaise. On setting out, however, Gilpin is unable to restrain his horse and gallops ahead all the way to Ware (ten miles beyond Edmonton) and back again. Obviously the name 'Gilpin' was not inappropriate for a 'runaway' star.

56   Struve, 'The Story of an Observatory,' p. 242

57   ibid., p. 243

58   WSA to EEB, December 9, 1917; VUA

59   E. B. Frost, 'The Total Solar Eclipse of 1918 June 8: A Reconnaisance,' *PA*, **26** (1918), 103–10:106

60   EEB, observing notebook; YOA

61   Calvert, 'Remiscences,' p.30

62   ibid.

63   E. B. Frost, MS draft; YOA

64   EEB, observing notebook; YOA

65   E. E. Barnard., 'Some Peculiarities of Nova Aquilae III of 1918,' *Ap J*, **49** (1919), 199–202

66   EEB to W. H. Wright, January 7, 1920; SLO

67   EEB to HC, June 13, 1895; SLO

68   E. B. Frost, MS draft; YOA

69   Norman Klein to EEB, June 16, 1914; EEB to Norman Klein, June 18, 1914; VUA

70   EEB to Charles Brown of Brown Brothers, photographers, September 28, 1916; VUA

71   EEB to C. P. Bollman, August 31, 1915; VUA

72   EEB, undated letter, paradox file; YOA

73   EEB to Joseph Jastrow, February 28, 1920; VUA

74   EEB to Julius Stone, January 25, 1919; VUA

75   EEB to Rev. W. E. Glanville, June 15, 1921; VUA

76   ibid.

24

Ad astra

1

Barnard received a terrible blow on March 11, 1921 when he received word of Burnham's death. An even greater blow followed on May 16, when Rhoda, who had been in poor health for years, suffered a stroke. She died five days later. The loss devastated Barnard. Soon after he returned from the funeral in Nashville, where she was buried next to his mother, he wrote to Frederick Slocum of the Van Vleck Observatory: 'I am sad beyond measure. She loved me and cared for me more than I knew or could appreciate, but it all comes back to me now and I am heartbroken.'[1] To W. H. Wright of the Lick Observatory he wrote, 'I have felt so unhappy and broken in spirit, that it seemed impossible to write . . . I am thinking of Mrs. Barnard all the time and it seems so cruel that she should be taken away.'[2] He had always hoped to return to California and die there, but now he knew that it was too late. 'We hoped some day to go back to California and build a home on the Ranch,' he confided to Campbell, 'but that hope is all gone now and I could not think of the Ranch without her.'[3]

Depressed by Rhoda's death and wasted by diabetes, Barnard had lost much of his former drive. Nevertheless, he worked on through sheer habit and sense of duty. His niece, Mary Calvert, took care of the household and helped him fix up his observing books every morning after he observed. He continued to take his regular turns at the 40-inch, and Frost, who considered it 'not the least important part of my duties as director to see that he did not overdo,' was not very successful. In addition, Barnard regularly worked with the Bruce telescope. 'He only needed some very long exposures on certain regions to clear up some problems about the dark markings,' Frost wrote to Aitken at Lick, 'but he would nevertheless begin, in spite of my entreaties, when the sky was not such as to promise proper return for his labor.'[4] As always, Barnard had to do the work himself. 'He never felt that he could risk the loss of a plate by having a less experienced person take a hand in the exposure,' Frost continued. 'I have often tried to persuade him to let Miss Calvert guide for part of the time, and she was always ready, and the other men here would have been very pleased to have had the honor of assisting him.'[5] Again Frost appealed in vain.

Despite his advancing age and ill-health, his eye remained legendary for acuity. He made the splendid discovery of the nebula around Nova Persei in December 1916, at a time when it had been looked for unsuccessfully by others with the Mt. Wilson 60-inch. Adams wrote to him, 'I want to congratulate you on your ability to discover this nebula visually. No one here has been able to see it with the 60-inch, and Mr. Hale was saying it is one more case of your astonishing capacity to observe what the ordinary man cannot.'[6] He also kept up his measures of the fifth satellite of Jupiter. After his death, it went unobserved for another twenty years – until Van Biesbroeck, using as an occulter for the bright planet a glass plate one half of which was thinly aluminized so as to allow the image of the planet to be faintly visible through it, managed to capture it again with the 40-inch refractor. Van Biesbroeck described just how difficult it was for him to see it:

> Unfavorable atmospheric conditions must account for my failure to see the satellite in March and April this spring. Finally at the western elongation May 22 . . . occasional glimpses of the satellite were obtained under moderately favorable conditions. It appeared much fainter than the magnitude 13 assigned to it by Barnard, and measuring the position relative to the planet proved impossible . . . When we consider the large number of elongations observed by Barnard at certain oppositions where he held the satellite often for over two hours one wonders if possibly the brightness of the satellite undergoes fluctuations.[7]

Among his other work, he made careful observations of the phenomena of the edgewise ring of Saturn between November 1920 and August 1921, the results of which largely corroborated those of 1907–8.[8] He also obtained accurate positions of a recently discovered asteroid, in which he had a sentimental interest – 907 Rhoda, which had been named by its discoverer, Max Wolf, in honor of Barnard's late wife.

He apologized to Slocum in the summer of 1921 for not coming east for the dedication of the new Van Vleck observatory. 'There are two things that prevent my coming,' he explained. 'I am not able to stand the long railway journey. I am very busy trying to get things finished up before it is too late – I am no Mathusela.'[9] Above all he hoped to finish the *Atlas of the Milky Way*, despite his depression and weariness with life, which led him to remark to Aitken that 'it was a very lonely world, and that he hoped he would not have to stay in it much longer.'[10] To Fox he sighed: 'Oh! I am so terribly lonely.'[11] He was still thinking of Rhoda.

During the summer of 1922, S. A. Mitchell and his wife stayed in Barnard's home while Mitchell was doing some work at the observatory. 'He was then showing the effects of his hard night work,' Mitchell later recalled.

> As I had known him so long and so intimately, I, myself, having grown older and being the director of an observatory . . . tried to show him how foolish was his habit of staying up all night, when he was assigned to the forty-inch telescope, even though it might be pouring rain . . . His reply was characteristic of him: 'I have never yet shirked my duty and been found wanting, and I shall not begin now.'[12]

That fall he refused yet another invitation to come to California. 'It is fated otherwise,' he told W. H. Wright. 'I am tied here by work that has been delayed for years. Until that is done, and I hope it may soon be finished, I cannot get away. I am afraid also the journey would be too much for me. I have not been well for a long time. I feel lonely and discouraged.'[13]

Edward Emerson Barnard in September 1922. This is probably the last photograph ever taken of him. Yerkes Observatory

2

Barnard visited Chicago for a meeting of a local astronomical club on December 16, 1922 – his sixty-fifth birthday. It was to prove to be his last visit to Chicago. On December 21, he tried to examine the fading Nova Persei with the 40-inch telescope. However, he was unable to get any kind of an image. On pointing the end of the telescope down in order to inspect the object glass, he found that it was completely covered with frost. The next night he was assigned to the instrument was December 23, but the sky was overcast. On Christmas morning, he was at Frost's house with the rest of the staff for about an hour, but excused himself early, due to distress from an acute inflammation of the bladder. The next day Frost drove to Milwaukee to consult his occulist. By the time he got back to the observatory he found that Barnard was laid up in bed. A physician, Dr. McDonald, was called out and succeeded in relieving Barnard's distress from the bladder inflammation by means of a catheter, but recognized that the real problem was an enlargement of the prostate, for which the remedy would be surgery. In Barnard's case, because of the diabetes, it was felt to be too dangerous. Nevertheless, the catheter cleared the bladder inflammation up within a week, and Frost noted optimistically that Barnard was now 'getting sleep and rest and [dressing up] to go down to breakfast, and [doing] quite a little writing for his Atlas.'[14]

In fact, by this time Barnard was a dying man. He had been greatly weakened from his long struggle with diabetes – in the last year or two of his life, he had become greatly emaciated. This was, at the time, a common complication in cases of untreated diabetes, and due to the passing of large amounts of sugar in the urine which led to an inexorable wasting despite the patient's having a voracious appetite. Through his bedroom window, Barnard made his last astronomical observation, of an occultation of Venus by the Moon, on the morning of January 13, 1923. He would never again return to his beloved telescope. He and Mary Calvert worked a few hours every afternoon on the *Atlas*, and he also tried to keep up his correspondence. At least at times he seems to have expected that the worst was behind him and that he would recover. 'I am still sick in bed,' he wrote to Campbell the day after the occultation, 'but hope to be out before long. It is slow and tedious.' He added: 'My greatest unhappiness [is] that I can do no observing at all.'[15] A week later he told Wright: 'I am sitting up this morning . . . My strength is returning but my main trouble is unchanged. I had hoped for a surgical operation [but the] physicians do not advise it under the circumstances. In the meantime I am subject to the catheter with the forlorn hope that things will come all right again . . . It is distressing to be away from my observations.'[16]

At the beginning of February, a Chicago specialist in metabolic diseases, Dr. Woodyatt, was called out to Williams Bay, and after examining Barnard expressed the hope that he could be taken to Presbyterian Hospital in Chicago, placed on a strict diet and given the then new treatment for diabetes – the pancreatic extract insulin, whose life-saving effects had first been demonstrated only a year before by Frederick Banting and Charles Best on a patient at the Toronto General Hospital. In this way, he thought, it might be possible to bring things to the point where an operation could be attempted. But it was too late. Barnard was by then suffering from other complications, including congestive heart failure. Frost and Ambrose Swasey, who had come out from Cleveland, called on him on Saturday, February 3. Barnard had been dozing and was not quite awake, but he was able to discuss old times with Swasey.[17] Thinking that things were stable, Frost decided to leave for Milwaukee that afternoon in order to give some lectures on Sunday; by the time he got back on Monday, Barnard had taken a turn for the worse.

Barnard knew that the end was approaching. He told Mary Calvert that 'he didn't mind dying, but was sorry not to finish his work.'[18] After about noon on Monday, February 5, his consciousness began to cloud up, and he was no longer coherent. He was attended through the night by Calvert, Oliver J. Lee of the Yerkes staff, and Lee's wife. Drs. Woodyatt and McDonald were also at his bedside, and remained with him throughout the night and the following day, when Barnard was given the new insulin treatment. But it was no use. His vital functions were now rapidly failing, and at 4 p.m. the doctors decided that nothing more could be done and gave him a hypodermic injection of morphine. His pulse ceased at 8 p.m. Frost, recording the time, added: '*ad astra*' – to the stars.

Barnard, worn out from disease and overwork, appeared in death 'pitifully wasted in body and . . . pitifully wasted indeed,' as his onetime colleague Philip Fox observed.[19]

The dour George Willis Ritchey noted that 'It is too bad that his ambition and temperament were such that he could not take life more easily as he grew older.'[20] In the end, his compulsion to observe had made him a martyr to his science. His body was laid out in state in the rotunda of the observatory. Among those who attended the brief services held the following day were Henry Gale and F. R. Moulton of the University of Chicago, Fox and Henry Crew from Northwestern University, and Joel Stebbins from the University of Wisconsin. That afternoon Mary Calvert, Parkhurst, and Lee caught the 4 p.m. train from Williams Bay, and accompanied Barnard's body back to Nashville, where he was laid to rest in Mt. Olivet Cemetery beside his wife and mother. The native son had come home at last.

His medals, including the Lalande gold medal (1892), the Arago gold medal (1893), and the Janssen gold medal (1900) of the French Academy of Sciences, the gold medal of the Royal Astronomical Society (1897), and the Bruce medal of the Astronomical Society of the Pacific (1917), were bequeathed to the Yerkes Observatory. So were his books and his residence – the latter as a memorial to his wife. During his lifetime he had published upwards of 900 articles. Of the works he left unfinished at his death, the most important were his *Atlas* and his studies of globular clusters. They were later completed by Mary Calvert and Frost.

3

His legacy remains. In true Horatio Alger fashion, the poor boy had made good. From the impoverished conditions of his boyhood in Nashville, he had risen against odds to become one of the most famous astronomers in the world, and had witnessed a remarkable transformation of human understanding of the universe to which he himself had contributed greatly. He left, as Swasey put it, 'a noble record for future generations.'[21]

His was an improbable success story, achieved at least in large part through sheer strength of character – what Alfred E. Howell, one of his acquaintances in the early days in Nashville, described as 'the immortal fire within himself'[22] – though not a little abetted by a phenomenally keen eye and the uncanny ability to get along without sleep. Today the odds against the self-made man are no doubt even longer than they were in Barnard's day. Nevertheless, the Barnard legend, like the Lincoln legend, continues to inspire all who start life without advantages, and suggests that struggle and hardship may themselves be formative. As the poet Keats once remarked, 'I must think that difficulties nerve the Spirit of a Man – they make our Prime Objects a Refuge as well as a Passion.'[23]

1 EEB to F. Slocum, June 10, 1921; VVO
2 EEB to W. H. Wright, June 16, 1921; SLO
3 EEB to WWC, June 16, 1921; SLO

4   EBF to RGA, February 19, 1923; YOA

5   ibid.

6   WSA to EEB, October 6, 1917; VUA

7   G. Van Biesbroeck, 'The Fifth Satellite of Jupiter,' *AJ*, **52** (1946), 114

8   E. E. Barnard, 'Observations of the Present Disappearance of the Rings of Saturn,' *PA*, **29** (1921), 147–9

9   EEB to F. Slocum, August 16, 1921; VVO

10   RGA to EBF, February 7, 1923; YOA

11   Philip Fox, 'Edward Emerson Barnard,' *PA*, **31** (1923), 195–200:196

12   S. A. Mitchell, 'Barnard at Yerkes Observatory and at the Sumatra Eclipse,' *JTAS*, 3:1 (1928), 27

13   EEB to W. H. Wright, November 3, 1922; SLO

14   EBF to GEH, February 8, 1923; YOA

15   EEB to WWC, January 14, 1923; SLO

16   EEB to W. H. Wright, January 21, 1923; SLO

17   EBF to GEH, February 8, 1923; YOA

18   EBF to RGA, February 19, 1923; YOA

19   Fox, 'Edward Emerson Barnard,' p. 195

20   G. W. Ritchey to EBF, February 24, 1923; YOA

21   Ambrose Swasey to EBF, February 8, 1923; YOA

22   Alfred E. Howell, *JTAS*, 3:1 (1928), 8

23   John Keats to Benjamin Robert Haydon, May 10, 1817; quoted in Walter Jackson Bate, *John Keats* (New York, Oxford University Press, 1966), p. 164

Index